Scaffolds and Implants for Bone Regeneration

Scaffolds and Implants for Bone Regeneration

Guest Editors

Kunyu Zhang
Qian Feng
Yongsheng Yu
Boguang Yang

Basel • Beijing • Wuhan • Barcelona • Belgrade • Novi Sad • Cluj • Manchester

Guest Editors

Kunyu Zhang
South China University of Technology
Guangzhou
China

Qian Feng
Chongqing University
Chongqing
China

Yongsheng Yu
Tongji University
Shanghai
China

Boguang Yang
The Chinese University of Hong Kong
Hong Kong
China

Editorial Office
MDPI AG
Grosspeteranlage 5
4052 Basel, Switzerland

This is a reprint of the Special Issue, published open access by the journal *Journal of Functional Biomaterials* (ISSN 2079-4983), freely accessible at: https://www.mdpi.com/journal/jfb/special_issues/implants_bone_regen.

For citation purposes, cite each article independently as indicated on the article page online and as indicated below:

Lastname, A.A.; Lastname, B.B. Article Title. *Journal Name* **Year**, *Volume Number*, Page Range.

ISBN 978-3-7258-2949-1 (Hbk)
ISBN 978-3-7258-2950-7 (PDF)
https://doi.org/10.3390/books978-3-7258-2950-7

© 2025 by the authors. Articles in this book are Open Access and distributed under the Creative Commons Attribution (CC BY) license. The book as a whole is distributed by MDPI under the terms and conditions of the Creative Commons Attribution-NonCommercial-NoDerivs (CC BY-NC-ND) license (https://creativecommons.org/licenses/by-nc-nd/4.0/).

Contents

About the Editors vii

Mitsuo Kotsu, Karol Alí Apaza Alccayhuaman, Mauro Ferri, Giovanna Iezzi, Adriano Piattelli, Natalia Fortich Mesa and Daniele Botticelli
Osseointegration at Implants Installed in Composite Bone: A Randomized Clinical Trial on Sinus Floor Elevation
Reprinted from: *J. Funct. Biomater.* **2022**, *13*, 22, https://doi.org/10.3390/jfb13010022 1

Marcus Jäger, Agnieszka Latosinska, Monika Herten, André Busch, Thomas Grupp and Andrea Sowislok
The Implant Proteome—The Right Surgical Glue to Fix Titanium Implants In Situ
Reprinted from: *J. Funct. Biomater.* **2022**, *13*, 44, https://doi.org/10.3390/jfb13020044 17

Woraporn Supphaprasitt, Lalita Charoenmuang, Nuttawut Thuaksuban, Prawichaya Sangsuwan, Narit Leepong, Danaiya Supakanjanakanti, et al.
A Three-Dimensional Printed Polycaprolactone–Biphasic-Calcium-Phosphate Scaffold Combined with Adipose-Derived Stem Cells Cultured in Xenogeneic Serum-Free Media for the Treatment of Bone Defects
Reprinted from: *J. Funct. Biomater.* **2022**, *13*, 93, https://doi.org/10.3390/jfb13030093 37

Qinghui Zhao and Shaorong Gao
Poly (Butylene Succinate)/Silicon Nitride Nanocomposite with Optimized Physicochemical Properties, Biocompatibility, Degradability, and Osteogenesis for Cranial Bone Repair
Reprinted from: *J. Funct. Biomater.* **2022**, *13*, 231, https://doi.org/10.3390/jfb13040231 64

Paweł Bąkowski, Kamilla Grzywacz, Agnieszka Prusińska, Kinga Ciemniewska-Gorzela, Justus Gille and Tomasz Piontek
Autologous Matrix-Induced Chondrogenesis (AMIC) for Focal Chondral Lesions of the Knee: A 2-Year Follow-Up of Clinical, Proprioceptive, and Isokinetic Evaluation
Reprinted from: *J. Funct. Biomater.* **2022**, *13*, 277, https://doi.org/10.3390/jfb13040277 81

Christina Polan, Christina Brenner, Monika Herten, Gero Hilken, Florian Grabellus, Heinz-Lothar Meyer, et al.
Increased UHMWPE Particle-Induced Osteolysis in Fetuin-A-Deficient Mice
Reprinted from: *J. Funct. Biomater.* **2023**, *14*, 30, https://doi.org/10.3390/jfb14010030 92

Michael Escobar, Oriol Careta, Nora Fernández Navas, Aleksandra Bartkowska, Ludovico Andrea Alberta, Jordina Fornell, et al.
Surface Modified β-Ti-18Mo-6Nb-5Ta (wt%) Alloy for Bone Implant Applications: Composite Characterization and Cytocompatibility Assessment
Reprinted from: *J. Funct. Biomater.* **2023**, *14*, 94, https://doi.org/10.3390/jfb14020094 108

Viviana R. Lopes, Ulrik Birgersson, Vivek Anand Manivel, Gry Hulsart-Billström, Sara Gallinetti, Conrado Aparicio and Jaan Hong
Human Whole Blood Interactions with Craniomaxillofacial Reconstruction Materials: Exploring In Vitro the Role of Blood Cascades and Leukocytes in Early Healing Events
Reprinted from: *J. Funct. Biomater.* **2023**, *14*, 361, https://doi.org/10.3390/jfb14070361 128

Hirotaka Mutsuzaki, Hidehiko Yashiro, Masayuki Kakehata, Ayako Oyane and Atsuo Ito
Femtosecond Laser Irradiation to Zirconia Prior to Calcium Phosphate Coating Enhances Osseointegration of Zirconia in Rabbits
Reprinted from: *J. Funct. Biomater.* **2024**, *15*, 42, https://doi.org/10.3390/jfb15020042 145

Pengwei Xiao, Caroline Schilling and Xiaodu Wang
Characterization of Trabecular Bone Microarchitecture and Mechanical Properties Using Bone Surface Curvature Distributions
Reprinted from: *J. Funct. Biomater.* **2024**, *15*, 239, https://doi.org/10.3390/jfb15080239 . . . . . . **162**

About the Editors

Kunyu Zhang

Kunyu Zhang is an associate professor at School of Biomedical Sciences and Engineering, South China University of Technology. He is enthusiastic about the design of functional hydrogels to reproduce the biochemical and biophysical signals existing in the natural cellular microenvironment and is particularly interested in their unique ability to regulate cellular behaviors and promote tissue regeneration.

Qian Feng

Qian Feng obtained her PhD in Biomedical Engineering from the Chinese University of Hong Kong. She is currently an associate professor at the Bioengineering College, Chongqing University. She works on the development of new dynamically crosslinking biomaterials, exploring the biological effects of these biomaterials on stem cells and immune cells, and investigating their applications in tissue repair.

Yongsheng Yu

Yongsheng Yu, who obtained his doctoral degree from the Chinese University of Hong Kong, is currently an associate professor at Chongqing Institute of Green and Intelligent Technology, Chinese Academy of Sciences. His research interests focus on regenerative medicine and cancer-targeting therapy. He has made several breakthroughs in developing selective covalent peptides and targeting delivery systems for disease therapy.

Boguang Yang

Dr. Boguang Yang graduated from Tianjin University and is currently an associate researcher in the Department of Biomedical Engineering at the Chinese University of Hong Kong. His research interests focus on organs-on-a-chip, 3D printing, and host–guest self-assembly. He has made breakthroughs in developing host–guest self-assembled crosslinked hydrogels with varying dynamics to regulate cell behavior.

Article

Osseointegration at Implants Installed in Composite Bone: A Randomized Clinical Trial on Sinus Floor Elevation

Mitsuo Kotsu [1], Karol Alí Apaza Alccayhuaman [2], Mauro Ferri [3], Giovanna Iezzi [4], Adriano Piattelli [4], Natalia Fortich Mesa [5] and Daniele Botticelli [1,*]

[1] ARDEC Academy, 47923 Rimini, Italy; dental_rescue@yahoo.co.jp
[2] Department of Oral Biology, Medical University of Vienna, 1090 Vienna, Austria; caroline7_k@hotmail.com
[3] ARDEC Foundation, Cartagena de Indias 130001, Colombia; medicina2000ctg@hotmail.com
[4] Department of Medical Oral and Biotechnological Sciences, University of Chieti-Pescara, 66100 Chieti, Italy; gio.iezzi@unich.it (G.I.); apiattelli@unich.it (A.P.)
[5] School of Dentistry, University Corporation Rafael Núñez, Cartagena de Indias 130001, Colombia; natalia.fortich@curnvirtual.edu.co
* Correspondence: daniele.botticelli@gmail.com

Citation: Kotsu, M.; Apaza Alccayhuaman, K.A.; Ferri, M.; Iezzi, G.; Piattelli, A.; Fortich Mesa, N.; Botticelli, D. Osseointegration at Implants Installed in Composite Bone: A Randomized Clinical Trial on Sinus Floor Elevation. *J. Funct. Biomater.* **2022**, *13*, 22. https://doi.org/10.3390/jfb13010022

Academic Editors: Kunyu Zhang, Qian Feng, Yongsheng Yu and Boguang Yang

Received: 29 December 2021
Accepted: 24 February 2022
Published: 28 February 2022

Publisher's Note: MDPI stays neutral with regard to jurisdictional claims in published maps and institutional affiliations.

Copyright: © 2022 by the authors. Licensee MDPI, Basel, Switzerland. This article is an open access article distributed under the terms and conditions of the Creative Commons Attribution (CC BY) license (https://creativecommons.org/licenses/by/4.0/).

Abstract: Osseointegration of implants installed in conjunction with sinus floor elevation might be affected by the presence of residual graft. The implant surface characteristics and the protection of the access window using a collagen membrane might influence the osseointegration. To evaluate these factors, sinus floor elevation was performed in patients using a natural bovine bone grafting material. The access windows were either covered with a collagen membrane made of porcine corium (Mb group) or left uncovered (No-Mb group) and, after six months, two mini-implants with either a moderate rough or turned surfaces were installed. After 3 months, biopsies containing the mini-implants were retrieved, processed histologically, and analyzed. Twenty patients, ten in each group, were included in the study. The two mini-implants were retrieved from fourteen patients, six belonging to the Mb group, and eight to the No-Mb group. No statistically significant differences were found in osseointegration between groups. However, statistically significant differences were found between the two surfaces. It was concluded that implants with a moderately rough surface installed in a composite bone presented much higher osseointegration compared to those with a turned surface. The present study failed to show an effect of the use of a collagen membrane on the access window.

Keywords: maxillary sinus; biomaterial; sinus augmentation; collagen membrane; access window; antrostomy; osteotomy

1. Introduction

Sinus floor elevation through a lateral access is a well-documented procedure used to increase bone volume in the posterior segments of the maxilla [1]. This approach includes the elevation of the sinus mucosa and the immediate placement of biomaterial [2,3], devices [4–6], implants alone [7,8], or in conjunction with biomaterial [9], aiming to maintain over time the elevated volume and allow bone growth within the subantral space [10–14]. The use of a membrane to cover the lateral bone window has been suggested to improve implant success [1] and might decrease both the dislodgment of the biomaterial through the access window [15,16] and the post-surgical morbidity [16]. Nevertheless, a systematic review with meta-analysis [17] failed to find effects on bone formation placing a membrane on the access window.

The implant surface instead might influence osseointegration. In an experimental study in dogs in which the osseointegration of a moderately rough surface was compared with a turned surface, better outcomes were observed at the former compared to the latter surface [18]. Even though good long-term results can be achieved also with turned surfaces [19], in a systematic review it was concluded that the best survival rate of implants installed in combination with sinus floor elevation was obtained by implants with a rough surface [1].

Nevertheless, experimental studies showed a higher progression of peri-implantitis at rough compared to turned surfaces [20]. However, systematic reviews concluded that the surface did not seem to affect the incidence of peri-implantitis [21,22]. In a retrospective study in patients with a history of periodontitis, a hybrid surface, i.e., presenting a turned surface limited to the coronal part and the remaining portion of the implant with a rough surface, showed less marginal bone loss compared to a conventional rough surface [23]. However, no clinical, radiographic, and microbiological differences were found between hybrid and traditional implants in a randomized clinical trial (RCT) in patients with history of periodontitis [24]. Even though a turned surface presented high clinical results when installed in pristine alveolar bone [19], the conditions for osseointegration might be compromised by the presence of regenerated composite bone, composed of newly formed bone and residual graft particles. In a human study after sinus floor elevation, biopsies taken from the elevated regions and from pristine zones were evaluated [25]. Both groups presented ~46% of vital bone. It has to be considered that immediate and delayed implants present different behaviors after installation in composite bone. In an experimental study in dogs, circumferential marginal defects were immediately filled with deproteinized bovine bone matrix [26]. Only few particles were found in contact to the implant surface after 4 months of healing. Instead, at implants installed after 6 months of healing after sinus floor elevation, up to ~16% of the implant surface was found in contact to graft particles, reducing the space available for osseointegration by up to 32% [27,28]. It should be considered that human biopsies harvested from the distal segments of the maxilla after 6 weeks of healing resulted in ~46–47% of osseointegration [29]. It might be argued that in a delayed mode the graft particles are stuck into newly formed bone, so that osteotomy preparation and implants might impact with the graft. Instead, when an immediate mode is applied, new bone has the chance to be formed between the implant and graft surfaces separating the particles from the implant.

Under such conditions, implant surface quality and osteoconductivity might acquire great importance.

Hence, the aim of the present study was to evaluate the osseointegration of different surfaces installed into composite bone. Moreover, the influence of the use of a collagen membrane on the access window was also assessed.

2. Materials and Methods

2.1. Ethical Statement

The protocol of the present randomized controlled trial (RCT) was approved by the Ethical Committee of the University Corporation Rafael Núñez, Cartagena de Indias, Colombia (protocol #02-2015; 19 May 2015). The study was carried out at the same university. The Declaration of Helsinki on medical protocols and ethics were adopted. All participants signed informed consent after being thoroughly notified about procedures and possible complications. The CONSORT checklist was followed to structure the article. The present study reports the histological finding while, in a previous RCT article, tomographic evaluations of the dimensional changes of the augmented space after sinus floor elevation were reported [30]. The RCT was registered at ClinicalTrials.gov with the following identifier code: NCT03899688.

2.2. Study Population

The inclusion criteria were the following: (i) the presence of an edentulous zone in the posterior segment of the maxilla presenting a height of the sinus floor ≤ 4 mm; (ii) requesting a fix prosthetic rehabilitation on implants in that region; (iii) ≥ 21 years of age; (iv); and (V) not being pregnant. The following excluding criteria were adopted: (i) no contraindications for oral surgical procedures; (ii) under chemotherapic or radiotherapic treatment; (iii) presence of an acute or a chronic sinusitis; and (iv) previous bone augmentation procedures in the region. Smokers of >10 cigarettes per day and patients under bisphosphonates treatment were also excluded.

2.3. Devices and Biomaterials

Two custom-made titanium screw-shaped mini-implants (Sweden & Martina, Due Carrare, Padua, Italy), 2.4 mm in diameter and 8 mm long, with either a moderately rough (ZirTi® surface, Sweden & Martina, Due Carrare, Padua, Italy) [31] or a turned surface, were used (Figure 1).

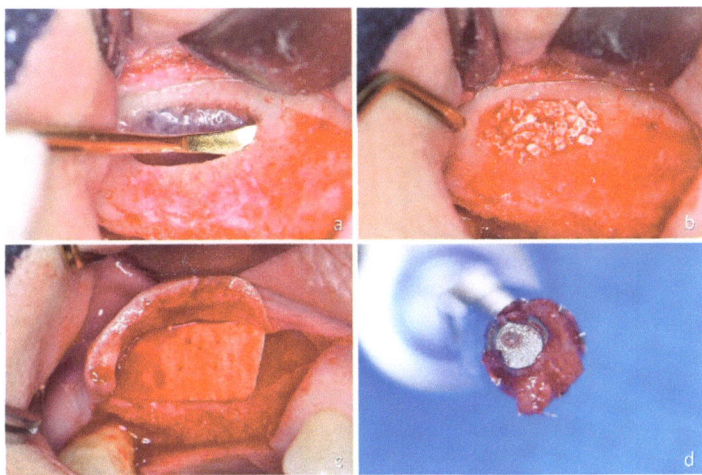

Figure 1. (**a**) Osteotomy and sinus mucosa elevation; (**b**) graft within the elevated space; (**c**) collagen membrane covering the access window; and (**d**) apical view of the biopsy: observe the eccentric position of the implant.

Cerabone granulate 1.0–2.0 mm (Botiss Biomaterials GmbH, Zossen, Germany) was used as filler material. It is composed of a ceramic made of hydroxyapatite (pentacalcium hydroxide trisphosphate) obtained from bovine cancellous bone at a high-temperature (>1200 °C). It has macroporosities with a range of 100–1500 µm in dimensions.

The collagen membrane used to protect the access window was a Collprotect membrane (Botiss Biomaterials GmbH) obtained from porcine corium.

2.4. Sample Size

The sample size for the tomographic evaluations was reported in a previous article [30]. For the present article, the data from a previous study performed on dogs by the same group were used [18], and in which a statistically significant difference was obtained using 6 animals. A sample of 9 subjects in each group was calculated to be sufficient in a one-tail test to disclose differences between the two surfaces in bone-to-implant contact, with a power 0.8, an α error of 0.05, and an effect size of 0.96.

2.5. Study Design and Allocation Concealment

This was a triple-blind study because the participants, the surgeon and the assessor of the outcome were not informed about allocation treatment. The surgeon was informed after the preparation of the two osteotomies of the recipient sites. Two mini-implants were placed in the edentulous distal segment of the maxilla in the elevated region. The position (distal or mesial) was randomly allocated. The implants were installed by an expert surgeon (MF) while the randomization of the mini-implant position was performed by another author (DB). The randomization was performed electronically by an author not involved in the mini-implant installation and biopsy retrieval (DB). The treatment assignments were kept in opaque sealed envelopes that were opened after the preparation of the two osteotomies of the recipient sites.

2.6. Clinical Procedures

Detailed descriptions of the surgical procedures were reported in a previous article [30]. Briefly, lateral bone windows were prepared using a sonic-air surgical instrument (Sonosurgery® TKD, Calenzano, FI, Italy), the sinus mucosa was elevated (Figure 1a), and a graft was used to fill the subantral space (Figure 1b). A collagen membrane was placed to cover the access window at the control sites (Figure 1c) while no membrane was used at the test sites. After 6 months of healing, two mini-implants were installed and retrieved after 3 months of submerged healing. A trephine (GA33M, Bontempi Strumenti Chirurgici, San Giovanni in Marignano, RN, Italy), 3.5 mm and 4 mm of internal and external diameter, respectively, was used, adopting an eccentric method to retrieve biopsies containing the mini-implants (Figure 1d) [32]. Standard implants were subsequently installed in the same position.

2.7. Histological Preparation of the Biopsies

The biopsies were not removed from the trephines to avoid damages and were immediately fixed in 10% buffered formalin, followed by dehydration in an ascending series of alcohol, inclusion in resin (Technovit® 7200 VLC; Kulzer, Wehrheim, Germany), and polymerization. Histological slide of ~30 μm of width were prepared following the longitudinal axis of the mini-implant and stained with acid fuchsine and toluidine blue.

2.8. Histomorphometric Evaluation

The histomorphometric evaluation were performed by a well-trained author (KAAA) blinded about allocations of the two mini-implants and an intra-rate agreement K > 0.90 was achieved. High-definition scanned photomicrographs (×200) of each histological slide were taken at an Eclipse Ci microscope (Nikon Corporation, Tokyo, Japan) equipped with a motorized stage (EK14 Nikon Corporation, Tokyo, Japan). The software NIS-Elements D 5.11.01 (Laboratory Imaging, Nikon Corporation, Tokyo, Japan) was used for histomorphometric measurements.

All measurements were performed from the most coronal contact of the bone to the implant surface to the apex. New bone, pre-existing bone (old bone and bone particles), residual graft, interpenetrating bone network (IBN; new bone penetrating the biomaterial), soft tissues (bone marrow, vessels) in contact to the implant surface (histometric linear measurements) and within 400 μm from the implant surface (morphometric measurements) were assessed.

For the morphometric measurements, a point counting method was applied [33], using a lattice with squares of 50 microns.

2.9. Data Analysis

Mean values are reported within the text while mean values and standard deviations as well as the 25th, 50th (median), and 75th percentiles are illustrated in the tables. The primary variable was new bone for both linear and morphometric evaluations. The other variables were considered as secondary variable.

Prism 9.1.1 (GraphPad Software, LLC, San Diego, CA, USA) was used for statistical analyses. The Shapiro–Wilk test was used to verify the normal distribution and either a paired t test or a Wilcoxon test was used to evaluate differences between rough and turned surface groups while an unpaired t test or a Mann–Whitney test was used to analyze differences between collagen membrane and no membrane groups. The level of significance was set at α 0.05. Pooled data with relation to the surface characteristics were also evaluated.

3. Results

3.1. Clinical Outcomes

Twenty patients were initially included in the study. Two sinus mucosa perforations, one in each group, occurred during the surgical procedures. Both were protected with a collagen membrane. No complications were reported or observed during the healing period. Further clinical and radiographic information were reported elsewhere [30]. After 6 months, in one patient of the membrane group, insufficient hard tissue was found to install both mini-implants so that the patient was excluded from the histological analysis. After a further 3 months, at the time of biopsies removal, in five patients the mini-implants were not integrated. Hence, both mini-implants were finally retrieved from fourteen patients, six patients for the membrane group ($n = 6$) and eight patients for the no-membrane group (Table 1; $n = 8$; Figure 2).

Table 1. Demographic data.

	Number	Age	Smokers	Mb	No-Mb
Females	10	53.1 ± 9.3	10 No	5	5
Males	4	59.0 ± 12.8	4 No	3	1

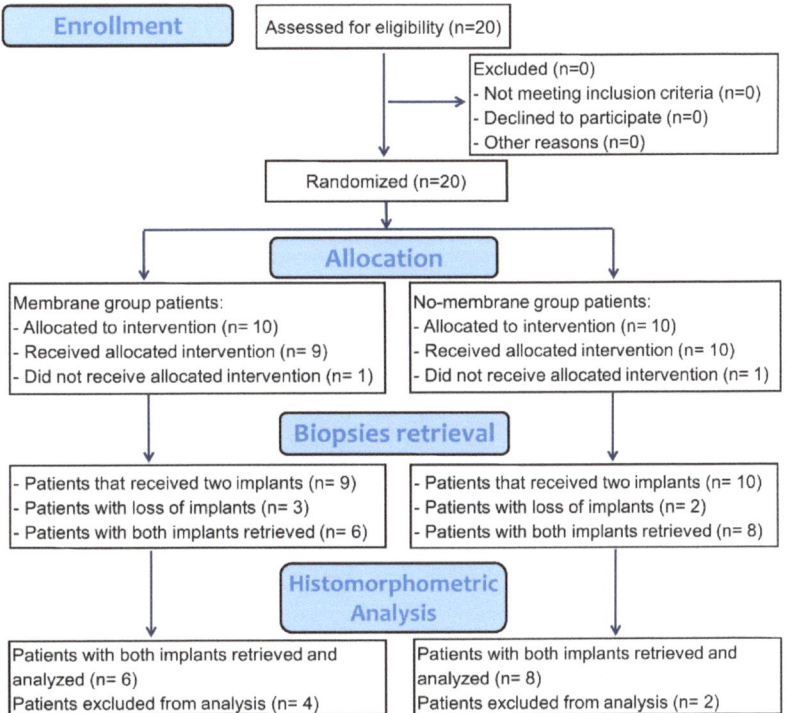

Figure 2. Consort flow diagram.

3.2. Histometric Evaluations—Tissues in Contact with the Implant Surface

All biopsies were retrieved applying the eccentric method (Figure 3).

Figure 3. Retrieved biopsy. Note the eccentric position on the mini-implant within the trephine.

The mini-implants presented new bone around and in contact to the surface (Figure 4a) while, in other regions, large amounts of biomaterial were still present (Figure 4b).

Figure 4. (**a**), New bone anchored to the implant surface. (**b**) Large amounts of biomaterial were still present.

In several instances, the biomaterial was found overlaying the new bone, taking on a foggy appearance (Figure 5a–d). In such cases, that new bone was assuming a different feature compared to new bone outside the biomaterials, as if the two tissues were interpenetrating each other (interpenetrating bone network; IBN).

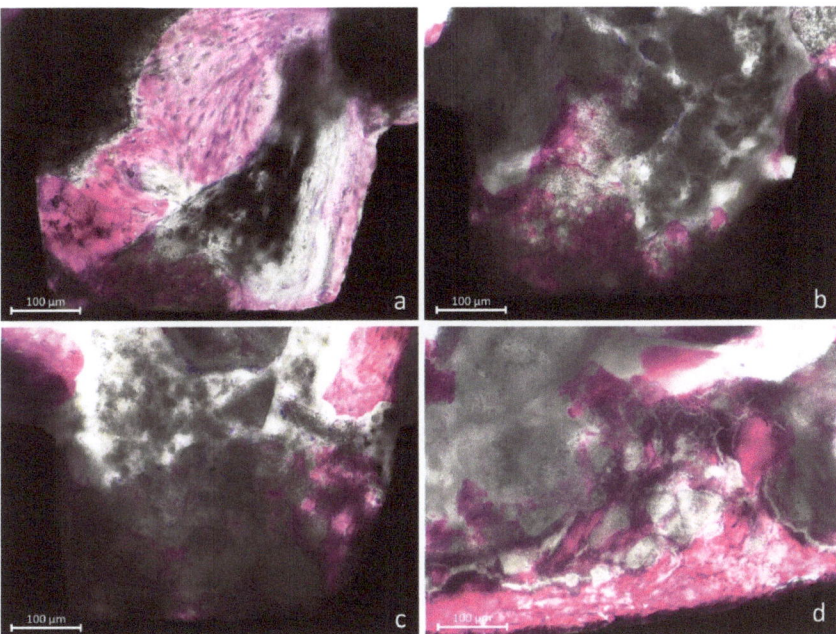

Figure 5. (**a–d**) Images showing new bone formed around and within the graft residues (interpenetrating bone network; IBN).

High light intensity was provided to better identify this structure (Figure 6a–d).

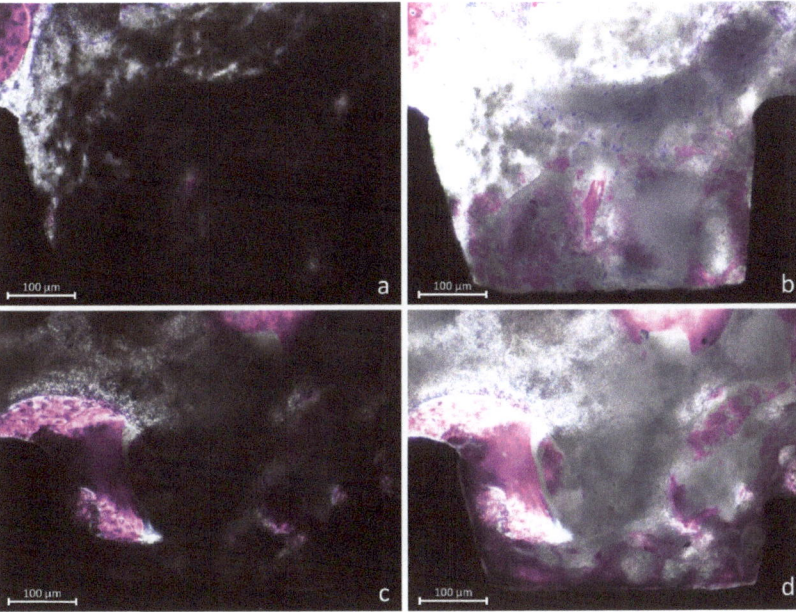

Figure 6. (**a–d**) Photomicrographs representing new bone and interpenetrating bone network (IBN). (**a**,**c**) Dark mode, at which a normal light exposure was adopted. (**b**,**d**) Overexposed images that better revealed the structure of the IBN.

Bone particles were sometimes identified (Figure 7a) as well granules of biomaterial (Figure 7b).

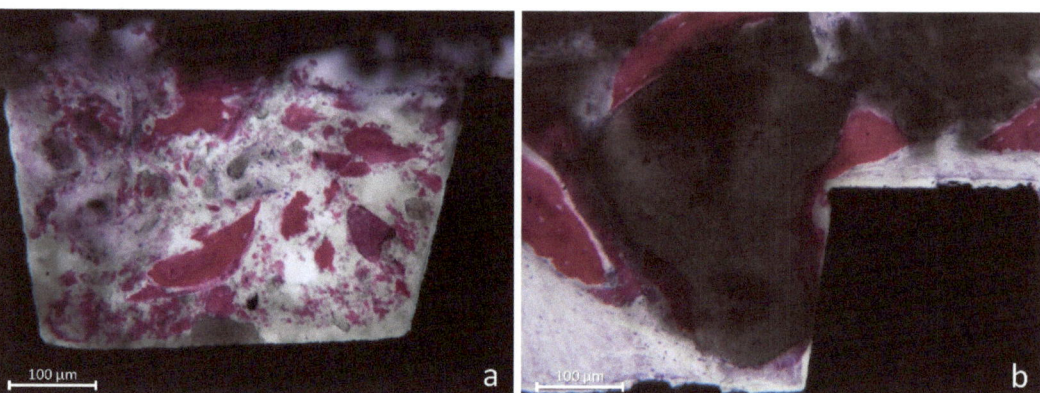

Figure 7. (**a**) Some bone particles not yet resorbed or included in new bone. (**b**) Granules of Cerabone® surrounded by newly formed bone.

In the coronal segment of the mini-implant, old bone was still visible and was in some cases anchored to the implant surface (Figure 8).

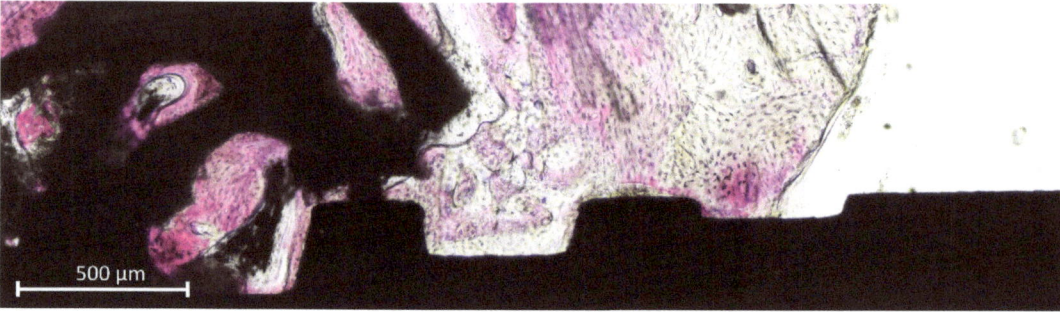

Figure 8. Old pre-existing bone at the coronal margin of the implant.

The mean percentage of new bone in contact with the implant surface was higher at the ZirTi compared to the turned surfaces in both membrane (28.9% and 11.0%, respectively; $p = 0.030$; Table 2) and no-membrane groups (30.5% and 9.2%, respectively; $p = 0.008$; Table 3).

The difference between the membrane and no-membrane groups was not statistically significant for both ZirTi ($p = 0.852$) and turned ($p = 0.636$) surfaces.

The interpenetrating bone network (IBN) was in the membrane group 13.5% and 16.6% at the ZirTi and turned surfaces, respectively. In the no-membrane group, the respective fractions were 7.0% and 6.1%. In the membrane group, the sum between new bone and IBN yielded 42.4% of total bone for the ZirTi surface and 27.6 % for the turned surface ($p = 0.258$). In the no-membrane group, the respective percentages were 37.5% and 15.3 % ($p = 0.001$).

Table 2. Membrane group (*n* = 6). Tissues in contact to the implant surface expressed in percentages (%). SD, standard deviation. IBN, interpenetrating bone network; 25%, first percentile; 75%, third percentile.

		New Bone	IBN	Total Bone	Old Bone	Graft	Soft Tissues
ZIRTI	Mean ± SD	28.9 ± 14.5	13.5 ± 8.0	42.4 ± 17.7	1.6 ± 3.8	25.2 ± 15.2	30.8 ± 17.3
	Median (25%; 75%)	25.2 (24.3; 34.1)	14.6 (7.7; 20.1)	35.5 (30.3; 48.1)	0.0 (0.0; 0.0)	24.0 (15.7; 29.7)	37.3 (15.9; 43.3)
TURNED	Mean ± SD	11.0 ± 5.7	16.6 ± 15.1	27.6 ± 14.5	1.2 ± 2.1	27.2 ± 17.7	43.9 ± 26.0
	Median (25%; 75%)	12.9 (8.0; 14.2)	11.2 (5.7; 27.6)	23.1 (17.0; 33.3)	0.0 (0.0; 1.7)	29.7 (15.7; 38.3)	38.4 (34.2; 59.6)
p-value ZirTi vs. Turned		0.030	0.750	0.258	>0.999	0.828	0.305
p-value Mb vs. No-Mb ZirTi		0.852	0.108	0.612	0.469	0.579	0.507
p-value Mb vs. No-Mb Turned		0.636	0.308	0.103	0.618	0.755	0.282

Table 3. No-membrane group (*n* = 8). Tissues in contact to the implant surface expressed in percentages (%). SD, standard deviation. IBN, interpenetrating bone network; 25%, first percentile; 75%, third percentile.

		New Bone	IBN	Total Bone	Old Bone	Graft	Soft Tissues
ZIRTI	Mean ± SD	30.5 ± 14.9	7.0 ± 8.1	37.5 ± 17.3	3.0 ± 3.6	30.6 ± 20.6	28.9 ± 12.6
	Median (25%; 75%)	27.1 (19.1; 34.7)	3.8 (1.6; 9.0)	32.4 (26.3; 43.5)	1.3 (0.0; 5.5)	35.6 (12.1; 46.0)	27.9 (23.0; 30.1)
TURNED	Mean ± SD	9.2 ± 7.3	6.1 ± 6.1	15.3 ± 8.1	2.4 ± 4.2	23.4 ± 24.3	58.9 ± 23.4
	Median (25%; 75%)	6.5 (5.1; 13.1)	5.8 (0.5; 9.7)	13.9 (8.7; 21.7)	0.3 (0.0; 3.1)	14.4 (10.6; 28.2)	70.6 (47.9; 75.4)
p-value ZirTi vs. Turned		0.008	0.672	0.001	0.625	0.461	0.016

Small amounts of old bone (mean ≤ 3%) were observed while large remnants of non-resorbed graft were present in contact with the implant surface, the means ranging between 23.4% and 30.6%. Soft tissues were present in high percentages, ranging from 28.9% to 58.9%. The highest values were observed at the turned compared to the ZirTi surfaces. However, the difference was statistically significant only in the no-membrane group (p = 0.016).

The pooled data (Table 4) revealed that ZirTi surface yielded a higher amount of new bone (29.8%) and total bone (39.6%) compared to the turned surface (10% and 20.6%, respectively). Similar amounts of IBN, old bone and graft percentages were found at the two surfaces while statistically high percentages of soft tissues were detected at the turned compared to the ZirTi surfaces.

Table 4. Pooled data of membrane and no-membrane groups (*n* = 14). Tissues in contact to the implant surface expressed in percentages (%). SD, standard deviation. IBN, interpenetrating bone network; 25%, first percentile; 75%, third percentile.

		New Bone	IBN	Total Bone	Old Bone	Graft	Soft Tissues
ZIRTI	Mean ± SD	29.8 ± 14.2	9.8 ± 8.4	39.6 ± 17.0	2.4 ± 3.7	28.3 ± 18.1	29.7 ± 14.2
	Median (25%; 75%)	26.0 (20.9; 34.9)	5.9 (2.4; 16.7)	32.9 (28.9; 48.1)	0.0 (0.0; 4.2)	28.5 (13.4; 43.9)	29.0 (19.4; 38.4)
TURNED	Mean ± SD	10.0 ± 6.5	10.6 ± 11.7	20.6 ± 12.5	1.9 ± 3.4	25.0 ± 21.0	52.5 ± 24.8
	Median (25%; 75%)	9.2 (5.4; 14.2)	9.3 (2.3; 11.5)	17.1 (12.5; 27.0)	0.0 (0.0; 2.7)	17.5 (12.0; 38.3)	58.7 (37.6; 74.5)
p-value ZirTi vs. Turned		0.000	0.594	0.003	0.813	0.580	0.004

3.3. Morphometric Evaluations

A similar density of new bone was found around both ZirTi and Turned surfaces in both membrane and no-membrane groups, the means ranging between 19.9% and 23.7% (Tables 5–7). IBN means ranged between 11.4% to 6.3% and the total bone from 28.9% and 31.3%. Graft remnants were still present in a high proportion, ranging between 28.9% and 46.2%. No statistically significant differences were found for all variables above mentioned between surfaces and between membrane and no-membrane groups. Small amounts of old bone were detected while softs tissues ranged between 24.5% and 39.0%.

Table 5. Membrane group (*n* = 6). Tissues density around the implant surface expressed in percentages (%). SD, standard deviation. IBN, interpenetrating bone network; 25%, first percentile; 75%, third percentile.

		New Bone	IBN	Total Bone	Old Bone	Graft	Soft Tissues
ZIRTI	Mean ± SD	21.8 ± 4.8	7.4 ± 4.0	29.2 ± 7.0	0.1 ± 0.2	46.2 ± 4.6	24.5 ± 7.5
	Median (25%; 75%)	22.9 (18.1; 23.7)	6.6 (5.0; 8.5)	27.3 (25.2; 28.2)	0.0 (0.0; 0.0)	45.3 (42.9; 50.1)	26.9 (21.2; 30.4)
TURNED	Mean ± SD	19.9 ± 8.9	11.4 ± 9.0	31.3 ± 6.2	3.2 ± 4.5	36.6 ± 11.3	28.9 ± 5.4
	Median (25%; 75%)	21.2 (18.5; 22.0)	8.4 (6.6; 11.2)	31.6 (28.9; 33.7)	0.7 (0.0; 5.5)	39.5 (34.5; 43.5)	31.0 (24.8; 32.7)
p-value ZirTi vs. Turned		0.552	0.438	0.563	0.250	0.048	0.282
p-value Mb vs. No-Mb ZirTi		0.662	0.878	>0.9999	0.021	0.342	0.883
p-value Mb vs. No-Mb Turned		0.573	0.282	0.534	0.505	0.308	0.037

Table 6. No-membrane group (*n* = 8). Tissues density around the implant surface expressed in percentages (%). SD, standard deviation. IBN, interpenetrating bone network; 25%, first percentile; 75%, third percentile.

		New Bone	IBN	Total Bone	Old Bone	Graft	Soft Tissues
ZIRTI	Mean ± SD	23.7 ± 10.3	7.1 ± 3.9	30.7 ± 10.0	4.9 ± 6.3	39.1 ± 19.4	25.4 ± 13.6
	Median (25%; 75%)	20.3 (17.1; 29.3)	6.7 (3.4; 11.0)	28.1 (22.7; 37.5)	2.7 (0.4; 6.7)	44.3 (25.7; 51.4)	27.5 (12.5; 31.4)
TURNED	Mean ± SD	22.6 ± 8.2	6.3 ± 3.2	28.9 ± 7.6	3.2 ± 3.7	28.9 ± 15.7	39.0 ± 10.3
	Median (25%; 75%)	22.1 (19.3; 25.9)	6.2 (5.4; 7.4)	31.3 (25.3; 33.6)	2.2 (0.7; 3.9)	26.2 (19.7; 32.7)	41.1 (36.3; 44.9)
p-value ZirTi vs. Turned		0.771	0.558	0.550	0.375	0.318	0.121

Table 7. Pooled data of membrane and no-membrane groups (*n* = 14). Tissues density around the implant surface expressed in percentages (%). SD, standard deviation. IBN, interpenetrating bone network; 25%, first percentile; 75%, third percentile.

		New Bone	IBN	Total Bone	Old Bone	Graft	Soft Tissues
ZIRTI	Mean ± SD	22.9 ± 8.1	7.2 ± 3.8	30.1 ± 8.6	2.8 ± 5.2	42.1 ± 15.0	25.0 ± 11.0
	Median (25%; 75%)	22.5 (17.0; 27.4)	6.7 (3.9; 10.4)	27.6 (24.4; 34.1)	0.2 (0.0; 6.7)	45.3 (41.2; 51.2)	27.5 (14.8; 30.6)
TURNED	Mean ± SD	21.5 ± 8.3	8.5 ± 6.6	29.9 ± 6.9	3.2 ± 3.9	32.2 ± 14.1	34.7 ± 9.8
	Median (25%; 75%)	21.3 (18.5; 23.9)	6.5 (5.8; 9.6)	31.3 (27.8; 33.7)	1.4 (0.1; 4.9)	32.2 (21.1; 39.6)	33.6 (27.1; 41.6)
p-value ZirTi vs. Turned		0.547	0.726	0.987	0.846	0.092	0.061

4. Discussion

The mini-implants retrieved were osseointegrated into newly formed bone. The different characteristics of the implant surface played an important role in osseointegration, generating a statistically significant higher amount of newly formed bone at the moderately rough compared to the turned surface. However, no differences could be detected between the membrane and no-membrane groups.

A total of ten mini-implants were found not integrated, independently from the surface characteristics. This is not in agreement with other RCTs that included a similar design with mini-implant installed after 6 months from sinus floor elevation and retrieved after a further 3 months [27,28]. In those studies, different dimensions and positions of the access window were included as variables, and a collagenated cortico-cancellous porcine bone was used as filler. Only implants with a moderate surface were used. Four mini-implants were lost in one study [27] while none in the other study [28].

Nevertheless, the xenogeneic graft used in the present study has been used in several studies that reported optimal results both in clinical [34–38] and animal [39–42] studies.

Even though in the present study the loss of implants was similar for both surfaces, the grade of osseointegration was statistically significantly higher at the moderately rough compared to the turned surface. It should be considered that, when an implant is installed in a standard alveolar crest, new bone can be formed from multiple sources, both in the cortical and marrow regions. A strong cellular reaction can be observed after 5 days of healing within the marrow compartment around the body of the implant [43]. New bone is subsequently formed, creating a bone barrier around the implant and on its surface showing an attempt to isolate the implant body from the marrow compartment. In the cortical region, the old pre-existing marginal bone around the implant is resorbed over time and substituted by newly formed bone, mainly through basic multicellular units (BMUs) [43]. Under such conditions of multiple bone sources, also a turned surface might work properly. Indeed, in an experiment in dogs, both surfaces were integrated 4 months after the installation in a healed alveolar bone [18] presenting osseointegration fractions of 56.3% and 50.6% at the moderately rough and turned surfaces, respectively. However, the presence of residual graft particles in composite bone limits the number of multiple bone sources [27,28], and in such a case, the degree of osteoconductivity of the implant surface might play an important role.

The importance of osteoconductivity properties has been elucidated in an experiment in dogs [44]. In that study, circumferential marginal defects with a depth of 5 mm and a horizontal gap of 1.25 were created around implants presenting either a moderately rough or a turned surface. Collagen membranes were used to protect the defects. Both submerged and not submerged healing were studied. After 4 months of healing, the marginal gain at the moderately rough surface was >4 mm while, at the turned surface, residual defects of about 3.4–3.6 mm in depth were still present at both submerged and not submerged implants. Residual defects were also observed in another study in which commercial turned implants were used [45]. Marginal defects, 5 mm in depth but with horizontal gaps of different dimensions, were tested. It was shown that the larger the marginal defect at installation, the deeper the residual defect after 12 weeks of healing. Moreover, it was shown that, due to the small horizontal dimensions, the residual marginal defects were not detectable at a clinical evaluation.

In other experimental studies, only moderately rough surfaces were adopted, and marginal defects were prepared. It was shown that, in the presence of marginal defects, the new bone was formed from the lateral bone walls during the first month of healing and the lateral growth stopped at ~0.4 mm from the implant surface [46,47]. During the same period, osseointegration started from the base of the defect and proceeded coronally to gain the closure of the defect in few months [46,48]. This period of healing is longer compared to that needed for the healing of artificial defects and extraction sockets [49–52]. Similar marginal defects, but larger in dimensions compared to those artificial described above, are obtained at implants installed simultaneously to sinus floor elevation performed by lateral or transcrestal accesses. In that case, new bone apposition on the implant surface starts from the sinus floor proceeding towards the apex [12,53], and reaches the implant apex, but only if the conditions for the growth are maintained over time [54].

However, in the present study, the mini-implants were inserted 6 months after sinus floor elevation. It might be argued that bone regeneration in that area should have already created similar conditions to that of a pristine alveolar bone. However, high amounts of residual grafts were still present after 9 months from the first surgery, providing different characteristics to the regenerated grafted bone (composite bone) compared to the pristine alveolar bone. Like in the present study, other similar RCTs showed a contact of the graft to the surface at implants installed after 6 months from sinus floor elevation and retrieved after 3 more months [27,28]. The histological analyses revealed 0.6–15.9% of graft in contact to the implant surface. This contact of the biomaterial to the surface reduced the available space for bone apposition as well as the number of bone sources compared to a pristine alveolar bone. In addition, it has to be considered that bone density was similar around both moderately rough and turned surfaces so that bone sources availability should be considered similar for the two different surfaces. Under such conditions, the osteoconductivity of the surfaces acquires an important role and the turned surface might be at a disadvantage compared to the moderately rough surface. This condition resulted in a much lower BIC% at the former compared to the latter surface than it was expected [18]. In fact, in the present study, the difference in pooled BIC% between the two groups of surfaces was ~19% while the difference found in another study between two similar surfaces at implants installed in pristine bone was 5.7% [18].

In the present study, the term "interpenetrating bone network (IBN)" was used. This term was first used in an experimental study in which a biphasic biomaterial, composed of 60% hydroxyapatite (HA) and 40% of beta-tricalcium phosphate (β-TCP), was used as filler material for sinus augmentation in rabbits [55]. Providing a higher light intensity at the optical microscope, it was possible to identify new bone overlapping or within the graft residues (Figure 6a–d). This was shown also in another previously published article on sinus floor elevation in sheep in which a biphasic biomaterial was also used as filler, again composed of HA 60% and β-TCP 40% [56]. The structure of the IBN recalled the structure of an "interpenetrating polymer network" [57] and, for this reason, the term "interpenetrating bone network" was adopted.

The foggy-like appearance of the biomaterial and the presence of old bone particles might depose for damage of the newly formed composite bone within the elevated area that occurred during the recipient site preparation for implant installation. Moreover, the cellular reaction that follows this event [45] triggered new bone formation and further degradation of the biomaterial.

No statistically significant differences were seen for osseointegration of mini-implants and bone density between membrane and no-membrane groups for both surfaces evaluated. Nevertheless, a systematic review concluded that better results might be obtained in implant survival after 3 years using a rough surface and a membrane coverage of the access window [1]. The results from the present study agree with the former but disagree with the second assumption. It has to be considered that, even though histological studies reported higher amounts of new bone density using membrane either in PTFE [58] or collagen [59–61], in those studies the biopsies were taken through the access windows so that the data does not represent correctly the region of interest. Data supporting this assumption were reported in a histological study in humans in which biopsies were taken after 9 months after sinus floor elevation [62]. Statistically higher fractions of mineralized bone were found at the biopsies taken from the alveolar crest (40.1%) compared to those taken from the lateral window (26.0%), even though the osteotomy was protected with a collagen membrane at the time of sinus floor elevation. Moreover, in a systematic review with meta-analysis [17], it was concluded that a membrane placed on the access window does not influence the proportion of bone formed within the elevated space. This outcome was also supported by the data from another experimental study on sinus floor augmentation in rabbits in which the percentages of new bone within the grafted sinuses after 8 weeks of healing were 24.9% and 24.5% for membrane and no-membrane groups, respectively [63].

The main limitation of the present study is related to the low numbers of patients included and of the biopsies retrieved. Nevertheless, the importance of the implant surface osteoconductivity has been clearly shown. Another limitation is the biomaterial used that might have influenced bone formation within the sinus cavity so that the results should not be inferred with other fillers. Studies comparing the present with biomaterial devoid of a similar property of interpenetration should be performed. RCTs using moderate rough and turned surfaces should be performed to compare the healing at implants installed in the pristine or composite alveolar bone. The results from the present study suggest that the osteoconductivity properties of the surface should be considered when the implant is installed in the composite bone because the residual graft might interfere with the osseointegration processes. In the present study, a bovine cancellous bone and a porcine corium collagen membrane were used as biomaterial. Other biomaterials that proved their capacity of supporting tissue regeneration should be used and evaluated [64–67].

The presence of graft material in contact with the implant surface suggests that composite bone might result in critical regions in which no integration of the implant surface occurs. This condition suggests the need of further investigations to identify the best biomaterials able to reduce this phenomenon and the capacity of the implant surface to favorite bone apposition also in the presence of composite bone.

5. Conclusions

It might be concluded that implants with a moderately rough surface installed in a composite bone presented much higher osseointegration compared to those with a turned surface. The present study failed to show the effect of the use of a collagen membrane on the access window.

Author Contributions: Conceptualization, M.K., D.B.; methodology, M.F., D.B.; validation, G.I., A.P.; formal analysis, K.A.A.A., D.B.; investigation, M.F., G.I., D.B.; resources, D.B., G.I., A.P.; data curation, D.B.; writing—original draft preparation, M.K., D.B.; writing—review and editing, G.I., A.P., D.B.; supervision, M.F., N.F.M., D.B.; project administration, M.F., N.F.M., D.B.; funding acquisition, D.B. All authors have read and agreed to the published version of the manuscript.

Funding: This research was funded by ARDEC Academy, Rimini, Italy, and Sweden & Martina, Due Carrare, Padua, Italy.

Institutional Review Board Statement: The study was conducted in accordance with the Declaration of Helsinki and approved by the Ethical Committee of the University Corporation Rafael Núñez, Cartagena de Indias, Colombia (protocol #02-2015; 19 May 2015).

Informed Consent Statement: Informed consent was obtained from all subjects involved in the study.

Data Availability Statement: Data are available on reasonable request.

Acknowledgments: The biomaterial was provided free of charge by Straumann, Milan, Italy. The mini-implants were provided free of charge by Sweden & Martina, Due Carrare, Padua, Italy.

Conflicts of Interest: The authors declare no conflict of interest.

References

1. Pjetursson, B.E.; Tan, W.C.; Zwahlen, M.; Lang, N.P. A systematic review of the success of sinus floor elevation and survival of implants inserted in combination with sinus floor elevation. *J. Clin. Periodontol.* **2008**, *35* (Suppl. 8), 216–240. [CrossRef] [PubMed]
2. Kawakami, S.; Lang, N.P.; Iida, T.; Ferri, M.; Apaza Alccayhuaman, K.A.; Botticelli, D. Influence of the position of the antrostomy in sinus floor elevation assessed with cone-beam computed tomography: A randomized clinical trial. *J. Investig. Clin. Dent.* **2018**, *9*, e12362. [CrossRef] [PubMed]
3. Kawakami, S.; Lang, N.P.; Ferri, M.; Apaza Alccayhuaman, K.A.; Botticelli, D. Influence of the height of the antrostomy in sinus floor elevation assessed by cone beam computed tomography- a randomized clinical trial. *Int. J. Oral Maxillofac. Implant.* **2019**, *34*, 223–232. [CrossRef] [PubMed]
4. Cricchio, G.; Palma, V.C.; Faria, P.E.; de Olivera, J.A.; Lundgren, S.; Sennerby, L.; Salata, L.A. Histological outcomes on the development of new space-making devices for maxillary sinus floor augmentation. *Clin. Implant. Dent. Relat. Res.* **2011**, *13*, 224–230. [CrossRef]
5. Schweikert, M.; Botticelli, D.; de Oliveira, J.A.; Scala, A.; Salata, L.A.; Lang, N.P. Use of a titanium device in lateral sinus floor elevation: An experimental study in monkeys. *Clin. Oral Implant. Res.* **2012**, *23*, 100–105. [CrossRef]
6. Johansson, L.Å.; Isaksson, S.; Adolfsson, E.; Lindh, C.; Sennerby, L. Bone regeneration using a hollow hydroxyapatite space-maintaining device for maxillary sinus floor augmentation—A clinical pilot study. *Clin. Implant. Dent. Relat. Res.* **2012**, *14*, 575–584. [CrossRef] [PubMed]
7. Omori, Y.; Botticelli, D.; Ferri, M.; Delgado-Ruiz, R.; Ferreira Balan, V.; Porfirio Xavier, S. Argon Bioactivation of Implants Installed Simultaneously to Maxillary Sinus Lifting without Graft. An Experimental Study in Rabbits. *Dent. J.* **2021**, *9*, 105. [CrossRef]
8. Ye, M.; Liu, W.; Cheng, S.; Yan, L. Outcomes of implants placed after osteotome sinus floor elevation without bone grafts: A systematic review and meta-analysis of single-arm studies. *Int. J. Implant. Dent.* **2021**, *7*, 72. [CrossRef] [PubMed]
9. Ekhlasmandkermani, M.; Amid, R.; Kadkhodazadeh, M.; Hajizadeh, F.; Abed, P.F.; Kheiri, L.; Kheiri, A. Sinus floor elevation and simultaneous implant placement in fresh extraction sockets: A systematic review of clinical data. *J. Korean Assoc. Oral Maxillofac. Surg.* **2021**, *47*, 411–426. [CrossRef]
10. Jensen, T.; Schou, S.; Svendsen, P.A.; Forman, J.L.; Gundersen, H.J.; Terheyden, H.; Holmstrup, P. Volumetric changes of the graft after maxillary sinus floor augmentation with Bio-Oss and autogenous bone in different ratios: A radiographic study in minipigs. *Clin. Oral Implant. Res.* **2012**, *23*, 902–910. [CrossRef]

11. Busenlechner, D.; Huber, C.D.; Vasak, C.; Dobsak, A.; Gruber, R.; Watzek, G. Sinus augmentation analysis revised: The gradient of graft consolidation. *Clin. Oral Implant. Res.* **2009**, *20*, 1078–1083. [CrossRef]
12. Scala, A.; Botticelli, D.; Faeda, R.S.; Garcia Rangel, I., Jr.; Américo de Oliveira, J.; Lang, N.P. Lack of influence of the Schneiderian membrane in forming new bone apical to implants simultaneously installed with sinus floor elevation: An experimental study in monkeys. *Clin. Oral Implant. Res.* **2012**, *23*, 175–181. [CrossRef] [PubMed]
13. Caneva, M.; Lang, N.P.; Garcia Rangel, I.J.; Ferreira, S.; Caneva, M.; De Santis, E.; Botticelli, D. Sinus mucosa elevation using Bio-Oss(®) or Gingistat(®) collagen sponge: An experimental study in rabbits. *Clin. Oral Implant. Res.* **2017**, *28*, e21–e30. [CrossRef] [PubMed]
14. Iida, T.; Carneiro Martins Neto, E.; Botticelli, D.; Apaza Alccayhuaman, K.A.; Lang, N.P.; Xavier, S.P. Influence of a collagen membrane positioned subjacent the sinus mucosa following the elevation of the maxillary sinus. A histomorphometric study in rabbits. *Clin. Oral Implant. Res.* **2017**, *28*, 1567–1576. [CrossRef]
15. Nosaka, Y.; Nosaka, H.; Arai, Y. Complications of postoperative swelling of the maxillary sinus membrane after sinus floor augmentation. *J. Oral Sci. Rehabil.* **2015**, *1*, 26–33.
16. Ohayon, L.; Taschieri, S.; Friedmann, A.; Del Fabbro, M. Bone Graft Displacement after Maxillary Sinus Floor Aug-mentation With or Without Covering Barrier Membrane: A Retrospective Computed Tomographic Image Evaluation. *Int. J. Oral Maxillofac. Implant.* **2019**, *34*, 681–691. [CrossRef]
17. Suárez-López Del Amo, F.; Ortega-Oller, I.; Catena, A.; Monje, A.; Khoshkam, V.; Torrecillas-Martínez, L.; Wang, H.L.; Galindo-Moreno, P. Effect of barrier membranes on the outcomes of maxillary sinus floor augmentation: A meta-analysis of histomorphometric outcomes. *Int. J. Oral Maxillofac. Implant.* **2015**, *30*, 607–618. [CrossRef]
18. Caroprese, M.; Lang, N.P.; Baffone, G.M.; Ricci, S.; Caneva, M.; Botticelli, D. Histomorphometric analysis of bone healing at implants with turned or rough surfaces: An experimental study in the dog. *J. Oral Sci. Rehabil.* **2016**, *2*, 74–79.
19. Wennerberg, A.; Albrektsson, T.; Chrcanovic, B. Long-term clinical outcome of implants with different surface modifications. *Eur. J. Oral Implantol.* **2018**, *11*, S123–S136.
20. Garaicoa-Pazmino, C.; Lin, G.H.; Alkandery, A.; Parra-Carrasquer, C.; Suárez-López Del Amo, F. Influence of implant surface characteristics on the initiation, progression and treatment outcomes of peri-implantitis: A systematic review and meta-analysis based on animal model studies. *Int. J. Oral Implant.* **2021**, *14*, 367–382.
21. Saulacic, N.; Schaller, B. Prevalence of Peri-Implantitis in Implants with Turned and Rough Surfaces: A Systematic Review. *J. Oral Maxillofac. Res.* **2019**, *10*, e1. [CrossRef] [PubMed]
22. Stavropoulos, A.; Bertl, K.; Winning, L.; Polyzois, I. What is the influence of implant surface characteristics and/or implant material on the incidence and progression of peri-implantitis? A systematic literature review. *Clin. Oral Implant. Res.* **2021**, *32* (Suppl. 21), 203–229. [CrossRef] [PubMed]
23. Gallego, L.; Sicilia, A.; Sicilia, P.; Mallo, C.; Cuesta, S.; Sanz, M. A retrospective study on the crestal bone loss as-sociated with different implant surfaces in chronic periodontitis patients under maintenance. *Clin. Oral Implant. Res.* **2018**, *29*, 557–567. [CrossRef] [PubMed]
24. Serrano, B.; Sanz-Sánchez, I.; Serrano, K.; Montero, E.; Sanz, M. One-year outcomes of dental implants with a hybrid surface macro-design placed in patients with history of periodontitis: A randomized clinical trial. *J. Clin. Periodontol.* **2022**, *49*, 90–100. [CrossRef]
25. Galindo-Moreno, P.; Moreno-Riestra, I.; Avila, G.; Fernández-Barbero, J.E.; Mesa, F.; Aguilar, M.; Wang, H.L.; O'Valle, F. Histomorphometric comparison of maxillary pristine bone and composite bone graft biopsies obtained after sinus augmentation. *Clin. Oral Implant. Res.* **2010**, *21*, 122–128. [CrossRef]
26. Botticelli, D.; Berglundh, T.; Lindhe, J. The influence of a biomaterial on the closure of a marginal hard tissue defect adjacent to implants. An experimental study in the dog. *Clin. Oral Implant. Res.* **2004**, *15*, 285–292. [CrossRef]
27. Hirota, A.; Iezzi, G.; Piattelli, A.; Ferri, M.; Tanaka, K.; Apaza Alccayhuaman, K.A.; Botticelli, D. Influence of the position of the antrostomy in sinus floor elevation on the healing of mini-implants: A randomized clinical trial. *Oral Maxillofac. Surg.* **2020**, *24*, 299–308. [CrossRef]
28. Imai, H.; Iezzi, G.; Piattelli, A.; Ferri, M.; Apaza Alccayhuaman, K.A.; Botticelli, D. Influence of the Dimensions of the Antrostomy on Osseointegration of Mini-implants Placed in the Grafted Region after Sinus Floor Elevation: A Randomized Clinical Trial. *Int. J. Oral Maxillofac. Implants* **2020**, *35*, 591–598. [CrossRef]
29. Sakuma, S.; Piattelli, A.; Baldi, N.; Ferri, M.; Iezzi, G.; Botticelli, D. Bone Healing at Implants Placed in Sites Prepared Either with a Sonic Device or Drills: A Split-Mouth Histomorphometric Randomized Controlled Trial. *Int. J. Oral Maxillofac. Implant.* **2020**, *35*, 187–195. [CrossRef]
30. Imai, H.; Lang, N.P.; Ferri, M.; Hirota, A.; Apaza Alccayhuaman, K.A.; Botticelli, D. Tomographic Assessment on the In-fluence of the Use of a Collagen Membrane on Dimensional Variations to Protect the Antrostomy After Maxillary Sinus Floor Augmentation: A Randomized Clinical Trial. *Int. J. Oral Maxillofac. Implant.* **2020**, *35*, 350–356. [CrossRef]
31. Caneva, M.; Lang, N.P.; Calvo Guirado, J.L.; Spriano, S.; Iezzi, G.; Botticelli, D. Bone healing at bicortically installed implants with different surface configurations. An experimental study in rabbits. *Clin. Oral Implant. Res.* **2015**, *26*, 293–299. [CrossRef]
32. Ferri, M.; Lang, N.P.; Angarita Alfonso, E.E.; Bedoya Quintero, I.D.; Burgos, E.M.; Botticelli, D. Use of sonic in-struments for implant biopsy retrieval. *Clin. Oral Implant. Res.* **2015**, *26*, 1237–1243. [CrossRef] [PubMed]

33. Schroeder, H.E.; Münzel-Pedrazzoli, S. Correlated morphometric and biochemical analysis of gingival tissue. Mor-phometric model, tissue sampling and test of stereologic procedures. *J. Microsc.* **1973**, *99*, 301–329. [CrossRef]
34. Riachi, F.; Naaman, N.; Tabarani, C.; Aboelsaad, N.; Aboushelib, M.N.; Berberi, A.; Salameh, Z. Influence of material properties on rate of resorption of two bone graft materials after sinus lift using radiographic assessment. *Int. J. Dent.* **2012**, *2012*, 737262. [CrossRef]
35. Mahesh, L.; Mascarenhas, G.; Bhasin, M.T.; Guirado, C.; Juneja, S. Histological evaluation of two different anorganic bovine bone matrixes in lateral wall sinus elevation procedure: A retrospective study. *Natl. J. Maxillofac. Surg.* **2020**, *11*, 258–262. [CrossRef] [PubMed]
36. Zahedpasha, A.; Ghassemi, A.; Bijani, A.; Haghanifar, S.; Majidi, M.S.; Ghorbani, Z.M. Comparison of Bone Formation After Sinus Membrane Lifting without Graft or Using Bone Substitute "Histologic and Radiographic Evaluation". *J. Oral Maxillofac. Surg.* **2021**, *79*, 1246–1254. [CrossRef]
37. Tawil, G.; Barbeck, M.; Unger, R.; Tawil, P.; Witte, F. Sinus Floor Elevation Using the Lateral Approach and Window Repositioning and a Xenogeneic Bone Substitute as a Grafting Material: A Histologic, Histomorphometric, and Radio-graphic Analysis. *Int. J. Oral Maxillofac. Implant.* **2018**, *33*, 1089–1096. [CrossRef]
38. Perić Kačarević, Z.; Kavehei, F.; Houshmand, A.; Franke, J.; Smeets, R.; Rimashevskiy, D.; Wenisch, S.; Schnettler, R.; Jung, O.; Barbeck, M. Purification processes of xenogeneic bone substitutes and their impact on tissue reactions and regeneration. *Int. J. Artif. Organs* **2018**, *41*, 789–800. [CrossRef]
39. Laschke, M.W.; Witt, K.; Pohlemann, T.; Menger, M.D. Injectable nanocrystalline hydroxyapatite paste for bone substitution: In vivo analysis of biocompatibility and vascularization. *J. Biomed. Mater. Res. B Appl. Biomater.* **2007**, *82*, 494–505. [CrossRef] [PubMed]
40. Huber, F.X.; Berger, I.; McArthur, N.; Huber, C.; Kock, H.P.; Hillmeier, J.; Meeder, P.J. Evaluation of a novel nanocrystalline hydroxyapatite paste and a solid hydroxyapatite ceramic for the treatment of critical size bone defects (CSD) in rabbits. *J. Mater. Sci. Mater. Med.* **2008**, *19*, 33–38. [CrossRef]
41. Catros, S.; Sandgren, R.; Pippenger, B.E.; Fricain, J.C.; Herber, V.; El Chaar, E. A Novel Xenograft Bone Substitute Supports Stable Bone Formation in Circumferential Defects around Dental Implants in Minipigs. *Int. J. Oral Maxillofac. Implant.* **2020**, *35*, 1122–1131. [CrossRef] [PubMed]
42. Shakir, M.; Jolly, R.; Khan, A.A.; Ahmed, S.S.; Alam, S.; Rauf, M.A.; Owais, M.; Farooqi, M.A. Resol based chi-tosan/nano-hydroxyapatite nanoensemble for effective bone tissue engineering. *Carbohydr. Polym.* **2018**, *179*, 317–327. [CrossRef] [PubMed]
43. Rossi, F.; Lang, N.P.; De Santis, E.; Morelli, F.; Favero, G.; Botticelli, D. Bone-healing pattern at the surface of titanium implants: An experimental study in the dog. *Clin. Oral Implant. Res.* **2014**, *25*, 124–131. [CrossRef]
44. Botticelli, D.; Berglundh, T.; Persson, L.G.; Lindhe, J. Bone regeneration at implants with turned or rough surfaces in self-contained defects. An experimental study in the dog. *J. Clin. Periodontol.* **2005**, *32*, 448–455. [CrossRef] [PubMed]
45. Akimoto, K.; Becker, W.; Persson, R.; Baker, D.A.; Rohrer, M.D.; O'Neal, R.B. Evaluation of titanium implants placed into simulated extraction sockets: A study in dogs. *Int. J. Oral Maxillofac. Implant.* **1999**, *14*, 351–360.
46. Botticelli, D.; Berglundh, T.; Buser, D.; Lindhe, J. Appositional bone formation in marginal defects at implants. *Clin. Oral Implant. Res.* **2003**, *14*, 1–9. [CrossRef]
47. Rossi, F.; Botticelli, D.; Pantani, F.; Pereira, F.P.; Salata, L.A.; Lang, N.P. Bone healing pattern in surgically created cir-cumferential defects around submerged implants: An experimental study in dog. *Clin. Oral Implant. Res.* **2012**, *23*, 41–48. [CrossRef]
48. Botticelli, D.; Berglundh, T.; Buser, D.; Lindhe, J. The jumping distance revisited: An experimental study in the dog. *Clin. Oral Implant. Res.* **2003**, *14*, 35–42. [CrossRef]
49. Carmagnola, D.; Berglundh, T.; Lindhe, J. The effect of a fibrin glue on the integration of Bio-Oss with bone tissue. An experimental study in Labrador dogs. *J. Clin. Periodontol.* **2002**, *29*, 377–383. [CrossRef]
50. Cardaropoli, G.; Araújo, M.; Lindhe, J. Dynamics of bone tissue formation in tooth extraction sites. An experimental study in dogs. *J. Clin. Periodontol.* **2003**, *30*, 809–818. [CrossRef] [PubMed]
51. Araújo, M.G.; Lindhe, J. Dimensional ridge alterations following tooth extraction. An experimental study in the dog. *J. Clin. Periodontol.* **2005**, *32*, 212–218. [CrossRef] [PubMed]
52. Scala, A.; Lang, N.P.; Schweikert, M.T.; de Oliveira, J.A.; Rangel-Garcia, I., Jr.; Botticelli, D. Sequential healing of open extraction sockets. An experimental study in monkeys. *Clin. Oral Implant. Res.* **2014**, *25*, 288–295. [CrossRef] [PubMed]
53. Masuda, K.; Silva, E.R.; Apaza Alccayhuaman, K.A.; Botticelli, D.; Xavier, S.P. Histologic and Micro-CT Analyses at Implants Placed Immediately after Maxillary Sinus Elevation Using Large or Small Xenograft Granules: An Experimental Study in Rabbits. *Int. J. Oral Maxillofac. Implant.* **2020**, *35*, 739–748. [CrossRef]
54. De Santis, E.; Lang, N.P.; Ferreira, S.; Rangel Garcia, I., Jr.; Caneva, M.; Botticelli, D. Healing at implants installed con-currently to maxillary sinus floor elevation with Bio-Oss® or autologous bone grafts. A histo-morphometric study in rabbits. *Clin. Oral Implant. Res.* **2017**, *28*, 503–511. [CrossRef]
55. Tanaka, K.; Botticelli, D.; Canullo, L.; Baba, S.; Xavier, S.P. New bone ingrowth into β-TCP/HA graft activated with argon plasma: A histomorphometric study on sinus lifting in rabbits. *Int. J. Implant. Dent.* **2020**, *6*, 36. [CrossRef] [PubMed]
56. Perini, A.; Ferrante, G.; Sivolella, S.; Velez, J.U.; Bengazi, F.; Botticelli, D. Bone plate repositioned over the antrostomy after sinus floor elevation: An experimental study in sheep. *Int. J. Implant. Dent.* **2020**, *6*, 11. [CrossRef]
57. IUPAC. *Compendium of Chemical Terminology*, 2nd ed.; Blackwell Scientific Publications: Oxford, UK, 1997. [CrossRef]

58. Tarnow, D.P.; Wallace, S.S.; Froum, S.J.; Rohrer, M.D.; Cho, S.C. Histologic and clinical comparison of bilateral sinus floor elevations with and without barrier membrane placement in 12 patients: Part 3 of an ongoing prospective study. *Int. J. Periodontics Restor. Dent.* **2000**, *20*, 117–125.
59. Wallace, S.S.; Froum, S.J.; Cho, S.C.; Elian, N.; Monteiro, D.; Kim, B.S.; Tarnow, D.P. Sinus augmentation utilizing anor-ganic bovine bone (Bio-Oss) with absorbable and nonabsorbable membranes placed over the lateral window: Histo-morphometric and clinical analyses. *Int. J. Periodontics Restor. Dent.* **2005**, *25*, 551–559.
60. Choi, K.S.; Kan, J.Y.; Boyne, P.J.; Goodacre, C.J.; Lozada, J.L.; Rungcharassaeng, K. The effects of resorbable membrane on human maxillary sinus graft: A pilot study. *Int. J. Oral Maxillofac. Implant.* **2009**, *24*, 73–80.
61. Barone, A.; Ricci, M.; Covani, U.; Nannmark, U.; Azarmehr, I.; Calvo-Guirado, J.L. Maxillary sinus augmentation using prehydrated corticocancellous porcine bone: Hystomorphometric evaluation after 6 months. *Clin. Implant. Dent. Relat. Res.* **2013**, *14*, 373–379. [CrossRef]
62. Tanaka, K.; Iezzi, G.; Piattelli, A.; Ferri, M.; Mesa, N.F.; Apaza Alccayhuaman, K.A.; Botticelli, D. Sinus Floor Elevation and Antrostomy Healing: A Histomorphometric Clinical Study in Humans. *Implant. Dent.* **2019**, *28*, 537–542. [CrossRef]
63. Perini, A.; Viña-Almunia, J.; Carda, C.; Martín de Llano, J.J.; Botticelli, D.; Peñarrocha-Diago, M. Influence of the Use of a Collagen Membrane Placed on the Bone Window after Sinus Floor Augmentation-An Experimental Study in Rabbits. *Dent. J.* **2021**, *9*, 131. [CrossRef]
64. Zahid, M.Z.; Rahman, S.A.; Alam, M.K.; Pohchi, A.; Jinno, M.; Sugita, Y.; Maeda, H. Prospective 3D Assessment of CORAGRAF and Bio-Oss as Bone Substitutes in Maxillary Sinus Augmentation for Implant Placement. *J. Hard Tissue Biol.* **2015**, *24*, 43–48. [CrossRef]
65. Yousaf Athar, Siti Lailatul Akmar Zainuddin, Zurairah Berahim, Akram Hassan, Aamina Sagheer, Mohammad Khursheed Alam Bovine Pericardium: A Highly Versatile Graft Material. *Int. Med. J.* **2014**, *21*, 321–324.
66. Al-Zoubi, I.A.; Patil, S.R.; Kato, I.; Sugita, Y.; Maeda, H.; Alam, M.K. 3D CBCT Assessment of Incidental Maxillary Sinus Abnormalities in a Saudi Arabian Population. *J. Hard Tissue Biol.* **2017**, *26*, 369–372. [CrossRef]
67. Athar, Y.; Zainuddin, S.L.A.; Berahim, Z.; Hassan, A.; Sagheer, A.; Alam, M.K. Bovine Pericardium Membrane and Periodontal Guided Tissue Regeneration: A SEM Study. *Int. Med. J.* **2014**, *21*, 325–327.

Article

The Implant Proteome—The Right Surgical Glue to Fix Titanium Implants In Situ

Marcus Jäger [1,2,*], Agnieszka Latosinska [3], Monika Herten [4], André Busch [1], Thomas Grupp [5,6] and Andrea Sowislok [2]

1. Department of Orthopedics, Trauma & Reconstructive Surgery, St. Marien Hospital Mülheim an der Ruhr, D-45468 Mülheim, Germany; a.busch@kk-essen.de
2. Orthopedics and Trauma Surgery, University of Duisburg-Essen, Hufelandstrasse 55, D-45147 Essen, Germany; andrea.sowislok@uni-due.de
3. Mosaiques Diagnostics GmbH, Rotenburger Straße 20, D-30659 Hannover, Germany; latosinska@mosaiques-diagnostics.com
4. Department of Trauma, Hand and Reconstructive Surgery, University Hospital Essen, University of Duisburg-Essen, Hufelandstrasse 55, D-45147 Essen, Germany; monika.herten@uk-essen.de
5. Aesculap AG, Research & Development, D-78532 Tuttlingen, Germany; thomas.grupp@aesculap.de
6. Department of Orthopedic and Trauma Surgery, Musculoskeletal University Center Munich, Ludwig Maximilians University Munich, Marchioninistraße 15, D-81377 Munich, Germany
* Correspondence: m.jaeger@contilia.de

Abstract: Titanium implants are frequently applied to the bone in orthopedic and trauma surgery. Although these biomaterials are characterized by excellent implant survivorship and clinical outcomes, there are almost no data available on the initial protein layer binding to the implant surface in situ. This study aims to investigate the composition of the initial protein layer on endoprosthetic surfaces as a key initiating step in osseointegration. In patients qualified for total hip arthroplasty, the implants are inserted into the femoral canal, fixed and subsequently explanted after 2 and 5 min. The proteins adsorbed to the surface (the implant proteome) are analyzed by liquid chromatography–tandem mass spectrometry (LC-MS/MS). A statistical analysis of the proteins' alteration with longer incubation times reveals a slight change in their abundance according to the Vroman effect. The pathways involved in the extracellular matrix organization of bone, sterile inflammation and the beginning of an immunogenic response governed by neutrophils are significantly enriched based on the analysis of the implant proteome. Those are generally not changed with longer incubation times. In summary, proteins relevant for osseointegration are already adsorbed within 2 min in situ. A deeper understanding of the in situ protein–implant interactions in patients may contribute to optimizing implant surfaces in orthopedic and trauma surgery.

Keywords: total hip arthroplasty; protein adsorption; proteomics; osseointegration; titanium; bone regeneration; biocompatibility; host-implant response

Citation: Jäger, M.; Latosinska, A.; Herten, M.; Busch, A.; Grupp, T.; Sowislok, A. The Implant Proteome—The Right Surgical Glue to Fix Titanium Implants In Situ. *J. Funct. Biomater.* **2022**, *13*, 44. https://doi.org/10.3390/jfb13020044

Academic Editors: Kunyu Zhang, Qian Feng, Yongsheng Yu and Boguang Yang

Received: 14 March 2022
Accepted: 14 April 2022
Published: 15 April 2022

Publisher's Note: MDPI stays neutral with regard to jurisdictional claims in published maps and institutional affiliations.

Copyright: © 2022 by the authors. Licensee MDPI, Basel, Switzerland. This article is an open access article distributed under the terms and conditions of the Creative Commons Attribution (CC BY) license (https://creativecommons.org/licenses/by/4.0/).

1. Introduction

Cementless total hip replacement of titanium (Ti) alloys is a worldwide standardized surgical procedure that improves mobility and quality of life, predominantly in the elderly population. In addition, plates, nails, screws and wires made of Ti belong to the daily standard repertoire of orthopedic and trauma surgeons. Compared to other biomaterials, Ti alloys have shown excellent clinical results in most patients [1].

In contrast to an ample number of in vitro and in vivo studies investigating Ti osseointegration, along with numerous clinical studies, the local humoral and cellular interactions governing processes in the human *situs* remain unclear. In particular, initial protein adsorption is currently the subject of research. In this context, the in situ implant proteome was described in a pilot study for the first time in 2019 [2]. As a major result, not only extracellular but also intracellular proteins were detected on the implant's surface (Ti-6Al-4V, cpTi). Predominantly, high amounts of cell-free hemoglobin were documented. Moreover, the implant proteome did not reflect a substantial contribution from the plasma proteome, as hypothesized by other investigators [2,3]. Other studies have demonstrated that not only osteoblastic precursors but also immunocompetent cells, such as macrophages and granulocytes, migrate early to the implantation site [4,5]. However, it is unclear if proteins adhered to the implant will have chemotactic effects on immune cells or if proteins from the immune system will adhere to the implant's surface itself. According to the Vroman effect, the most mobile proteins generally adsorb first to the biomaterial, and those are subsequently replaced with less mobile proteins with a higher surface affinity [6,7]. The adsorption of proteins to the implant's surface in situ has not been fully explored. Thus, it is still unclear which proteins adhere to the implant's surface and how their abundance changes over time and on different surface structures. Therefore, in this study, we aimed to characterize the protein adsorption onto the Ti implant's surface towards answering the following two emerging questions: (1) is there a difference in the protein adsorption on the implant's surface over time (from 2 to 5 min), and (2) if adsorption varies between differently structured Ti surfaces (rough and smooth). Our data provide a deeper insight into the understanding of local bone/bone-marrow-implant interactions in the human situs, allow a better description of the osseointegration process at the molecular level, and may lead to further improvement and design of orthopedic implants in the future.

2. Materials and Methods

2.1. Implants

A commercial Ti-6Al-4V femoral stem (BiCONTACT™ stem, Braun Aesculap, Tuttlingen, Germany) was used. This meta- and diaphyseal press-fit implant has been successfully applied for more than 20 years and has shown excellent clinical long-term results [8,9]. The implant contains two different surface structures [10]. Briefly, the proximal part is rough as it is covered with a 0.35 mm thick plasma-sprayed layer of commercial pure Ti (cpTi) (Plasmapore™, porosity 35%, pore size 50–200 μm), while the distal part of the implant is smooth as it was treated with glass bead blasting.

2.2. Probands, Surgical Technique and Implant Retrieval

Probands: Twelve adult patients in advanced stages of osteoarthritis scheduled for total hip replacement were included. All patients had given their informed consent. Exclusion criteria were septic conditions, active neoplasm or other consuming diseases (autoimmune diseases) or coagulopathy. The study has been approved by the Ethical Commission of the University Duisburg-Essen (AZ 17-7844-BO).

Surgical technique and implant retrieval: An antero-lateral approach to the right hip joint was performed (Harding–Bauer approach) [11,12] by one surgeon (M.J.). Following femoral neck resection, the femoral canal was prepared by two different rasps following the manufacturer's manual (A-/B-Osteoprofiler). Here, instead of removing the spongy bone, local bone was preserved and compressed by controlled mallet stokes, allowing a maximum of stability. Afterwards, an original BiCONTACT™ stem was implanted and controlled by fluoroscopy image intensifier. After an in situ time of either 2 or 5 min the stems were explanted via no touch technique. The stems were washed twice for 1 min with saline, packed in a sterile plastic bag and quick-frozen in liquid nitrogen for transport and subsequent storage at $-80\ °C$. In total, 13 implants were analyzed.

Removal and Collection of the Adsorbed Protein Layer from Implants: Protein removal from both implant surfaces (rough and smooth) followed a standard protocol using the following three different elution buffer solutions at room temperature: (i) buffer A (4% SDS, 1 M DTT, 0.1 M Tris-HCl pH 7.6), (ii) buffer B (4% SDS, 1 M DTT, 1 M NaCl, 0.1 M Tris-HCl pH 7.6) and (iii) buffer C (4% SDS, 0.1 M DTT, 1 M NaCl, 0.1 M Tris-HCl pH 7.6), as described in detail in [2].

2.3. Protein Quantification

Before protein quantification, the eluted protein samples of the hip implant were concentrated by TCA precipitation, as described previously [2]. For eluates of the smooth surface, the protein content was concentrated three-fold, while the eluates of the rough surface remained unconcentrated. Protein quantification was performed using the Lowry method [13] for eluates of the rough surface as described before [2]. For eluates of the smooth surface, a modified micro BCA assay [14] (Thermo Fischer Scientific, Rockford, IL, USA) was used. In brief, 50 µL of the protein sample was mixed with 250 µL of the BCA reagent (4.8 mL solution B + 200 µL solution C + 5 mL solution A) and incubated at 60 °C for 1 h. Afterward, the samples were measured at 570 nm in a plate reader (Multiskan Ascent, Thermo-Fischer, Rockford, IL, USA). The protein concentrations were measured in triplicates at three different dilutions (1:2, 1:3, 1:4).

2.4. Proteome Analysis

In total, 1 mL aliquots of the lyophilized samples were re-suspended in 10 mM dithioerythritol (DTE) solution. After initial SDS-PAGE analysis, samples from the smooth surface were further concentrated using Amicon filters (10 kDa cut-off) so that a similar number of proteins could be loaded onto the second SDS-PAGE. Here, 20 µL of each sample was loaded onto the gel and prepared according to the GeLC-MS method, as described in [15]. LC-MS/MS analysis was performed as described previously [2,16]. Briefly, liquid chromatography was performed with 5 µL protein digest loaded onto a 0.1 × 20 mm 5 µm C18 nano trap column (Dionex Ultimate 3000 RSLS nanoflow) and subsequently separated on an Acclaim PepMap C18 nano column 75 µm × 50 cm (Dionex, Sunnyvale, CA, USA) using an 8 h gradient. The LC was connected to an Orbitrap Velos FTMS (Thermo Finnigan, San Jose, CA, USA) via a Proxeon nanospray.

Evaluation of proteomics data: Protein identification was performed with Proteome Discoverer 1.4 (Thermo Scientific) using the SEQUEST search engine, as described previously [2]. In brief, a protein search was performed against the Swiss-Prot human protein database [17]. The following search parameters were applied: (i) precursor mass tolerance: 10 ppm and fragment mass tolerance: 0.05 Da; (ii) full tryptic digestion; (iii) maximum missed cleavage sites: 2; (iv) static modifications: carbamidomethylation of cysteine; (v) dynamic modifications: oxidation of methionine and proline. After completing the analysis of individual RAW MS files, proteomics data were exported at the peptide level using the following filters: (i) peptide rank up to 5; (ii) mass deviation (ΔM) ± 5 ppm; (iii) peptide grouping was enabled, and protein grouping was disabled. Data were further evaluated using the clustering approach, together with previously acquired RAW data [2], as described before with some modifications [2]. Briefly, after the calibration of the retention time against the one sample selected as reference, peptides from different proteomics runs were grouped ("clustered") based on the predefined window of mass (±5 ppm) and retention time (±5% of the peptides' retention time). Each cluster contains a group of peptides for which the sequence (belonging to the cluster) with the highest sum of Xcorr values was reported as reference. When the same Xcorr sum was reported, the frequency of sequence was considered as a second criterion. The sequences were selected among those with Xcorr above 1.9 and without proline hydroxylation on the non-collagen proteins. Quantification of proteomics data was based on the peak area that uses the precursor ions for estimation of relative abundance. Individual peptide peak areas were normalized using the following part per million (ppm) normalization: normalized peak area = (peptide peak area/total peak area in a sample) × 1,000,000. Protein abundance in each sample was calculated as the sum of all normalized peptide areas for a given protein, as described previously [2].

Statistical analysis: Statistical analysis of the proteomics data was performed using the non-parametric Wilcoxon test, which was shown to be more appropriate for the proteomics data [18]. In addition, data normality was assessed using Kolmogorov–Smirnov test, supporting that in most cases data are not normally distributed. Paired statistics were performed only when comparing proteins between the rough and smooth surfaces at different exposure times. The following comparisons were made: (1) 2 min and 5 min exposure at the rough surface, (2) 2 min and 5 min exposure at the smooth surface, (3) smooth and rough surface at 2 min exposure and (4) smooth and rough surface at 5 min exposure. Differentially abundant proteins were defined separately in each comparison based on the following criteria: proteins identified in at least 60% of samples in at least one group, for which nominal significant difference in protein abundance between conditions was observed (unadjusted $p < 0.05$). Linear regression was applied to assess the relationship between protein abundance across different conditions.

2.5. Annotation of Proteomics Data and Bioinformatics Analysis

Identified proteins were extensively annotated using different resources and previously published data. Specifically, for information on the plasma proteome, data were retrieved from Plasma Proteome Database [19] and literature. The latter included a list of 945 proteins (≥ 10 peptide spectrum matches and ≥ 2 peptides) described by Farrah et al. [20], 1175 proteins described by Anderson et al. [21] and 713 protein groups (≥ 2 peptides) identified by the report from Geyer et al. [22]. The implant proteome was also compared with the red blood cell proteome comprised of 2309 protein groups (≥ 2 peptides) [23]. The Human Protein Atlas [24] served as a source for information on the subcellular locations (including also secretome locations), while information about human extracellular matrix (ECM) proteins was retrieved from MatrisomeDB [25,26]. Information regarding protein length, mass and function was retrieved from the Uniprot database [17].

Pathway enrichment analysis was conducted using the online tool ShinyGO v0.66 [27,28] with a curated Reactome database used as an ontology source. Significantly enriched pathways were defined based on the p-value cut-off (false discovery rate (FDR)) below 0.05.

3. Results

3.1. Patients Characteristic

Implants from 12 patients, mainly females, with advanced osteoarthritis, with an average age of 73.5 were investigated. An overview of the demographic and clinical data of the study cohort is displayed in Table 1. Among these 12 patients, 1 patient provided 2 implants, and each of the other 11 patients provided 1 implant. Thus, 13 implants in total were included in the study and led to the generation of 26 implant eluates (13 from rough and 13 from smooth implant surfaces), for which proteomics analysis was conducted. Twenty-six proteomics datasets covered 20 newly acquired and 6 previously generated datasets [2]. A graphical representation of the study design is presented in Figure 1.

Table 1. Baseline demographic and clinical data. Continuous data are shown as mean and standard deviation and categorical data as number and percentage (%).

Variable		Value
Number of patients		12
Age (years)		73.5 ± 7.34
Male		3 (25%)
BMI		27.6 ± 3.01
Surgery site	left	5 (42%)
	right	7 (58%)
Surgery time (min)		88.5 ± 7.56
Hospital time (days)		11.4 ± 3.18
Comorbidities		
Hypertension		6 (50%)
ACVB (aorto-coronary-vein-bypass)		2 (16.6%)
Hypothyreosis		2 (16.6%)
Rheumatoid arthritis		2 (16.6%)
Adipositas		1 (8.3%)
Diabetes mellitus		1 (8.3%)

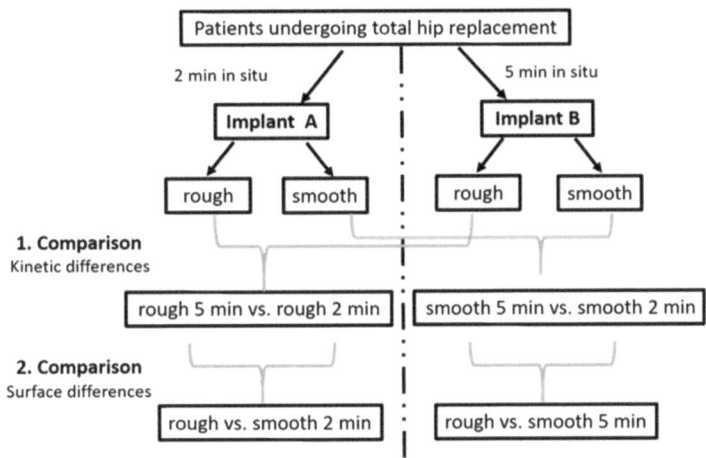

Figure 1. Visualization of our study design with clarification of pairwise comparisons.

The time of surgery was comparable in all patients. The pre- and post-operative serum and blood parameters were uneventful and comparable (Table 2).

Table 2. Patients' pre- and post-operative serum parameters (*n* = 12). These parameters were assessed one day before surgery (−1), on the day of surgery (0) and two days after surgery (2). Abbreviation: pTT–partial thromboplastin time.

Variable	Time Point	Mean	Standard Deviation	Normal Values
Sodium (mmol/L)	−1	139.8	4.2	136–145
	0	138.9	3.7	
	2	138.4	3.7	
Potassium (mmol/L)	−1	4.6	0.3	3.5–5.1
	0	4.2	0.3	
	2	4.0	0.3	
Leucocytes (1/nL)	−1	7.2	1.7	4.0–11.0
	0	11.8	2.8	
	2	9.6	2.9	
Hemoglobin (g/dL)	−1	14.0	1.2	11.6–16.3
	0	11.8	1.9	
	2	10.2	1.5	
Thrombocytes (1/nL)	−1	255.8	63.3	140–320
	0	213.7	38.2	
	2	195.2	40.5	
Orotein (g/L)	−1	67.1	5.9	64–83
	0	58.3	8.1	
	2	54.0	3.4	
Quick (%)	−1	101.4	23.9	70–130
	0	91.0	12.9	
	2	96.5	16.8	
pTT (s)	−1	26.6	3.8	24–32.2
	0	26.1	2.4	
	−1	28.2	1.9	
Fibrinogen (mg/dL)	−1	324.3	66.2	180–350
	0	312.8	63.8	
	2	584.6	196.4	

3.2. Characterization of the In Situ Implant Proteome

Proteomic profiles of 26 implant eluates were analyzed, including 14 (7 per surface) obtained after 2 min and 12 eluates (6 per surface) obtained after 5 min in situ. To investigate the proteomic changes between different exposure times and surfaces, the proteomic profiles were separated into the following four groups (conditions): (1) rough surface at 2 min exposure ($n = 7$), (2) smooth surface at 2 min exposure ($n = 7$), (3) rough surface at 5 min exposure ($n = 6$) and (4) smooth surface at 5 min exposure ($n = 6$). An overview of the analyzed samples is provided in Table 3.

Table 3. Mean elution volumes, protein concentrations and total protein content with standard deviations of implant eluates from different surfaces and different time points.

Surface/Condition	Sample Volume (ml)	Protein Concentration (µg/mL)	Total Protein (µg)
Rough/2 min	3.5 ± 0.8	66.0 ± 30.0	231.2 ± 133.0
Rough/5 min	3.2 ± 0.3	69.2 ± 25.0	221.3 ± 93.5
Smooth/2 min	3.4 ± 0.5	13.3 ± 8.6	40.9 ± 18.6
Smooth/5 min	2.9 ± 0.6	5.1 ± 2.4	15.3 ± 8.6

Only proteins identified based on at least two peptides in the entire dataset of 26 eluates were considered for subsequent investigation. In total, 1367 proteins after 2 min and 1687 proteins after 5 min of exposure in situ were identified on average per eluate from the rough surface. The average number of proteins identified per eluate from the smooth surface was 847 after 2 min and 1476 after 5 min in situ (Figure 2A). In general, more proteins were identified after a 5-min exposure as well as on the rough surface, which is consistent with the protein concentrations of the sample (Table 3). There was a strong relationship between the averages of the relative abundance of each of the identified proteins after 2 and 5 min in situ (Figure 2B, $p < 0.0001$) and between the smooth and rough surfaces (Figure 2C, $p < 0.0001$), indicating a good consistency of the results.

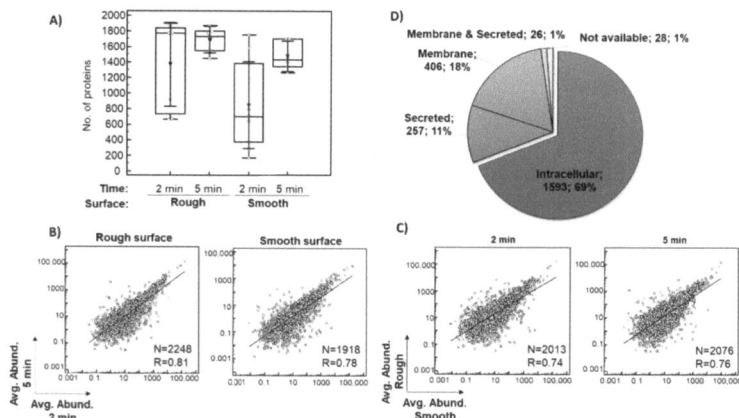

Figure 2. Overview of the implant proteomics data. The box plot shows the distribution of the number of proteins identified (≥ 2 peptides) per sample across the analyzed conditions, with an average value indicated with triangle (**A**). Scatter diagrams with regression lines showing the relationship between averages of the relative abundance of each of the identified proteins for eluates collected after 2 and 5 min exposure in situ (**B**), as well as eluates from smooth and rough surfaces are presented (**C**). A pie chart shows the subcellular localization of the identified proteins based on the data provided in Human Protein Atlas (**D**). Abbreviations: Avg. Abund.–ppm normalized protein abundance, R–coefficient of correlations, N–number of proteins included in the analysis (common proteins between the investigated conditions).

The proteomic analysis resulted in the identification of a total of 2310 proteins based on at least two peptides (Supplementary Table S1). Among these 2310 proteins (Figure 2D), 69% (1593) were of intra-cellular origin, 18% (406) were assigned to the membrane fraction, while 11% (257) were classified as secreted, with around 50% being secreted into the blood. Moreover, 8% (183) of the implant proteomes belonged to the group of ECM proteins (core ECM or ECM-associated proteins). Depending on the plasma reference set investigated, between 4% and 18% of implant proteins were annotated as plasma proteins, while 25% of all proteins belonged to the red blood cell proteome (Supplementary Table S1). A similar distribution of proteins was observed for each analyzed condition. The 10 most abundant proteins within this classification are listed in Supplementary File A and are among the 200 proteins with the highest peptide counts. Table 4 lists the top 10 most abundant proteins in the pooled analysis of 26 samples. Highly abundant intracellular/membrane proteins, e.g., include different hemoglobin subunits, actin cytoplasmic 1, myosin-7, carbonic anhydrases, spectrin beta chain and peroxiredoxin 1. Proteins secreted into the blood included serum albumin, alpha-1-antichymotrypsin, fibrinogen chains, alpha-1-antitrypsin, serotransferrin, alpha-2-macroglobulin, apolipoproteins and complement C3, while ECM proteins were represented by different types of collagen.

Table 4. Shortlist of the 10 most abundant proteins eluted from the implant surfaces based on the compiled dataset (n = 26). These proteins were selected among the top 200 proteins identified with the highest number of peptides (based on all datasets). For each localization (intracellular/membrane, secreted to blood, ECM), proteins were ranked based on the average abundance in the respective group. Abbreviations: Avg. Abund.–ppm normalized protein abundance, ECM–extracellular matrix, #—number.

Localization	Name	# Peptides	Avg. Abund.
Intracellular/Membrane	Hemoglobin subunit beta	25	139,291.93
	Hemoglobin subunit alpha	25	42,991.18
	Protein AHNAK2	10	5187.38
	Keratin, type II cytoskeletal 1	21	4755.28
	Hemoglobin subunit delta	12	4421.55
	Actin, cytoplasmic 1	16	3895.84
	Carbonic anhydrase 1	19	3776.61
	Spectrin beta chain, erythrocytic	40	2911.55
	Myosin-7	99	2564.4
	Peroxiredoxin-2	16	2186.3
Secreted to blood	Serum albumin	99	70,046.62
	Alpha-1-antichymotrypsin	12	7424.58
	Fibrinogen beta chain	28	2523.37
	Alpha-1-antitrypsin	27	2361.18
	Fibrinogen gamma chain	26	2286.78
	Serotransferrin	49	2063.98
	Alpha-2-macroglobulin	56	1660.91
	Apolipoprotein B-100	98	1433.53
	Fibrinogen alpha chain	29	1398.07
	Complement C3	85	1241.86
ECM	Collagen alpha-1(II) chain	67	6790.18
	Collagen alpha-1(XXIV) chain	32	5961.83
	Collagen alpha-2(I) chain	63	3161.86
	Collagen alpha-1(XXII) chain	44	3020.04
	Collagen alpha-6(IV) chain	24	2429.28
	Collagen alpha-1(VII) chain	56	1996.28
	Collagen alpha-2(XI) chain	52	1264.12
	Collagen alpha-1(III) chain	73	1181.17
	Collagen alpha-1(V) chain	34	1154.63
	Collagen alpha-3(VI) chain	72	869.6

The pathway enrichment analysis of the above-mentioned top 200 proteins was performed to place proteomic findings in their biological context and indicate pathways represented by the proteins that adsorb to the implant surface. The 25 most significantly enriched pathways were related to ECM organization (e.g., pathways involved in collagen formation, ECM proteoglycans, integrin cell surface and non-integrin membrane-ECM interactions), and hemostasis (platelet activation, signaling and aggregation). In addition, the pathways related to signal transduction (signaling by receptor tyrosine kinases), metabolism of proteins and the immune system (neutrophil degranulation) were identified as listed in Table 5.

Table 5. List of the top 25 significantly enriched pathways derived from analysis of the implant proteome. Based on their location in the pathway hierarchy, pathways belonging to the same "arental pathway" were grouped, and those being highest in the hierarchy were highlighted in bold. Proteins associated with these pathways are presented in Supplementary Table S2. Abbreviation: FDR–false discovery rate.

Parental Pathway	Enriched Pathway	Total Number of Proteins in the Pathway	No. of Assigned Proteins	FDR
	Extracellular matrix organization	418	54	7.72×10^{-47}
Extracellular matrix organization	Collagen chain trimerization	44	40	6.68×10^{-78}
	Collagen biosynthesis and modifying enzymes	67	41	3.81×10^{-67}
	Collagen formation	90	43	3.34×10^{-64}
	Assembly of collagen fibrils and other multimeric structures	61	33	3.09×10^{-51}
	Anchoring fibril formation	15	9	2.18×10^{-14}
	Collagen degradation	89	34	2.73×10^{-46}
	Degradation of the extracellular matrix	188	39	1.43×10^{-41}
	Integrin cell surface interactions	109	33	5.15×10^{-41}
	ECM proteoglycans	90	24	5.12×10^{-28}
	Non-integrin membrane-ECM interactions	73	18	2.42×10^{-20}
Hemostasis	**Hemostasis**	738	39	6.04×10^{-19}
	Platelet degranulation	128	25	1.35×10^{-25}
	Response to elevated platelet cytosolic Ca2+	133	25	3.44×10^{-25}
	Platelet activation signaling and aggregation	295	27	7.54×10^{-19}
Developmental Biology	**NCAM signaling for neurite out-growth**	74	20	1.59×10^{-23}
	NCAM1 interactions	43	18	3.84×10^{-25}
Metabolism of proteins	**Regulation of Insulin-like Growth Factor IGF transport and uptake by Insulin-like Growth Factor Binding Proteins IGFBPs**	124	17	1.15×10^{-14}
	Post-translational protein phosphorylation	107	16	1.90×10^{-14}
Signal Transduction	**Signaling by Receptor Tyrosine Kinases**	555	29	7.95×10^{-14}
	Signaling by PDGF	58	20	7.84×10^{-26}
	MET activates PTK2 signaling	30	11	1.50×10^{-14}
	MET promotes cell motility	41	11	6.61×10^{-13}
Immune System	**Immune System**	2610	60	1.97×10^{-12}
	Neutrophil degranulation	495	26	1.97×10^{-12}

3.3. Changes in Protein Adsorption on Titanium Implants

3.3.1. Proteomic Differences between 2 and 5 Min In Situ

Pairwise statistical comparisons revealed 113 significantly different proteins when comparing eluates from the rough implant side after 2 and 5 min in situ, whereas 315 proteins were significantly different on the smooth surface. For most of these proteins, the abundance increased after 5 min (Figure 3, left panel). Among the 200 proteins with the highest peptide number, only a few proteins on the rough surface were significantly changed, whereas the most striking changes were observed on the smooth surface (Table 6, Supplementary Table S1). The proteins with increased abundance after 5 min in situ on the smooth surface included some of the intracellular and ECM proteins with high abundance (Table 4), including hemoglobin subunit beta, collagen alpha-1(XXII) chain and collagen alpha-1(III) chain. Many of these differentially abundant proteins also showed statistical significance when comparing surface properties (Table 6).

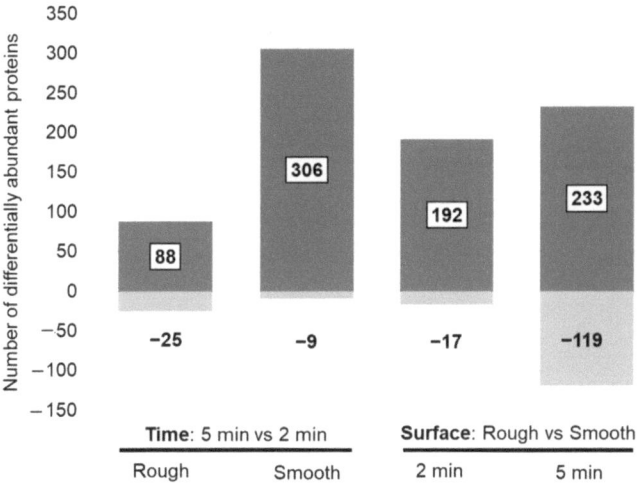

Figure 3. Number of differentially abundant proteins when comparing protein changes between two exposure times on a defined surface (**left panel**) and between surface properties at a defined time point (**right panel**). Discrimination for proteins with an increased (dark grey = up) and decreased (light grey = down) abundance in the "case condition" per comparison is given.

Table 6. Selected proteins with significantly different abundance between 2 and 5 min exposure in situ. Differentially abundant proteins among the top 200 proteins identified with the highest number of peptides (based on all datasets) are presented. For these proteins, the respective results are also provided for the comparison between the different implant surfaces. Significant changes ($p < 0.05$) in the respective comparisons are highlighted in bold. The fold change was calculated by dividing the average abundance of the respective proteins from the case versus the control group. Abbreviations: #—number.

Name	# Peptides	Fold Change			
		Rough Surface: 5 Min vs. 2 Min	Smooth Surface: 5 Min vs. 2 Min	2 Min Exposure: Rough vs. Smooth	5 Min Exposure: Rough vs. Smooth
Intracellular, Membrane					
Probable E3 ubiquitin-protein ligase HECTD4	8	2.56	**8.76**	**4.22**	1.24
Aldehyde dehydrogenase, mitochondrial	10	0.37	**8.33**	**26.44**	1.16
Glutathione S-transferase omega-1	8	0.39	**6.53**	2.18	**0.13**
Glycerol-3-phosphate dehydrogenase [NAD(+)], cytoplasmic	9	0.89	**5.05**	**9.33**	1.65
Heat shock protein HSP 90-beta	10	0.27	**2.57**	**4.49**	**0.48**
Hemoglobin subunit gamma-2	10	1.58	**2.22**	**2.35**	1.67
Ankyrin-1	33	2.08	**2.17**	1.19	1.15
Hemoglobin subunit beta	25	0.93	**0.54**	0.95	**1.65**
L-lactate dehydrogenase B chain	12	0.99	**0.34**	1.36	**3.91**
Myeloperoxidase	19	0.40	**0.29**	**2.03**	**2.83**
Rab GDP dissociation inhibitor beta	8	**0.07**	**0.06**	**4.07**	**4.63**

Table 6. Cont.

Name	# Peptides	Fold Change			
		Rough Surface: 5 Min vs. 2 Min	Smooth Surface: 5 Min vs. 2 Min	2 Min Exposure: Rough vs. Smooth	5 Min Exposure: Rough vs. Smooth
		Secreted to blood			
Complement C5	11	1.36	**5.60**	**9.18**	2.23
Plasminogen	15	1.34	**4.83**	**4.17**	1.15
Coagulation factor XIII A chain	10	1.06	**3.56**	0.76	**0.23**
Histidine-rich glycoprotein	9	0.80	**3.32**	0.67	**0.16**
Antithrombin-III	15	0.94	**2.16**	1.48	**0.65**
Leukocyte elastase inhibitor	10	0.71	**2.42**	**2.52**	0.74
		ECM			
Mucin-19	10	1.86	**9.41**	**11.43**	2.26
Filaggrin	8	1.81	**7.15**	**8.80**	2.22
Collagen alpha-1(III) chain	73	0.45	**6.01**	**6.89**	0.52
Collagen alpha-1(XXI) chain	11	1.91	**3.41**	1.37	0.77
Collagen alpha-1(XVIII) chain	17	**2.40**	**3.30**	1.10	0.80
Collagen alpha-2(IX) chain	15	**3.80**	**3.26**	1.56	1.82
Collagen alpha-1(XXII) chain	44	0.97	**2.87**	**2.92**	0.99
Collagen alpha-3(V) chain	35	1.70	**2.32**	1.88	1.37
Collagen alpha-1(XI) chain	41	1.20	**2.27**	0.99	**0.52**
Collagen alpha-6(VI) chain	20	0.80	**0.69**	1.18	1.37
Collagen alpha-4(IV) chain	45	**1.95**	1.42	0.97	1.33

The pathway enrichment analysis of all significantly different proteins (5 vs. 2 min in situ) revealed seven enriched pathways on the smooth surface. The proteins belonging to these significantly enriched pathways were mainly related to ECM organization, hemostasis and NCAM signaling (Table 7). For the rough surface, no significantly enriched pathways were found.

Table 7. List of the seven significantly enriched pathways derived from the analysis of differentially abundant proteins between 5 and 2 min exposure in situ on the smooth surface. Based on the location of the pathway in the pathway hierarchy, pathways belonging to the same "Parental pathway" were grouped, and those being highest in the hierarchy were highlighted in bold. Proteins associated with these pathways are presented in Supplementary Table S2. Abbreviation: FDR–false discovery rate.

Parental Pathway	Enriched Pathway	Total Number of Proteins in the Pathway	No. of Assigned Proteins	FDR
Extracellular matrix organization	**Collagen formation**	90	9	1.14×10^{-3}
	Collagen chain trimerization	44	8	1.11×10^{-4}
	Collagen biosynthesis and modifying enzymes	67	9	1.37×10^{-4}
	Assembly of collagen fibrils and other multimeric structures	61	6	2.94×10^{-2}
Hemostasis	**Common pathway of fibrin clot formation**	22	4	2.94×10^{-2}
Developmental Biology	**NCAM signaling for neurite out-growth**	74	7	1.47×10^{-2}
	NCAM1 interactions	43	5	3.56×10^{-2}

3.3.2. Proteomic Differences between the Smooth and the Rough Surfaces

Pairwise statistic comparisons (rough vs. smooth) revealed 209 differentially abundant proteins after 2 min and 352 proteins after 5 min in situ. In general, protein abundance was higher on the rough surface (Figure 2). Several of these proteins, belonging to the top 200 proteins with the highest peptide number, were among the top 10 highly abundant intracellular/membrane, secreted and ECM proteins (Table 8). After 2 min in situ, the proteins with increased abundance on the rough surface were the following: hemoglobin subunit delta, myosin-7 (intracellular/membrane), alpha-1-antitrypsin, alpha-2-macroglobulin, complement C3, apolipoprotein A-I (secreted to blood), collagen alpha-1(II) chain, collagen alpha-1(XXII) chain, collagen alpha-6(IV) chain and collagen alpha-1(III) chain (ECM proteins). After 5 min in situ, hemoglobin subunits (beta, alpha, delta), actin, carbonic anhydrase 1 and 2, myosin-7, peroxiredoxin-2, vimentin (intracellular/membrane), alpha-1-antitrypsin, fibrinogen gamma chain, apolipoprotein B-100 and A-I, haptoglobin (secreted to blood) were increased on the rough surface. Furthermore, 16 of these differentially abundant proteins (Table 8) experienced a significant change in abundance after 2 min and 5 min in situ, with most showing increased abundance on the rough surface. These proteins included tropomyosin beta chain, myeloperoxidase, malate dehydrogenase, hemoglobin subunit delta, histone H4 and myosin-7 (belonging to intracellular proteins), complement factor B, alpha-1-antitrypsin, apolipoprotein A-I, angiotensinogen, complement C4-B (proteins secreted into the blood), collagen alpha-1(XV) chain, collagen alpha-1(XIII) chain and decorin (ECM proteins).

Table 8. Selected proteins with significantly changed abundance between rough and smooth surfaces at two different exposure times in situ. Differentially abundant proteins among the top 200 proteins identified with the highest number of peptides (based on all datasets) are presented. For these proteins, the respective results are also given for the comparison between the different exposure times. Significant changes ($p < 0.05$) in the respective comparisons are highlighted in bold. The fold change was calculated by dividing the average abundance of the respective proteins from the case versus the control group. Abbreviations: #—number.

Name	# Peptides	Fold Change			
		2 Min Exposure: Rough versus Smooth	5 Min Exposure: Rough versus Smooth	Rough Surface: 5 Min vs. 2 Min	Smooth Surface: 5 Min vs. 2 Min
		Intracellular, Membrane			
Nuclear receptor corepressor 2	8	3.31	2.40	0.99	1.36
14-3-3 protein epsilon	10	2.99	3.45	1.04	0.90
Endoplasmic reticulum chaperone BiP	10	4.97	1.42	0.34	1.18
Transitional endoplasmic reticulum ATPase	13	3.87	1.82	0.93	1.96
Eosinophil peroxidase	9	8.63	1.85	0.41	1.91
Four and a half LIM domains protein 1	8	5.03	4.08	0.63	0.78
Rab GDP dissociation inhibitor beta	8	4.07	4.63	0.07	0.06
Filamin-A	33	2.97	1.28	1.32	3.05
Protein piccolo	8	20.26	1.51	0.36	4.85
Glycerol-3-phosphate dehydrogenase [NAD(+)], cytoplasmic	9	9.33	1.65	0.89	**5.05**
Alcohol dehydrogenase 1B	8	8.93	6.96	1.37	1.76
Aldehyde dehydrogenase, mitochondrial	10	26.44	1.16	0.37	**8.33**
Protein bassoon	9	2.82	1.44	0.58	1.14
Tropomyosin beta chain	17	**3.20**	**6.02**	0.37	0.20
Myeloperoxidase	19	**2.03**	**2.83**	0.40	**0.29**
Malate dehydrogenase, cytoplasmic	8	**6.28**	**6.93**	0.67	0.60
Hemoglobin subunit delta	12	**2.84**	**2.58**	0.86	0.95
Histone H4	10	**2.43**	**2.52**	1.38	1.33
Keratin, type II cytoskeletal 2 epidermal	25	0.12	0.09	0.80	1.00
Myosin-7	99	**3.40**	**4.10**	0.92	0.76
Tubulin alpha-1B chain	9	0.61	0.38	0.87	1.39
Isocitrate dehydrogenase [NADP], mitochondrial	10	14.95	Only Rough	0.44	Only Rough
ADP/ATP translocase 1	8	5.88	**22.48**	0.72	0.19
Protein 4.1	10	13.95	**15.43**	1.98	1.79
Prelamin-A/C	17	3.98	**13.29**	1.62	0.49
ATP synthase subunit alpha, mitochondrial	12	2.52	**10.00**	0.50	0.13
Alpha-crystallin B chain	8	21.87	7.80	1.33	3.73
Myosin light chain 1/3, skeletal muscle isoform	9	1.92	7.74	2.07	0.51
Myosin light chain 3	9	2.13	6.33	0.52	0.17
Coronin-1A	8	10.30	6.11	0.27	0.46
Triosephosphate isomerase	12	6.40	6.06	0.84	0.89
L-lactate dehydrogenase A chain	9	3.04	4.56	1.51	1.01
L-lactate dehydrogenase B chain	12	1.36	3.91	0.99	**0.34**

Table 8. Cont.

Name	# Peptides	Fold Change			
		2 Min Exposure: Rough versus Smooth	5 Min Exposure: Rough versus Smooth	Rough Surface: 5 Min vs. 2 Min	Smooth Surface: 5 Min vs. 2 Min
Glyceraldehyde-3-phosphate dehydrogenase	12	0.95	**3.79**	0.89	0.22
Carbonic anhydrase 2	16	1.38	**3.75**	0.60	0.22
Transketolase	14	3.14	**3.35**	1.42	1.32
Vinculin	8	3.24	**3.22**	1.27	1.28
Erythrocyte membrane protein band 4.2	10	2.05	**3.00**	0.52	0.36
Pyruvate kinase PKM	10	2.06	**2.85**	1.57	1.13
Alpha-enolase	14	1.05	**2.75**	0.60	0.23
Vimentin	24	0.81	**2.57**	0.21	0.07
Fructose-bisphosphate aldolase A	13	2.17	**2.50**	0.99	0.86
Hemoglobin subunit alpha	25	1.11	**2.46**	0.87	0.40
Phosphoglycerate kinase 1	9	1.97	**2.36**	1.40	1.16
Peroxiredoxin-2	16	1.95	**2.35**	1.03	0.85
Plectin	11	0.05	**2.11**	2.05	0.05
Actin, cytoplasmic 1	16	1.52	**2.05**	0.97	0.72
Carbonic anhydrase 1	19	1.78	**2.02**	0.71	0.62
Catalase	25	1.66	**1.98**	0.98	0.82
Hemoglobin subunit beta	25	0.95	**1.65**	0.93	**0.54**
Alpha-actinin-2	31	1.06	**1.52**	0.98	0.68
Keratin, type I cytoskeletal 9	14	0.09	**0.41**	1.91	0.41
Glutathione S-transferase omega-1	8	2.18	**0.13**	0.39	**6.53**
Secreted to blood					
Fibronectin	34	**3.60**	1.25	0.72	**2.07**
Annexin A2	15	**3.15**	7.27	1.63	0.71
Leukocyte elastase inhibitor	10	**2.52**	0.74	0.71	**2.42**
Plasminogen	15	**4.17**	1.15	1.34	**4.83**
Complement C5	11	**9.18**	2.23	1.36	**5.60**
Afamin	11	**8.08**	1.77	0.78	3.55
Alpha-2-macroglobulin	56	**2.21**	1.95	1.23	1.40
Complement C3	85	**2.59**	2.39	1.19	1.28
Complement factor B	17	**3.61**	4.34	0.68	0.57
Alpha-1-antitrypsin	27	**2.49**	3.18	1.27	1.00
Apolipoprotein A-I	26	**2.54**	1.98	0.99	1.28
Angiotensinogen	9	**2.73**	1.81	1.38	2.08
Complement C4-B	43	**4.36**	4.06	1.02	1.10
Antithrombin-III	15	1.48	0.65	0.94	**2.16**
Inter-alpha-trypsin inhibitor heavy chain H2	14	**4.05**	4.40	1.04	0.96
Plasma kallikrein	12	**4.04**	14.63	2.52	0.70
Neutrophil elastase	8	**3.25**	3.89	0.98	0.82
Haptoglobin	24	1.95	2.33	1.46	1.21
Histidine-rich glycoprotein	9	0.67	**0.16**	0.80	3.32
Inter-alpha-trypsin inhibitor heavy chain H1	9	**2.55**	2.55	0.54	0.54
Hemopexin	17	0.74	1.57	1.13	0.54
Coagulation factor XIII A chain	10	0.76	**0.23**	1.06	**3.56**
Apolipoprotein B-100	98	1.01	1.39	0.65	0.47
Fibrinogen gamma chain	26	0.73	**0.49**	0.86	1.28
ECM					
Collagen alpha-1(II) chain	67	1.76	1.34	1.21	1.59
Collagen alpha-1(III) chain	73	**6.89**	0.52	0.45	**6.01**
Collagen alpha-3(IV) chain	13	**7.07**	0.27	0.45	11.56
Collagen alpha-1(XXVIII) chain	17	**2.75**	1.99	0.81	1.12
Filaggrin	8	**8.80**	2.22	1.81	**7.15**
Collagen alpha-6(IV) chain	24	0.66	0.51	0.90	1.18
Collagen alpha-1(XXII) chain	44	**2.92**	0.99	0.97	**2.87**
Collagen alpha-1(XV) chain	9	**3.71**	2.62	0.72	1.03
Decorin	9	**10.76**	10.38	1.39	1.44
Collagen alpha-1(XIII) chain	27	**2.26**	2.04	0.75	0.83
Collagen alpha-1(XIV) chain	20	1.94	**2.60**	2.00	1.50
Lumican	8	1.47	**13.21**	2.18	0.24
Collagen alpha-1(XI) chain	41	0.99	0.52	1.20	**2.27**

The pathway enrichment analysis of all significantly different proteins (rough vs. smooth) revealed multiple enriched pathways for both time points (Table 9). After a shorter incubation time of 2 min, significantly enriched pathways mainly related to extracellular matrix organization (mainly involving collagens), the immune system (neutrophil degranulation, signaling by interleukins, complement cascade), hemostasis (platelet activation) and muscle contraction. After a prolonged incubation time of 5 min, the most significantly enriched pathways were related to metabolism (carbohydrate metabolism), the immune system (neutrophil degranulation, signaling by interleukins), hemostasis (platelet activation) and signaling (by MAPK family cascade).

Table 9. List of the top 25 significantly enriched pathways derived from the analysis of differentially abundant proteins between smooth and rough surfaces. The analysis covers pathways enriched after 2 min and 5 min exposure in situ. For each comparison, the top 25 most significant pathways (based on FDR values) were shortlisted. The table presents a compilation of the shortlisted pathways, with FDR values for the pathways belonging to the top 25 highlighted in bold. Based on the location of the pathway in the pathway hierarchy, pathways belonging to the same "Parental pathway" were grouped, and those being the highest in the hierarchy were highlighted in bold. Proteins associated with these pathways are presented in Supplementary Table S2. Abbreviation: FDR–false discovery rate.

Parental Pathway	Enriched Pathway	Total Number of Proteins in the Pathway	After 2 Min Exposure		After 5 Min Exposure	
			FDR	No. of Assigned Proteins	FDR	No. of Assigned Proteins
Extracellular matrix organization	**Extracellular matrix organization**	418	1.96×10^{-5}	17	2.20×10^{-3}	18
	Collagen chain trimerization	44	2.40×10^{-6}	8	3.34×10^{-2}	4
	Collagen biosynthesis and modifying enzymes	67	1.87×10^{-5}	8	-	-
	Collagen formation	90	1.87×10^{-5}	9	-	-
	Degradation of the extracellular matrix	188	1.87×10^{-5}	12	3.78×10^{-3}	11
	Non-integrin membrane-ECM interactions	73	2.37×10^{-4}	7	-	-
	Collagen degradation	89	7.86×10^{-4}	7	2.15×10^{-2}	6
	ECM proteoglycans	90	7.86×10^{-4}	7	-	-
	Assembly of collagen fibrils and other multimeric structures	61	7.89×10^{-4}	6	-	-
	Integrin cell surface interactions	109	9.73×10^{-3}	6	-	-
Hemostasis	**Hemostasis**	738	7.10×10^{-3}	17	3.16×10^{-6}	33
	Platelet degranulation	128	1.65×10^{-4}	9	1.16×10^{-6}	14
	Response to elevated platelet cytosolic Ca^{2+}	133	1.99×10^{-4}	9	1.72×10^{-6}	14
	Platelet activation signaling and aggregation	295	7.10×10^{-3}	10	1.50×10^{-4}	17
Metabolism of proteins	Regulation of Insulin-like Growth Factor IGF transport and uptake by Insulin-like Growth Factor Binding Proteins IGFBPs	124	4.27×10^{-3}	7	7.15×10^{-4}	10
Signal Transduction	MET activates PTK2 signaling	30	4.29×10^{-3}	4	-	-
	MAPK family signaling cascades	299	-	-	1.72×10^{-4}	17
	RAF/MAP kinase cascade	248	-	-	1.88×10^{-5}	17
	MAPK1/MAPK3 signaling	254	-	-	2.49×10^{-5}	17
Immune System	**Immune System**	2610	4.27×10^{-3}	41	1.22×10^{-5}	73
	Neutrophil degranulation	495	3.63×10^{-5}	18	9.16×10^{-11}	34
	Gene and protein expression by JAK-STAT signaling after Interleukin-12 stimulation	39	1.05×10^{-3}	5	2.41×10^{-2}	4
	Interleukin-12 signaling	56	4.29×10^{-3}	5	-	-
	Activation of C3 and C5	12	4.33×10^{-3}	3	-	-
	Innate Immune System	1313	2.15×10^{-2}	23	4.18×10^{-8}	52
	Signaling by Interleukins	706	4.41×10^{-2}	14	3.32×10^{-6}	32

Table 9. Cont.

Parental Pathway	Enriched Pathway	Total Number of Proteins in the Pathway	After 2 Min Exposure		After 5 Min Exposure	
			FDR	No. of Assigned Proteins	FDR	No. of Assigned Proteins
	Cytokine Signaling in Immune system	983	-	-	6.30×10^{-5}	36
	FLT3 Signaling	275	-	-	6.37×10^{-5}	17
	Other interleukin signaling	298	-	-	3.16×10^{-6}	20
Muscle contraction	Muscle contraction	216	4.29×10^{-3}	9	3.16×10^{-6}	17
	Smooth Muscle Contraction	42	1.41×10^{-3}	5	7.50×10^{-3}	5
	Striated Muscle Contraction	36	7.10×10^{-3}	4	6.01×10^{-11}	12
Transport of small molecules	Erythrocytes take up oxygen and release carbon dioxide	9	-	-	2.71×10^{-4}	4
Vesicle-mediated transport	Binding and Uptake of Ligands by Scavenger Receptors	112	4.15×10^{-2}	5	3.50×10^{-4}	10
Metabolism	Metabolism	2262	-	-	6.39×10^{-8}	73
	Metabolism of carbohydrates	312	-	-	8.30×10^{-12}	29
	Glucose metabolism	93	-	-	2.81×10^{-9}	15
	Gluconeogenesis	34	4.26×10^{-2}	3	1.16×10^{-8}	10
	Glycolysis	73	-	-	1.33×10^{-7}	12
Programmed Cell Death	Programmed Cell Death	193	-	-	8.29×10^{-5}	14
	Apoptosis	186	-	-	6.23×10^{-5}	14
	Activation of BH3-only proteins	32	5.02×10^{-3}	4	-	-

4. Discussion

The local humoral and cellular mechanisms in situ after the first intraoperative bone contact of an implant until load-stable osseointegration are complex and have not yet been fully elucidated in detail. They follow a strict sequence of protein adsorption, cellular migration, proliferation, biomineralization and subsequent bone remodeling processes [29,30]. In our previously published study, we were able to disprove the generally accepted hypothesis of immediate adsorption of plasma proteins to the implant surface in situ [2]. We reproduced these data with a significant correlation between the average protein abundance ($p < 0.0001$). However, in this study, we were interested in the kinetic changes of the initial protein layer and its function in osseointegration. According to the Vroman's effect, in which smaller proteins with higher concentrations adsorb to the surface first and are later replaced by larger proteins with higher binding affinities [31], we expected differences in the protein composition of eluates from a specific surface with increasing incubation time. Investigating the proteome of the implant eluates at two different time points revealed that both surfaces assimilated their protein portfolio over time, with the protein abundance being higher on the rough implant surface (Figure 2). To better understand the molecular mechanisms underlying osseointegration, proteomic results were placed in the context of biology, discovering pathways belonging to ECM organization, hemostasis, signal transduction, metabolism of proteins and the immune system. Most of these pathways were also detected when analyzing changes between smooth and rough surfaces, while only a few pathways (mainly those involved in ECM organization) were significantly enriched on the smooth surface when comparing the 2 and 5 min exposures in situ (Supplementary Table S2). These findings suggest a higher relevance of surface structure to the osseointegration process rather than that of time. The evaluation of the findings in the biological context is discussed in detail in the following sections.

4.1. Proteins of the Bone ECM and Their Distribution within the Implant Proteome

ECM is a basic component of tissues and organs and provides both structural and non-structural support that leads to osseointegration. This is supported by our findings showing a significant enrichment in the pathways involved in ECM organization, represented by collagenous and non-collagenous proteins. Since collagen accounts for 90% of the organic ECM of bone, collagen biosynthesis and collagen biochemistry may be key determinants of osseointegration [32]. Among bone ECM proteins, collagen type I, typically organized into collagen fibrils consisting of triple helices of polypeptides, is the most abundant. The collagen fibrils interact with other collagenous and non-collagenous proteins to assemble higher-order fibril bundles and fibers. The fiber diameter and fibrillogenesis are regulated by collagen type III and V, which are present in smaller amounts in the bone ECM [33]. Within the implant proteome, various collagen types were found to adsorb to the implant surface, with the abundance significantly affected by either incubation time (Table 6) or surface properties (Table 8). While the abundance of collagen type I was similar on both surfaces after a 2 min exposure in situ and did not change substantially with longer incubation time, the abundance of collagen type III was increased on the rough surface, followed by further enrichment on the smooth surface with longer incubation time (5 min). Collagen type V also accumulated on the smooth surface during prolonged exposure (Table 6). The non-collagenous proteins of bone ECM constitute a large group of diverse proteins that are non-stoichiometric with type I collagen but of great importance for bone physiology. Mutations in a number of these proteins result in bone abnormalities [34]. Within the implant proteome, some of the most common members of these proteins have been identified as having a higher sensitivity to surface properties than to incubation times. In the group of small leucine-rich proteoglycans, which are extracellular secreted proteins involved in all phases of bone formation, including cell proliferation, osteogenesis, mineral deposition, and bone remodeling [33], biglycan, asporin, decorin and lumican were found, with the latter two proteins being among the 200 most frequently identified proteins (Table 8). While biglycan showed strong binding to both surfaces, which was not significantly affected by the exposure time, the abundance of both asporin and lumican was increased on the rough surface compared to the smooth surface after prolonged incubation in situ. Along these lines, the decorin level was increased on the rough surface in comparison to the smooth surface, and its abundance was not affected by incubation time. Glycoproteins such as thrombospondins, fibronectin and vitronectin were also part of the implant proteome, with fibronectin belonging to the top 200 proteins identified with the highest number of peptides (Table 8). Thrombospondins are expressed by osteoblasts and are present at different stages of bone maturation and development [34]. The abundance of thrombospondin 1 and 2 was stable on both implant surfaces and did not increase with time, while thrombospondin 4 showed an increase after 5 min on the rough surface in comparison to the smooth surface. Fibronectin and vitronectin are plasma proteins that interact with other ECM proteins and are crucial for collagen-matrix assembly [35]. In the implant proteome, fibronectin that is produced during the early stages of bone formation and upregulated in osteoblasts [34] is increased after 2 min of incubation on the rough surface, but it does not significantly increase with time (Table 6). Whereas after a 5 min exposure in situ, the abundance of vitronectin, which is found at low levels in a mineralized matrix [34], is significantly reduced on the rough in comparison to the smooth surface. The enzymes involved in the posttranslational modification of collagen, or its enzymatic degradation may be of further importance when considering the organization of bone ECM. The formation of inter- and intra-molecular crosslinks of collagen is regulated by posttranslational hydroxylation and oxidation of specific lysine residues catalyzed by lysyl-hydroxylases and oxidases [32]. Some of these proteins, such as prolyl 4-hydroxylase subunit alpha-, lysyl-oxidase 1 and 2, were among the proteins adsorbing to the implant surface. Their abundance (although low) was stable and not affected by incubation time or surfaces. Another important class of proteins concerning the degradation of extracellular matrix components are the zinc-dependent matrix metalloproteases (MMPs) that are also involved in bone resorption by osteoclasts [36]. In our dataset, MMP21 does not show a significant increase over time, and its abundance was stable independent of the surface. MMP9 is increased on the rough surface, albeit only at the 2 min time point.

4.2. Hemostasis and Inflammation

Hemostasis and inflammation are interconnected physiological processes, with inflammation leading to hemostasis activation that, in turn, influences the inflammation. A pathway enrichment analysis revealed that the top 200 proteins that adsorb to the implant's surface (Table 5) are involved in hemostasis, the complement cascade and neutrophil degranulation. Those pathways were also found to be significantly enriched based on the analysis of differentially abundant proteins between the two implant surfaces (Table 9) but were not significantly different when comparing the two incubation times (Table 7). As defined by Alberktsson et al., osseointegration is a mild inflammatory response leading to an integrated implant with a bone-implant interface that remains in a state of equilibrium, susceptible to changes in the environment [37]. Disturbance in this foreign body equilibrium may lead to peri-implant bone loss through reactivation of inflammation, the formation of foreign body giant cells and the activation of osteoclastogenesis [38].

4.2.1. Proteins Potentially Involved in Hemostasis and Neutrophil Activity

Enhanced hemostasis is a natural reaction to prevent excessive blood loss and maintain blood flow to the rest of the body as an answer to the damage of blood vessels in the periosteum, endosteum and surrounding soft tissues during surgery [39]. Activated platelets are the first cells to respond to wound healing through the processes of adherence, aggregation and degranulation [40]. In vivo, the disruption of the continuity of the endothelial layer and the exposure of the underlying subendothelial matrix leads to the activation of platelets through the interactions between collagen, fibronectin, von Willebrand factor and various glycoprotein receptors on the platelets [41]. Interestingly, most of these proteins were of higher abundance on the rough surface after a 2 min exposure in situ, with collagen and fibronectin being among the top abundant 200 proteins (Table 8). The abundance of von Willebrand factor was reduced on the rough surface in comparison to the smooth surface after 2 and 5 min in situ, and its abundance was further increased on the rough surface with longer incubation (5 min). Besides their role in hemostasis, platelets were also found to be involved in the activation of the immune system through their surface receptors and their granules, which contain a plethora of biologically active products [42]. The physical interactions between neutrophils and platelets are triggered by the expression of p-selectin on activated platelets that can bind to p-selectin glycoprotein ligand-1 on the surface of neutrophils. Additionally, the secretion of CD40 ligand by platelets has been shown to upregulate integrin expression on neutrophils, and the secretion of serotonin and platelet factor 4 leads to neutrophil recruitment and adhesion [41,42]. We could not detect adhesion of p-selectin to the Ti implant but found platelet factor 4 adsorbed to the implant surface in situ, albeit at a low abundance with no significant changes concerning surface properties and exposure times.

As aforementioned, the pathway analysis also suggested the activation of neutrophils and their degranulation. Along these lines, the leucocyte concentration was found to increase after surgery in the blood levels of all patients (Table 2). Neutrophils, together with basophils and eosinophils, belong to polymorphonuclear leukocytes, constitute 40–65% of the white blood cell population and are hallmarks of acute inflammation [43,44]. Immediately following tissue injury, these cells are recruited and exert anti-microbial activity via degranulation, enzymatic release, phagocytosis of foreign substances and debris and the production of large DNA-based fiber networks called neutrophil extracellular traps [5]. Upon activation, neutrophils release toxic mediators including elastase, myeloperoxidase, cathepsins and defensins from their primary granules [45]. These proteins were also present in the implant proteome, with myeloperoxidase and elastase among the most striking differences when comparing surfaces (Table 8). Myeloperoxidase and cathepsin G were found significantly enriched on the rough surface (both after 2 and 5 min). Myeloperoxidase catalyzes the formation of reactive oxygen species, such as hypochlorous acid (HOCl) [46], while cathepsin G stimulates the production of cytokines and chemokines responsible for the activation and mobilization of immune cells to the site of pathogen or tissue damage [47].

4.2.2. Possible Neutrophil Recruitment through DAMPs and the Complement System

The proteomic analysis also indicated another possible mechanism for neutrophil recruitment. Because of surgery-induced tissue damage, necrotic cells release self-derived molecules that are either altered or relocated from their normal cellular compartment [48]. Those damage-associated molecular patterns (DAMPs) are known to activate neutrophils and trigger a sterile inflammatory response [49]. The DAMPs include extracellular ATP, formylated peptides and DNA of mitochondrial origin, nucleic acids, heat shock proteins, S100 proteins, high mobility group protein B1 and altered extracellular matrix components, such as hyaluronan [48,49]. The members of these protein families were also found in the implant proteome. Hyaluronan, high mobility group protein B1, numerous heat shock proteins and S100A8 and S100A11 were identified on the implant surface, with their abundance on the implant surface not being affected by the surface type or time. However, the 10 kDa heat shock protein was found to be significantly increased on the rough surface after 5 min in situ, and its abundance was also accumulated on the smooth surface with a longer incubation time. Heat shock protein HSP 90-beta was only increased with a longer incubation time on the smooth surface. In addition, the protein S100A9, known to induce neutrophil chemotaxis and adhesion [50], was found to be consistently elevated on the rough in comparison to the smooth surface after both 2 min and 5 min of exposure in situ.

Another immunologic recruitment mechanism for neutrophils is the activation of the complement system. Biomaterials, including Ti implants, are known to activate the complement cascade through the classical pathway, with further amplification through the alternative pathway [51]. Several complement factors are also detected in the implant proteome. Among them, complement factor B (CFB), C3, C4 and C5 were found significantly enriched on the rough and smooth surfaces after an exposure time of 2 min in situ (Table 8). In addition, the abundance of CFB and C4 was also increased on the rough surface after 5 min exposure, while C5 showed an increase in abundance on the smooth surface after 5 min in comparison to 2 min. Furthermore, the abundance of complement C1q subcomponent subunit B was also significantly increased with prolonged incubation on the smooth surface, while C8 showed an increase in the abundance on the rough surface after 5 min in situ. Interestingly, the CFB and C4 proteins are involved in the classical (C4) and alternative (CFB) complement activation pathways. However, the question of whether the complement system is activated cannot be answered with confidence based on the proteomics data.

Moreover, bone cells can produce complement factors and are in turn targets of activated complement [51]. Osteoblasts can produce C3 and C5, while osteoclasts can release activated C5a. The anaphylatoxins C3a and C5a are not only chemotactic for neutrophils but also for osteoblasts and their mesenchymal stem cell precursors [52,53]. In osteoblasts, they stimulate the release of inflammatory cytokines, including IL-6 and IL-8. Additionally, C5a can stimulate the secretion of osteoclastogenic factors [51,52]. This suggests that the activation of the complement system might also directly influence bone healing through its interaction with bone cells. Enhancing osteoclastogenesis might increase bone resorption while recruitment of osteoblasts and their precursor cells might favor bone formation [52].

4.3. Limitations of Our Study

The sample size for each group was low and did not allow adjustment for multiple testing. Similarly, an evaluation of differences in the protein adsorption between male and female patients could not be performed. Since females are more likely to have bone-related diseases (e.g., osteoporosis and osteoarthritis), the influence of gender on protein adsorption at the implant surface will be investigated in a follow-up study. Furthermore, we are aware that incorrect sequence assignment may occur on occasion, especially in the case of low abundant peptides where the spectral quality is lower. However, the risk of false protein identification is reduced by excluding proteins identified based on one peptide only. For the assignment of proteins to the plasma proteome, we used the Plasma Proteome Database downloaded in June 2018, for which the download option in the online database has been temporarily unavailable, so newer additions to the database were not considered. Although we examined the implant proteome at two different time points, our study documents only a snapshot of protein adsorption after a short period. However, longer incubation times in situ during implant surgery are not possible for ethical reasons. Since some animal studies indicate that low protein intake might affect the osseointegration of Ti implants [54], the dietary habits of our patients, which were not considered, might have impacted the results of our study.

5. Conclusions

Our study indicates that the proteins adsorbed to orthopedic implants may allow insights into the molecular mechanisms directly taking place in humans after biomaterial insertion. We were able to match the proteins identified in the eluates from the implant's surface to pathways implicated in osseointegration. Despite the short exposure time of 2 to 5 min in situ, we were able to comprehend remodeling of the bone extracellular matrix and the onset of an inflammatory response as key steps of osseointegration. The proteins adsorbed to the implant's surface may show a chemotactic effect on immune cells and other immune cell-derived proteins adhered to the implant's surface. The composition of the adsorbed proteins indicates that bone healing immediately begins at a molecular level after implantation.

Supplementary Materials: The following supporting information can be downloaded at: https://www.mdpi.com/article/10.3390/jfb13020044/s1, Supplementary Table S1: List of proteins identified with at least two peptides, Supplementary Table S2: List of significantly enriched pathways, Supplementary File A: Full version of the tables: Tables in this article have been decreased in complexity. Complete versions of these tables can be found here.

Author Contributions: M.J., M.H.: Conceptualization; A.S., A.L.: Data curation; A.L.: Formal analysis; M.J., T.G.: Funding acquisition; M.H., A.S.: Investigation; M.J., M.H., A.S. and A.B.: Methodology; M.J., A.B.: Project administration; M.J., M.H., A.B., T.G. and A.S.: Resources; A.L.: Software; M.J.: Supervision; M.H., A.B., M.J., A.L., T.G. and A.S.: Validation; A.S., A.L.: Visualization; A.S., A.L., M.H., M.J.: Writing—original draft; A.S., A.L., M.H., M.J., T.G. and A.B.: Writing—review and A.S.: editing. All authors have read and agreed to the published version of the manuscript.

Funding: This work was supported by B. Braun Aesculap AG, Germany.

Institutional Review Board Statement: The study was conducted in accordance with the Declaration of Helsinki and approved by the Ethics Committee of the University of Duisburg-Essen. (AZ 17-7844-BO).

Informed Consent Statement: Informed consent was obtained from all subjects involved in the study.

Data Availability Statement: The processed data required to reproduce these findings are available to download from the supporting information.

Acknowledgments: We thank Heike Rekasi of the Department of Orthopedics and Trauma Surgery, University Hospital Essen for her technical input. Finally, we also thank B. Braun Aesculap AG for their support and funding. We acknowledge support by the Open Access Publication Fund of the University of Duisburg-Essen, Germany.

Conflicts of Interest: A.L. is employed by Mosaiques Diagnostics GmbH. All other authors declare that they have no competing financial interests.

References

1. van Oldenrijk, J.; Scholtes, V.A.B.; van Beers, L.W.A.H.; Geerdink, C.H.; Niers, B.B.A.M.; Runne, W.; Bhandari, M.; Poolman, R.W. Better early functional outcome after short stem total hip arthroplasty? A prospective blinded randomised controlled multicentre trial comparing the Collum Femoris Preserving stem with a Zweymuller straight cementless stem total hip replacement for the treatment of primary osteoarthritis of the hip. *BMJ Open* **2017**, *7*, e014522. [CrossRef] [PubMed]
2. Jaeger, M.; Jennissen, H.P.; Haversath, M.; Busch, A.; Grupp, T.; Sowislok, A.; Herten, M. Intrasurgical Protein Layer on Titanium Arthroplasty Explants: From the Big Twelve to the Implant Proteome. *Proteom. Clin. Appl.* **2019**, *13*, e1800168. [CrossRef] [PubMed]
3. Sowislok, A.; Weischer, T.; Jennissen, H.P. The In Situ Human Dental Implantome: A First Appraisal. *Curr. Dir.Biomed. Eng.* **2021**, *7*, 827–830. [CrossRef]
4. Busch, A.; Jäger, M.; Mayer, C.; Sowislok, A. Functionalization of Synthetic Bone Substitutes. *Int. J. Mol. Sci.* **2021**, *22*, 4412. [CrossRef] [PubMed]
5. Abaricia, J.O.; Shah, A.H.; Musselman, R.M.; Olivares-Navarrete, R. Hydrophilic titanium surfaces reduce neutrophil inflammatory response and NETosis. *Biomater. Sci.* **2020**, *8*, 2289–2299. [CrossRef]
6. Vroman, L.; Adams, A.L. Findings with the recording ellipsometer suggesting rapid exchange of specific plasma proteins at liquid/solid interfaces. *Surf. Sci.* **1969**, *16*, 438–446. [CrossRef]
7. Vroman, L.; Adams, A.L. Identification of rapid changes at plasma-solid interfaces. *J. Biomed. Mater. Res.* **1969**, *3*, 43–67. [CrossRef]
8. Swamy, G.; Pace, A.; Quah, C.; Howard, P. The Bicontact cementless primary total hip arthroplasty: Long-term results. *Int. Orthop.* **2012**, *36*, 915–920. [CrossRef]
9. Ateschrang, A.; Weise, K.; Weller, S.; Stöckle, U.; Zwart, P.D.; Ochs, B.G. Long-term results using the straight tapered femoral cementless hip stem in total hip arthroplasty: A minimum of twenty-year follow-up. *J. Arthroplast.* **2014**, *29*, 1559–1565. [CrossRef]
10. Learmonth, I.D. (Ed.) *Interfaces in Total Hip Arthroplasty*; Springer: London, UK, 2000.
11. Bauer, R.; Kerschbaumer, F.; Poisel, S.; Oberthaler, W. The transgluteal approach to the hip joint. *Arch. Orthop. Trauma. Surg.* **1979**, *95*, 47–49. [CrossRef]
12. Hardinge, K. The direct lateral approach to the hip. *J. Bone Joint Surg. Br.* **1982**, *64*, 17–19. [CrossRef] [PubMed]
13. Lowry, O.; Rosebrough, N.; Farr, A.L.; Randall, R. Protein Measurement with the Folin Phenol Reagent. *J. Biol. Chem.* **1951**, *193*, 265–275. [CrossRef]
14. Hill, H.D.; Straka, J.G. Protein Determination Using Bicinchoninic Acid in the Presence of Sulfhydryl Reagents. *Anal. Biochem.* **1988**, *170*, 203–208. [CrossRef]
15. Makridakis, M.; Vlahou, A. GeLC-MS: A Sample Preparation Method for Proteomics Analysis of Minimal Amount of Tissue. *Methods Mol. Biol.* **2018**, *1788*, 165–175. [CrossRef]
16. Lygirou, V.; Latosinska, A.; Makridakis, M.; Mullen, W.; Delles, C.; Schanstra, J.P.; Zoidakis, J.; Pieske, B.; Mischak, H.; Vlahou, A. Plasma proteomic analysis reveals altered protein abundances in cardiovascular disease. *J. Transl. Med.* **2018**, *16*, 104. [CrossRef]
17. The UniProt Consortium. UniProt: The universal protein knowledgebase in 2021. *Nucleic Acids Res.* **2021**, *49*, D480–D489. [CrossRef]
18. Dakna, M.; Harris, K.; Kalousis, A.; Carpentier, S.; Kolch, W.; Schanstra, J.P.; Haubitz, M.; Vlahou, A.; Mischak, H.; Girolami, M. Addressing the challenge of defining valid proteomic biomarkers and classifiers. *BMC Bioinform.* **2010**, *11*, 594. [CrossRef]
19. Nanjappa, V.; Thomas, J.K.; Marimuthu, A.; Muthusamy, B.; Radhakrishnan, A.; Sharma, R.; Ahmad Khan, A.; Balakrishnan, L.; Sahasrabuddhe, N.A.; Kumar, S.; et al. Plasma Proteome Database as a resource for proteomics research: 2014 update. *Nucleic Acids Res.* **2014**, *42*, D959–D965. [CrossRef]
20. Farrah, T.; Deutsch, E.W.; Omenn, G.S.; Campbell, D.S.; Sun, Z.; Bletz, J.A.; Mallick, P.; Katz, J.E.; Malmström, J.; Ossola, R.; et al. A high-confidence human plasma proteome reference set with estimated concentrations in PeptideAtlas. *Mol. Cell. Proteom.* **2011**, *10*, M110.006353. [CrossRef]
21. Anderson, N.L.; Polanski, M.; Pieper, R.; Gatlin, T.; Tirumalai, R.S.; Conrads, T.P.; Veenstra, T.D.; Adkins, J.N.; Pounds, J.G.; Fagan, R.; et al. The human plasma proteome: A nonredundant list developed by combination of four separate sources. *Mol. Cell. Proteom.* **2004**, *3*, 311–326. [CrossRef]
22. Geyer, P.E.; Kulak, N.A.; Pichler, G.; Holdt, L.M.; Teupser, D.; Mann, M. Plasma Proteome Profiling to Assess Human Health and Disease. *Cell Syst.* **2016**, *2*, 185–195. [CrossRef] [PubMed]
23. Bryk, A.H.; Wiśniewski, J.R. Quantitative Analysis of Human Red Blood Cell Proteome. *J. Proteome Res.* **2017**, *16*, 2752–2761. [CrossRef] [PubMed]
24. The Human Protein Atlas. Available online: https://www.proteinatlas.org (accessed on 16 December 2021).
25. Shao, X.; Taha, I.N.; Clauser, K.R.; Gao, Y.T.; Naba, A. MatrisomeDB: The ECM-protein knowledge database. *Nucleic Acids Res.* **2020**, *48*, D1136–D1144. [CrossRef] [PubMed]
26. MatrisomeDB. Available online: http://matrisomedb.pepchem.org (accessed on 16 December 2021).
27. Ge, S.X.; Jung, D.; Yao, R. ShinyGO: A graphical gene-set enrichment tool for animals and plants. *Bioinformatics* **2020**, *36*, 2628–2629. [CrossRef] [PubMed]
28. ShinyGO. Available online: http://bioinformatics.sdstate.edu/go/ (accessed on 16 December 2021).
29. Brånemark, P.I. Osseointegration and its experimental background. *J. Prosthet. Dent.* **1983**, *50*, 399–410. [CrossRef]
30. Mavrogenis, A.F.; Dimitriou, R.; Parvizi, J.; Babis, G.C. Biology of implant osseointegration. *J. Musculoskelet Neuronal Interact* **2009**, *9*, 61–71.

31. Robert, A.; Latour, J.R. Biomaterials: Protein—Surface Interactions. In *Encyclopedia of Biomaterials and Biomedical Engineering*, 2nd ed.; Gary, E., Gary, W., Bowlin, L., Eds.; CRC Press: Boca Raton, FL, USA, 2008; pp. 270–284. ISBN 9781420078022.
32. Mendonça, D.B.S.; Miguez, P.A.; Mendonça, G.; Yamauchi, M.; Aragão, F.J.L.; Cooper, L.F. Titanium surface topography affects collagen biosynthesis of adherent cells. *Bone* **2011**, *49*, 463–472. [CrossRef]
33. Lin, X.; Patil, S.; Gao, Y.-G.; Qian, A. The Bone Extracellular Matrix in Bone Formation and Regeneration. *Front. Pharmacol.* **2020**, *11*, 757. [CrossRef]
34. Gehron Robey, P.G. Robey. Bone Matrix Proteoglycans and Glycoproteins. In *Principles of Bone Biology*, 2nd ed.; Bilezikian, J., Raisz, L., Rodan, G., Eds.; Academic Press: San Diego, CA, USA, 2002; pp. 225–237.
35. Sroga, G.E.; Vashishth, D. Effects of bone matrix proteins on fracture and fragility in osteoporosis. *Curr. Osteoporos. Rep.* **2012**, *10*, 141–150. [CrossRef]
36. Thalji, G.N.; Nares, S.; Cooper, L.F. Early molecular assessment of osseointegration in humans. *Clin. Oral Implant. Res.* **2014**, *25*, 1273–1285. [CrossRef]
37. Albrektsson, T.; Dahlin, C.; Jemt, T.; Sennerby, L.; Turri, A.; Wennerberg, A. Is marginal bone loss around oral implants the result of a provoked foreign body reaction? *Clin. Implant Dent. Relat. Res.* **2014**, *16*, 155–165. [CrossRef] [PubMed]
38. Trindade, R.; Albrektsson, T.; Tengvall, P.; Wennerberg, A. Foreign Body Reaction to Biomaterials: On Mechanisms for Buildup and Breakdown of Osseointegration. *Clin. Implant Dent. Relat. Res.* **2016**, *18*, 192–203. [CrossRef] [PubMed]
39. Shiu, H.T.; Goss, B.; Lutton, C.; Crawford, R.; Xiao, Y. Formation of blood clot on biomaterial implants influences bone healing. *Tissue Eng. Part B Rev.* **2014**, *20*, 697–712. [CrossRef]
40. Teller, P.; White, T.K. The Physiology of Wound Healing: Injury Through Maturation. *Perioper. Nurs. Clin.* **2011**, *6*, 159–170. [CrossRef]
41. Golebiewska, E.M.; Poole, A.W. Platelet secretion: From haemostasis to wound healing and beyond. *Blood Rev.* **2015**, *29*, 153–162. [CrossRef] [PubMed]
42. Zucoloto, A.Z.; Jenne, C.N. Platelet-Neutrophil Interplay: Insights Into Neutrophil Extracellular Trap (NET)-Driven Coagulation in Infection. *Front. Cardiovasc. Med.* **2019**, *6*, 85. [CrossRef]
43. Trindade, R.; Albrektsson, T.; Galli, S.; Prgomet, Z.; Tengvall, P.; Wennerberg, A. Osseointegration and foreign body reaction: Titanium implants activate the immune system and suppress bone resorption during the first 4 weeks after implantation. *Clin. Implant Dent. Relat. Res.* **2018**, *20*, 82–91. [CrossRef]
44. El Kholy, K.; Buser, D.; Wittneben, J.-G.; Bosshardt, D.D.; van Dyke, T.E.; Kowolik, M.J. Investigating the Response of Human Neutrophils to Hydrophilic and Hydrophobic Micro-Rough Titanium Surfaces. *Materials* **2020**, *13*, 3421. [CrossRef]
45. Lacy, P. Mechanisms of Degranulation in Neutrophils. *Allergy Asthma Clin. Immunol.* **2006**, *2*, 98–108. [CrossRef]
46. Aratani, Y. Myeloperoxidase: Its role for host defense, inflammation, and neutrophil function. *Arch. Biochem. Biophys.* **2018**, *640*, 47–52. [CrossRef]
47. Zamolodchikova, T.S.; Tolpygo, S.M.; Svirshchevskaya, E.V. Cathepsin G-Not Only Inflammation: The Immune Protease Can Regulate Normal Physiological Processes. *Front. Immunol.* **2020**, *11*, 411. [CrossRef] [PubMed]
48. Pittman, K.; Kubes, P. Damage-associated molecular patterns control neutrophil recruitment. *J. Innate Immun.* **2013**, *5*, 315–323. [CrossRef] [PubMed]
49. Wang, J. Neutrophils in tissue injury and repair. *Cell Tissue Res.* **2018**, *371*, 531–539. [CrossRef] [PubMed]
50. Ryckman, C.; Vandal, K.; Rouleau, P.; Talbot, M.; Tessier, P.A. Proinflammatory activities of S100: Proteins S100A8, S100A9, and S100A8/A9 induce neutrophil chemotaxis and adhesion. *J. Immunol.* **2003**, *170*, 3233–3242. [CrossRef] [PubMed]
51. Mödinger, Y.; Teixeira, G.Q.; Neidlinger-Wilke, C.; Ignatius, A. Role of the Complement System in the Response to Orthopedic Biomaterials. *Int. J. Mol. Sci.* **2018**, *19*, 3367. [CrossRef] [PubMed]
52. Schoengraf, P.; Lambris, J.D.; Recknagel, S.; Kreja, L.; Liedert, A.; Brenner, R.E.; Huber-Lang, M.; Ignatius, A. Does complement play a role in bone development and regeneration? *Immunobiology* **2013**, *218*, 1–9. [CrossRef] [PubMed]
53. Campos, V.; Melo RC, N.; Silva, L.P.; Aquino, E.N.; Castro, M.S.; Fontes, W. Characterization of neutrophil adhesion to different titanium surfaces. *Bull. Mater. Sci.* **2014**, *37*, 157–166. [CrossRef]
54. Dayer, R.; Brennan, T.C.; Rizzoli, R.; Ammann, P. PTH improves titanium implant fixation more than pamidronate or renutrition in osteopenic rats chronically fed a low protein diet. *Osteoporos. Int.* **2010**, *21*, 957–967. [CrossRef]

Journal of Functional Biomaterials

Article

A Three-Dimensional Printed Polycaprolactone–Biphasic-Calcium-Phosphate Scaffold Combined with Adipose-Derived Stem Cells Cultured in Xenogeneic Serum-Free Media for the Treatment of Bone Defects

Woraporn Supphaprasitt [1], Lalita Charoenmuang [1], Nuttawut Thuaksuban [1,*], Prawichaya Sangsuwan [2], Narit Leepong [1], Danaiya Supakanjanakanti [1], Surapong Vongvatcharanon [1], Trin Suwanrat [1] and Woraluk Srimanok [1]

1. Department of Oral and Maxillofacial Surgery, Faculty of Dentistry, Prince of Songkla University, Hatyai 90110, Thailand; worapornsup@gmail.com (W.S.); oom.lalita@yahoo.com (L.C.); narit.l@psu.ac.th (N.L.); danaiya.s@psu.ac.th (D.S.); surapong.v@psu.ac.th (S.V.); bomtep.b@gmail.com (T.S.); woraluk.ws@hotmail.com (W.S.)
2. Department of Molecular Biotechnology and Bioinformatics, Faculty of Science, Prince of Songkla University, Hatyai 90110, Thailand; sangsuwan.ji@gmail.com
* Correspondence: nuttawut.t@psu.ac.th; Tel.: +66-954592492

Citation: Supphaprasitt, W.; Charoenmuang, L.; Thuaksuban, N.; Sangsuwan, P.; Leepong, N.; Supakanjanakanti, D.; Vongvatcharanon, S.; Suwanrat, T.; Srimanok, W. A Three-Dimensional Printed Polycaprolactone–Biphasic-Calcium-Phosphate Scaffold Combined with Adipose-Derived Stem Cells Cultured in Xenogeneic Serum-Free Media for the Treatment of Bone Defects. *J. Funct. Biomater.* **2022**, *13*, 93. https://doi.org/10.3390/jfb13030093

Academic Editors: Kunyu Zhang, Qian Feng, Yongsheng Yu, Boguang Yang and Chunming Wang

Received: 5 June 2022
Accepted: 12 July 2022
Published: 15 July 2022

Publisher's Note: MDPI stays neutral with regard to jurisdictional claims in published maps and institutional affiliations.

Copyright: © 2022 by the authors. Licensee MDPI, Basel, Switzerland. This article is an open access article distributed under the terms and conditions of the Creative Commons Attribution (CC BY) license (https://creativecommons.org/licenses/by/4.0/).

Abstract: The efficacy of a three-dimensional printed polycaprolactone–biphasic-calcium-phosphate scaffold (PCL–BCP TDP scaffold) seeded with adipose-derived stem cells (ADSCs), which were cultured in xenogeneic serum-free media (XSFM) to enhance bone formation, was assessed in vitro and in animal models. The ADSCs were isolated from the buccal fat tissue of six patients using enzymatic digestion and the plastic adherence method. The proliferation and osteogenic differentiation of the cells cultured in XSFM when seeded on the scaffolds were assessed and compared with those of cells cultured in a medium containing fetal bovine serum (FBS). The cell–scaffold constructs were cultured in XSFM and were implanted into calvarial defects in thirty-six Wistar rats to assess new bone regeneration. The proliferation and osteogenic differentiation of the cells in the XSFM medium were notably better than that of the cells in the FBS medium. However, the efficacy of the constructs in enhancing new bone formation in the calvarial defects of rats was not statistically different to that achieved using the scaffolds alone. In conclusion, the PCL–BCP TDP scaffolds were biocompatible and suitable for use as an osteoconductive framework. The XSFM medium could support the proliferation and differentiation of ADSCs in vitro. However, the cell–scaffold constructs had no benefit in the enhancement of new bone formation in animal models.

Keywords: polycaprolactone; biphasic calcium phosphate; scaffold; adipose; stem cells

1. Introduction

Over the last decade, the concept of tissue engineering has been applied to bone grafting procedures. Combining bone substitute scaffolds with osteoprogenitor cells or stem cells and some osteo-inductive growth factors is an effective strategy for the enhancement of new bone formation. To our knowledge, the scaffold is still the most important factor. Our techniques of melt stretching and multilayer deposition (MSMD) and melt stretching and compression molding (MSCM) have been successfully used for the fabrication of the polycaprolactone (PCL)–ceramic bone substitute scaffolds [1–6]. Both techniques involve filament-based processing that uses PCL–ceramic filaments to fabricate three-dimensional (3D) scaffolds. PCL is a synthetic polyester that was approved by the Food and Drug Administration (FDA) as a medical and drug delivery device. Its biocompatibility has been widely demonstrated in several in vivo and clinical studies [7–10]. Quantities of the ceramic materials, including biphasic calcium phosphate (BCP) and hydroxyapatite (HA), of up to 30% by weight have been used as fillers in the PCL matrix. Based on our previous studies [1–6], scaffolds fabricated using both these techniques are biocompatible and can act as effective osteoconductive frameworks for new bone regeneration in animal models. The PCL–HA scaffold was proved to be a good carrier of bone morphogenetic protein-2 (BMP) [6]. The BMP-soaked scaffolds could be applied to guided bone regeneration models without using conventional xenogeneic bone graft particles. BCP consists of the stable phase of HA and the more soluble phase of β-tricalcium phosphate (β-TCP) [11,12]. The bioactivity of the PCL–BCP scaffolds is achieved because of their ability to release calcium and phosphate ions, which are essential substrates for the bone formation process [1,2]. However, the machines used in both techniques are prototypes with a low productive capacity and possible batch-to-batch variations. For marketing for wider clinical use, the consistency of the scaffold product lines can be improved using 3D printing technology. A major advantage of biomedical 3D printing is that clinical data from computed tomography can be transferred to the computer-printing software, enabling scaffolds to be designed in precise sizes and shapes to fit into the defects. In this study, PCL–BCP filaments were used in a filament-based 3D printing process for the fabrication of PCL–BCP scaffolds.

Several studies [4,5,13–19] reported promising results in the enhancement of new bone formation when PCL-based scaffolds were combined with osteogenic cells, including primary osteoblasts, mesenchymal stem cells (MSCs) from bone marrow, and dental pulp tissue. Fat tissue is another source of MSCs. Buccal fat pads are a suitable intra-oral source of ADSCs and provide a large amount of fat tissue. The tissue can be easily harvested during routine intra-oral surgical operations of maxillary third molar removal and orthognathic surgeries. The volumes of the fat available for isolating the stem cells are greater, compared with dental pulp and periodontal tissue. Several studies [20–24] demonstrated that adipose-derived stem cells (ADSCs) expressed immunophenotyping markers similar to bone marrow MSCs. Moreover, they could differentiate toward the lineages of different cell types, including osteoprogenitor cells. Broccaioli et al. [25] and Niada et al. [26] demonstrated that there was no difference in immunophenotype, proliferation, and multi-differentiation between ADSCs isolated from buccal fat pads and those from subcutaneous adipose tissue. In a clinical trial, Khojasteh et al. assessed the efficacy of ADSCs from buccal fat pads as an adjunct to autogenous iliac bone block grafting for the treatment of extensive alveolar ridge defects in eight patients and compared the results with those of the control group, which had no cells [27]. The results demonstrated greater new bone formation in the test group when compared with the control group (65.32% and 49.21%, respectively).

Currently, although the use of MSCs in cell-based therapy is accepted for several tissue engineering models and clinical trials, large amounts of the cells are required for each application. Therefore, the small numbers of stem cells isolated from the tissue need to be increased using culture procedures. In general, culturing MSCs in a fetal bovine serum (FBS)-containing medium is a standard protocol for increasing the cells. FBS is suitable for supporting proliferation and differentiation of the cells because it contains many essential nutrients, hormones, and growth factors. However, concerns remain regarding the risks of disease transmission from contamination by animal-originated pathogens and of immunologic reactions from unidentified bovine proteins. The xenogeneic proteins in FBS could be infused into cells because a wash step prior to cell infusion cannot remove their internalized xenogeneic antigens [28]. Moreover, undefined ingredients and batch-to-batch variations can affect the accuracy of research results and therapeutic outcomes [29]. Currently, xenogeneic serum-free media (XSFM) are considered likely to replace the use of FBS for clinical applications. Several studies [30–36] demonstrated that XSFM have the capacity to maintain the morphologies, the expression of phenotypic surface markers, and multipotent differentiation of MSCs. Therefore, culturing stem cells in XSFM is expected to become a standard protocol in the large-scale expansion of this technique for clinical use. In our study, the growth and osteogenic differentiation of ADSCs isolated from buccal fat pads when seeded on the PCL–BCP 3D printed (PCL–BCP TDP) scaffolds and cultured in XSFM were evaluated in vitro. In addition, the efficacy of the cell–scaffold constructs for the enhancement of new bone regeneration was assessed in rat models.

2. Materials and Methods

PCL pellets (Purasorb PC12, Mn 79,760, Viscometry 1.0–1.3 dL/g) were purchased from Corbion, the Netherlands. BCP particles (HA/β-TCP = 30/70%, particle size < 75 μm) were supplied by the National Metal and Materials Technology Center (MTEC), Pathumthani, Thailand.

2.1. Scaffold Fabrication

The PCL pellets and BCP particles were mixed in a ratio of PCL: BCP at 70:30 by weight in the chamber of a melting–extruding machine [1]. A homogenous PCL–BCP blend was obtained by heating and stirring at 120 °C, and then the blend was extruded through the nozzle tip of the machine to form the filament. The architectures of the PCL–30%BCP TDP scaffold were designed in a grid pattern with 500 μm^2 between the filament rows and at 0°, 45°, and 90° to each lay-down layer using 3D Slicer Software (ideaMaker version 4.1.0.4990, Raise-3D Technologies Inc., Irvine, CA, USA) (Figure 1). The printing parameters included layer height 0.25 mm, infill density 40%, and infill angle 0°/45°/90°. To fabricate the scaffold, the filament was loaded into the 3D printer (RAISE3D-E2, Raise-3D Technologies Inc., Irvine, CA, USA) and extruded through the 0.4 mm nozzle tip of the machine using an extruder temperature at 180 °C and printing speed at 30 mm/s. The scaffolds were placed in sterilization pouches and sterilized using ethylene oxide gas (ethylene oxide 100%, 37 °C, humidity 76%, 2 h) 2 weeks prior to the subsequent experiments.

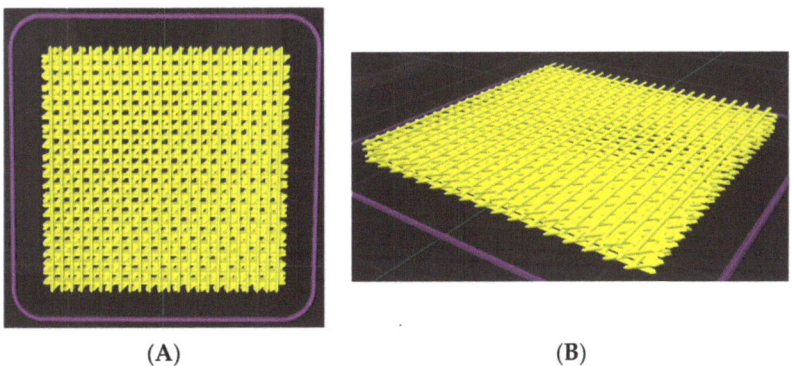

Figure 1. The preview architectures of the scaffold prior to the printing process; (**A**) top view and (**B**) perspective view.

2.2. Scaffold Morphologies and Structural Analysis

The microscopic architectures of the scaffold were evaluated using a stereomicroscope (Nikon, Tokyo, Japan) and a scanning electron microscope (SEM, JOEL Ltd., Tokyo, Japan). The scaffolds were stained with Alizarin Red S (AR, Sigma-Aldrich Inc., St. Louis, MO, USA) for detecting the areas where the BCP particles had deposited using a fluorescent microscope (ZEISS Axio Observer 7, Carl Zeiss, Oberkochen, Germany). The dispersion of the BCP particles in the PCL matrix of the scaffold filaments was assessed using field emission SEM (FE-SEM, Apreo, Thermo Fisher Scientific, Waltham, MA, USA) and energy-dispersive X-ray spectroscopy (EDX, Apero, Thermo Fisher Scientific, Waltham, MA, USA). Functional groups and the chemical interaction between the BCP particles and the PCL matrix were analyzed using Fourier transform infrared spectroscopy (FTIR, Bruker Vertex70, Billerica, MA, USA).

2.3. Mechanical Testing

The $10 \times 10 \times 5$ mm^3 scaffold specimens were immersed in phosphate buffer saline (PBS) and incubated at 37 °C for 24 h before the experiments. Compression tests were applied to the superior and lateral aspects of the specimens in the wet stage using a universal testing machine (Lloyd Instruments Ltd., West Sussex, UK) (n = 5/aspect) (Figure 2). For the superior aspect, each specimen was placed on the flat testing platform against the compressing probe (15.77 mm diameter; 5 kN load cell). Then, compression force was applied to its superior aspect from 0 to 300 N at a crosshead speed of 10 mm/min. For the lateral aspect, the lateral aspect of each specimen was placed against the compressing probe (15.77 mm diameter; 250 N load cell) and secured using a vice grip. Compression force was applied to its lateral aspect at a crosshead speed of 10 mm/min until the strain level reached 30%. The compressive strength of the scaffolds was analyzed using analysis software (NEXYGEN, Lloyd Instruments Ltd., Hampshire, UK).

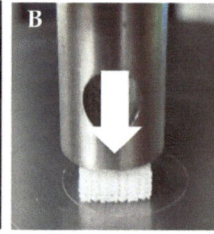

Figure 2. The compression forces (indicated by arrows) were applied to the superior aspect (**A**,**B**) and the lateral aspect (**C**,**D**) of the scaffolds.

2.4. Subject Enrollment

The experimental protocol was approved by the human research ethics committee, Faculty of Dentistry, Prince of Songkla University (EC6012-37-P-LR). The six volunteers were patients undergoing either surgical removal of impacted maxillary third molars or orthognathic surgeries for correction of skeletal discrepancies, in the Oral & Maxillofacial Surgery clinic, Dental Hospital, Faculty of Dentistry, Prince of Songkla University. The inclusion criteria for the surgical removal of the third molars were healthy men or women from 20 to 40 years old, weighing more than 50 kg. The inclusion criteria for the orthognathic surgeries were healthy men or women (ASA class I) from 20 to 40 years old, weighing more than 50 kg, and with a hematocrit level of at least 35%. The exclusion criteria included patients with a history of radiation of the head and neck region, diabetes, uncontrolled metabolic diseases, compromised immune system, blood-transmitted diseases, infection of surgical sites, postliposuction of buccal regions and pregnancy. All patients provided informed consent prior to the experiments.

2.5. Isolating ADSCs from Fat Tissue

For the removal of the third molars, triangular flap incisions were created extending to the buccal vestibule of maxillary first molars. After the impacted teeth were removed, blunt dissection was made through the buccinator muscle, and some parts of the buccal fat tissue were excised. For the orthognathic surgeries, the buccal fat tissue often leaked during Lefort I osteotomy of the maxilla or bilateral sagittal split ramus osteotomy (BSSRO) of the mandible without blunt dissection or force traction. Some parts of the tissue in the operation fields were excised. The fat tissue harvested from each patient was divided equally into 2 groups

1. XSF group: The fat issue was stored in XSFM (MesenCult™-XF, STEMCELL Technologies Inc, Vancouver, BC, Canada).

2. FBS group: The tissue was stored in Dulbecco's Modified Eagle Medium (DMEM, Gibco, Thermo Fisher Scientific, Waltham, MA, USA) supplemented with 10% FBS (Gibco, Thermo Fisher Scientific, Waltham, MA, USA).

The stem-cell-isolation procedure for both groups was performed within 2 h. In brief, the fat tissue was washed several times with sterile PBS to remove contaminating debris and red blood cells. The fat tissue was then minced into small pieces and enzymatically digested using 3 mg/mL type I collagenase (Gibco, USA) in PBS at 37 °C with gentle agitation for 60 min. The cell pellet was obtained by centrifugation at 1200 g for 10 min and then resuspended in the medium of each group. The solution was filtered through a 100 µm cell strainer (Corning, Merck KGaA, Darmstadt, Germany) and then plated into 6-well plates (Corning, Merck KGaA, Darmstadt, Germany). The plates were incubated in a humidified atmosphere with 5% CO_2 at 37 °C until the adherent cells reached 70–80% confluence, and then subculture was performed. The cells from passages 2 to 5 were used for the experiments (Figure 3).

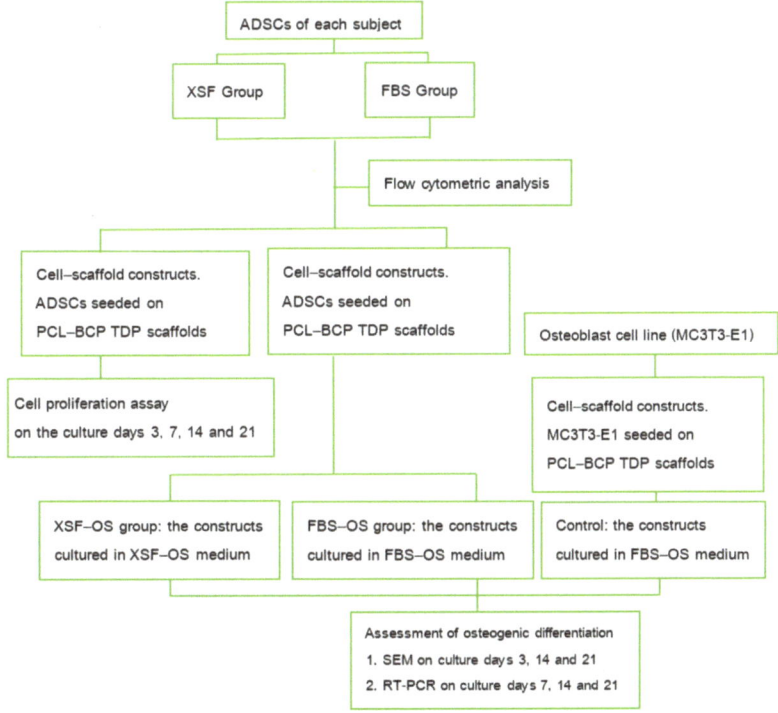

Figure 3. A schematic diagram showing an overview of the in vitro experiments.

2.6. Characterizing ADSCs

Flow Cytometry Analysis

The MSC immunophenotypes of the cells were defined following the protocol of the International Society for Cellular Therapy (ISCT) [37]. The analysis was performed using a fluorochrome-conjugated monoclonal antibody cocktail in the MSC Phenotyping Kit, human (Miltenyi Biotec, Bergisch Gladbach, Germany). In brief, 5×10^5 cells in passages 2–3 were incubated in antibodies against the surface antigens CD73, CD90 and CD105 as the positive markers and CD14, CD19, CD34 and CD45 as the negative markers. In addition, antibodies against CD 271 and CD 146 were included in the sequences. At least 10,000 events were acquired for each sample using a fluorescence-activated cell sorting instrument (FACSCalibur, BD Biosciences, Franklin Lakes, NJ, USA) and the data were analyzed using CELLQUEST software (version 4, BD Biosciences, Franklin Lakes, NJ, USA).

2.7. In Vitro Proliferation and Osteogenic Differentiation of ADSCs Seeded on the PCL–BCP TDP Scaffolds

2.7.1. Assessment of Cell Proliferation

Prior to cell seeding, the scaffolds were placed in 48-well plates (Corning, Merck KGaA, Darmstadt, Germany) and immersed in the medium of each group for 24 h. The cell–scaffold constructs were obtained by seeding 2×10^4 of the ADSCs of the XSF and FBS groups onto each scaffold (diameter 11 mm and height 2 mm) using the static seeding method. The constructs were left for 24 h in a humidified atmosphere with 5% CO_2 at 37 °C to allow cell attachment. Next, they were moved to new wells and 200 µL of each group's medium was added. The constructs were cultured in a humidified atmosphere with 5% CO_2 at 37 °C, and the media were changed every 3 days until the time of the test. At days 3, 7, 14 and 21 after seeding, the quantity of the viable cells in the constructs was measured using PrestoBlue reagent (Thermo Fisher Scientific Inc., Waltham, MA, USA) (n = 6/group/time point). The medium in each well was removed and the constructs were washed using phosphate buffer saline (PBS), then replaced with 180 µL of fresh media without FBS. Twenty microliters of PrestoBlue reagent was added to each well and the constructs were incubated in 5% CO_2 at 37 °C for 10 min while protected from direct light. After incubation, 100 µL of the medium in each well was transferred to a 96-well plate in duplicate and the absorbance of each well was read at 570 nm using a microplate reader (Thermo Fisher Scientific, Vantaa, Finland). The levels of OD were compared with a standard curve to infer the quantities of the cells. At each time point after the measurement, the constructs were refreshed with 200 µL of each group's medium and the culture was continued until the next time point.

2.7.2. Assessment of Cell Differentiation

At 21, 14 and 7 days prior to the experiment, the cell–scaffold constructs were obtained by seeding 1×10^6 of the ADSCs of the XSF and FBS groups onto each scaffold (diameter 11 mm and height 2 mm) using the static seeding method as previously described (n = 6/group/time point). They were left for 24 h in a humidified atmosphere with 5% CO_2 at 37 °C to allow cell attachment. Next, the constructs were moved to the new wells and 200 µL of osteogenic differentiation (OS) media was added as follows.

1. XSF–OS group: cultured in xenogeneic serum-free OS medium (MesenCult™ Osteogenic Differentiation Human, STEMCELL Technologies Inc., Vancouver, BC, Canada).
2. FBS–OS group: cultured in DMEM OS medium (DMEM supplemented with 10%FBS, 10 mM β-glycerophosphate (Sigma-Aldrich Inc., St. Louis, MO, USA), 10^{-7} M dexamethasone (Sigma, city, state, USA) and 50 µM ascorbic acid-2 phosphate (Sigma-Aldrich Inc., St. Louis, MO, USA).
3. Control group: The osteoblasts (MC3T3-E1 cell line, subclone 4, ATCC, Manassas, VA, USA) were seeded onto the scaffolds using the same ADSC protocol and cultured in the FBS–OS medium.

The constructs were cultured in a humidified atmosphere with 5% CO_2 at 37 °C, and the media were changed every 2 days until the time of the test. On the day of the experiment, a quantitative reverse transcription polymerase chain reaction (RT-qPCR) was performed to assess the osteogenic differentiation genes. In brief, the total RNA was extracted and purified using a PureLink™ RNA Mini Kit (Invitrogen, Thermo Fisher Scientific, Waltham, MA, USA) and then reverse-transcribed using a SuperScript III First-Strand Synthesis System (Invitrogen, Thermo Fisher Scientific, Waltham, MA, USA). RT-qPCR was performed using a SensiFAST™ SYBR® No-ROX Kit (Meridian Bioscience, London, United Kingdom), a LightCycler system (Roche Diagnostics, Mannheim, Germany), and specific primers for the osteoblast-related genes as listed in Table 1. The RT-PCR condition was performed via 45 cycles of denaturation, annealing and elongation. The expression of the genes at each time point was analyzed using the $2^{-\Delta\Delta CT}$ method and normalized with the expression of glyceraldehyde 3-phosphate dehydrogenase (GAPDH) housekeeping gene.

Table 1. The oligonucleotide primer sequences of the osteogenic differentiation genes.

Genes	Primer Sequence (5′-3′)	GenBank Accession No.
Collagen type 1 (Col-1)	R: ACCAGGTTCACCGCTGTTAC	NC_000017.11
	F: GTGCTAAAGGTGCCCAATGGT	
Bone sialoprotein (BSP)	R: AGGATAAAAGTAGGCATGCTTG	NC_000004.12
	F: ATGGCCTGTGCTTTCTCAATG	
Alkaline phosphatase (ALP)	R: GCGGCAGACTTTGGTTTC	NM_001127501
	F: CCACCAGCCCGTGACAGA	
Runt-related transcription factor 2 (RUNX-2)	R: TGCTTTGGTCTTGAAATCACA	NC_10472
	F: TCTTAGAACAAATTCTGCCCTTT	
Osteocalcin (OCN)	R: CTTTGTGTCCAAGCAGGAGG	NM_00582.2
	F: CTGAAAGCCGATGTGGTCAG	
Bone morphogenetic protein-2 (BMP-2)	R: AAGAGACATGTGAGGATTAGCAGGT	NM_007553
	F: GCTTCCGCCTGTTTGTGTTTG	
Osteopontin (OPN)	R: TGTGAGGTGATGTCCTCGTCTGT	NM_00582.2
	F: ACACATATTGATGGCCGAAGGTGA	
GAPDH	R: CCACCACCCTGTTGCTGTA	NM_001289745.1
	F: GCATCCTGGGCTACACTGA	

2.7.3. SEM

At days 3, 14 and 21 after seeding, the characteristics of the cells in the constructs of the XSF–OS and FBS–OS groups were descriptively assessed. The constructs were removed from the culture plates, rinsed with PBS and then fixed in 2.5% glutaraldehyde (Sigma-Aldrich Inc., St. Louis, MO, USA) in PBS for 2 h. The specimens were dehydrated in the 50–100% ethanol series and coated with gold–palladium. The constructs were observed using SEM.

2.8. Assessment of Efficacy of the Cell–Scaffold Constructs for Repairing Calvarial Defects in Rat Models

2.8.1. Preparing the Cell–Scaffold Constructs

The cell–scaffold constructs of the XSF–OS group were obtained by seeding 2×10^6 of the pooled ADSCs onto each scaffold (diameter 8 mm and height 1 mm) using the protocol previously described. The constructs were left to cultivate in the XSF–OS medium for 7 days prior to surgical implantation.

2.8.2. The Animals

Forty-eight adult male Wistar rats, weighing 300–400 g (Nomura Siam International, Bangkok, Thailand), were used for the experiment. This was approved by the animal experiment ethics committee of the Prince of Songkla University (MHESI 68014/860). The animals were anesthetized using ketamine (60 mg/kg) and xylazine (10 mg/kg), administered intraperitoneally. Next, a bi-cortical calvarial defect (8 mm in diameter) was created in the mid-sagittal plane of each animal. In Group A, the cell–scaffold construct was placed into the defect and covered with resorbable membrane (OssGuide, noncrosslinked collagen membrane, SK bioland Co., Ltd., Seoul, Korea). In Group B, the scaffold without the cells was placed into the defect and covered with the membrane. In Group C, the defects were filled with autogenous calvarial bone chips and covered with the membrane. In Group

D, the defect was left empty and covered with the membrane (Figure 4). The wounds were closed with 4/0 absorbable sutures (Vicryl, Ethicon LLC, London, UK). All animals received postoperative antibiotic prophylaxis with subcutaneous cephalexin, 10 mg/kg, and postoperative analgesic with subcutaneous buprenorphine, 0.1 mg/kg once daily for 3 days. At the time points of 2, 4 and 8 weeks after the operation, the animals were sacrificed using intraperitoneal administration of 120 mg/kg of overdose pentobarbital sodium. The calvarial specimens, including the 3 mm margin of normal bone surrounding the areas of the bone defects, were collected and then fixed in 10% formalin for microcomputed tomography (µ-CT) and histological assessment (n = 4/group/time point).

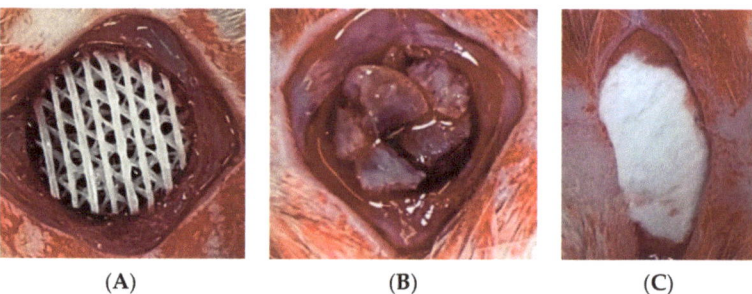

Figure 4. The pictures demonstrate the surgical sites of Group A (**A**) and Group C (**B**). The defects were covered with collagen membrane before suturing (**C**).

2.8.3. µ-CT Analysis

The specimens were scanned using a µ-CT machine (µ-CT 35, SCANCO Medical AG, Wangen-Brüttisellen, Switzerland) in a direction parallel to the coronal aspect of the calvariums, at settings of 55 kVp, 72 mA and 4 W. The gray-scale threshold values were adjusted to discriminate between new bone and the ceramic particles in the scaffolds. New bone formation within each implant site was measured as the new bone volume fraction (VF) using analysis software (µ-CT 35 Version 4.1, SCANCO Medical AG, Wangen-Brüttisellen, Switzerland) with the following formula:

$$\text{New bone VF} = [\text{New bone volume} \div \text{Total defect volume}] \times 100$$

2.8.4. Histologic Processing and Histological Assessment

The specimens were decalcified in 14% ethylenediaminetetraacetic acid (EDTA) and embedded in paraffin. Serial 5 µm thick sections were cut at positions 500 µm from the midline of each specimen. The sections were stained with hematoxylin and eosin (H&E) (2 sections/specimen). The stained sections were scanned using a slide scanner (Aperio, Leica Biosystems, Deer Park, IL, USA) to create image files. The microscopic features were assessed descriptively.

2.9. Statistical Analysis

The chemical and mechanical properties of the scaffolds, characteristics of the cell–scaffold constructs and histological features were descriptively evaluated. The measurable parameters, which included the number of viable cells in the constructs, the levels of the osteogenic genes and the new bone VF, were analyzed using statistical analysis software (SPSS, version 14, IBM Corporation, Armonk, NY, USA). One-way analysis of variance (ANOVA) followed by a Tukey HSD test were applied to assess the differences between the groups and time points. The level of statistical significance was set at $p < 0.05$.

3. Results

3.1. Scaffold Morphologies

The architectures of the scaffolds are demonstrated in Figure 5. The SEM images demonstrated irregular surfaces of the scaffolds that had a few of the BCP particles depositing (Figure 6A–C). The AR staining indicated that the BCP particles were distributed throughout the surfaces of the scaffolds (Figure 6D).

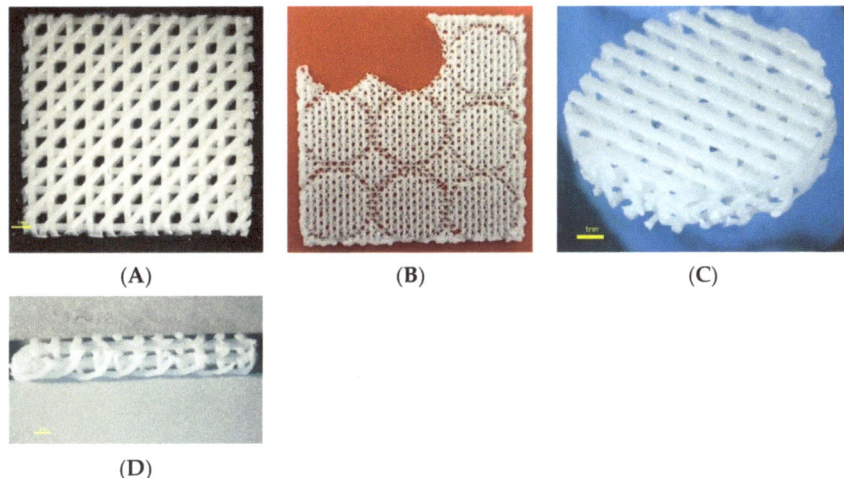

Figure 5. The architecture of the PCL–BCP TDP scaffold; (**A**) top view, (**B**) the scaffold was cut into round-shaped specimens for the experiments, (**C**) magnified picture of the scaffold specimen and (**D**) magnified picture of the lateral aspect of the scaffold. The scale bars represent 1 mm.

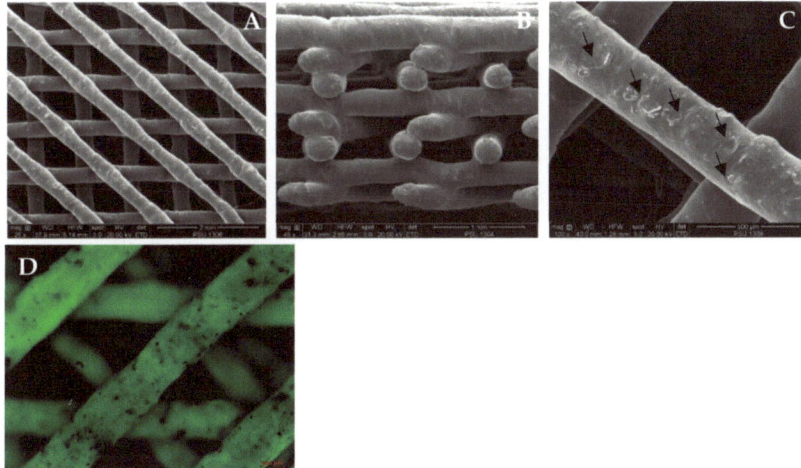

Figure 6. The SEM pictures demonstrate the architectures of the PCL–BCP TDP scaffolds. (**A**) top view, (**B**) lateral view and (**C**): image shows the BCP particles depositing on the surfaces of the scaffold (arrows). (**D**): The AR-stained BCP particles are seen as black spots throughout the surfaces of the scaffolds in the fluorescent microscope image.

3.2. Structural Analysis of the Scaffolds

The structural bands of the PCL–BCP composite were demonstrated via the FTIR spectra (Figure 7). The spectrum of the composite was slightly different from the PCL and BCP spectra, which indicated that there was no chemical bond between the materials. The FE-SEM demonstrated immiscible dispersion of the BCP particles throughout the PCL matrix and several voids inside the filaments (Figure 8A,B). The EDX analysis indicated calcium–phosphate crystals on the scaffold surfaces (Figure 8C).

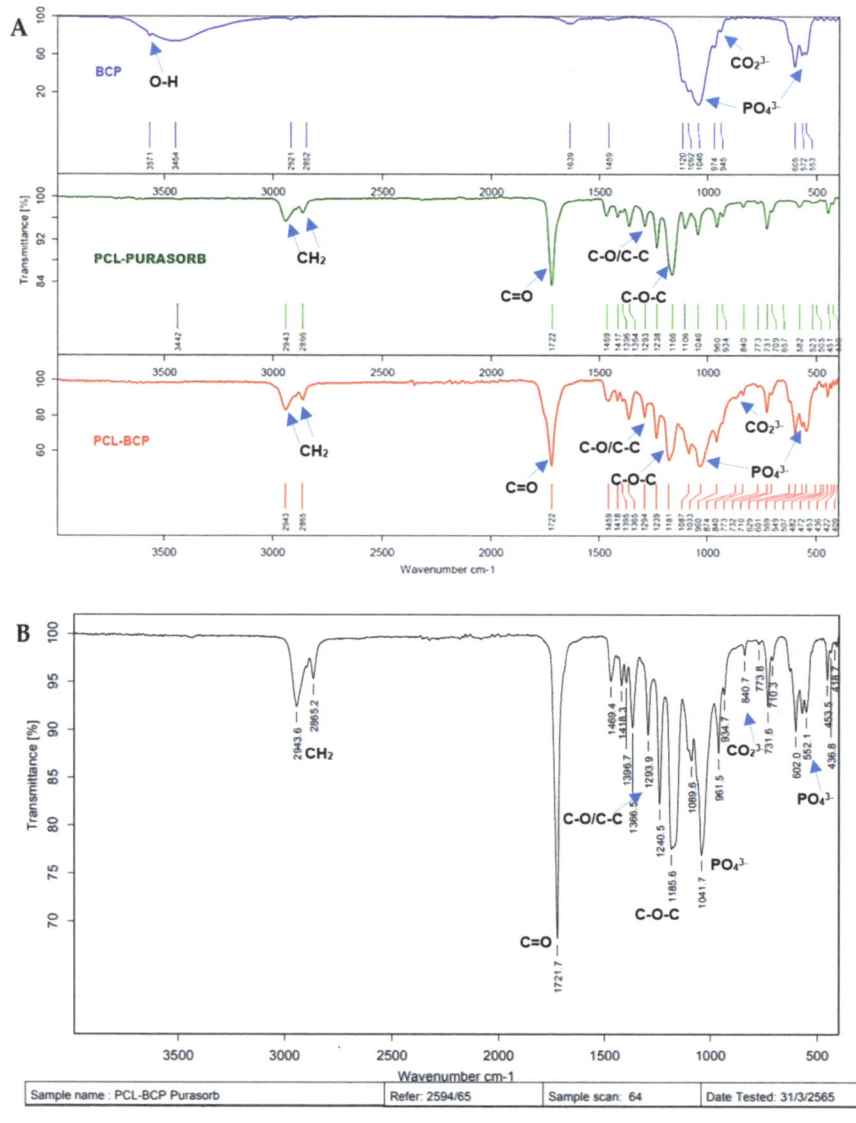

Figure 7. FTIR spectra for BCP, PCL and PCL-30%BCP composite. (**A**) The bands of the PCL–BCP composite showed no change when compared with the peaks of BCP and PCL raw materials. (**B**) The bands of the PCL-BCP composite were not changed when compared to those of each material.

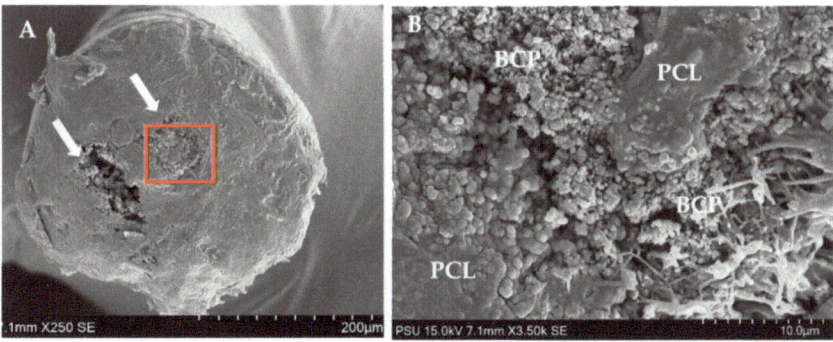
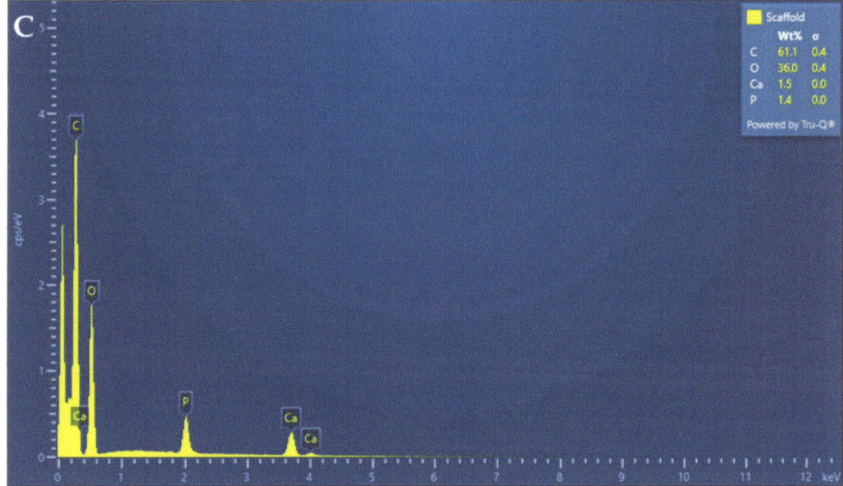

Figure 8. (**A**) The cross-sectional SEM image shows large voids inside the scaffold filament (indicated by arrows). (**B**) The magnified image of the box demonstrates immiscible blending of the BCP crystals and the PCL matrix in the box. (**C**) EDX analysis shows the high peaks of calcium (Ca) and phosphate (P) on the scaffold surfaces.

3.3. Mechanical Properties

The mechanical properties of the scaffolds are shown in Table 2. The scaffolds could successfully withstand compression forces from the superior and lateral directions. They recovered to their initial height without distortion after the forces had been applied.

Table 2. The mechanical properties of the scaffolds.

Mechanical Properties	Superior Aspect	Lateral Aspect
Compressive strength (MPa)	2.98 ± 0.01	2.18 ± 0.09
Strain at maximum load (%)	36.53 ± 1.8	30
Young's modulus (MPa)	13.68 ± 1.01	34.75 ± 3.52
Maximum load (N)	300	108.7 ± 4.06

3.4. Demographic Data

Two males and four females, at an average age of 25.17 ± 5.64 years old, were enrolled in the study. There were two cases involving third molar removal and four cases involving orthognathic surgeries (Table 3). All patients tolerated the operation well without postoperative complications. The average fat volume was 4.17 ± 0.98 milliliters.

Table 3. Demographic data of the subjects.

Case	Age	Gender	Fat Volume (mL)	Operations
1	24	Male	5	BSSRO setback
2	21	Female	3	Surgical removal #18
3	23	Female	3	Surgical removal #28
4	36	Female	5	BSSRO advancement
5	21	Male	4	BSSRO setback
6	26	Female	5	BSSRO setback

3.5. Cell Morphologies

After plating the cell suspension of the XSF and FBS groups, adherent cells could be detected within 7 days, and they gradually proliferated over time. The cells of both groups had spindle-shaped fibroblast-like morphology (Figure 9). By observation, the cells in the XSF group were more spindle-shaped and grew at higher density compared with those in the FBS group.

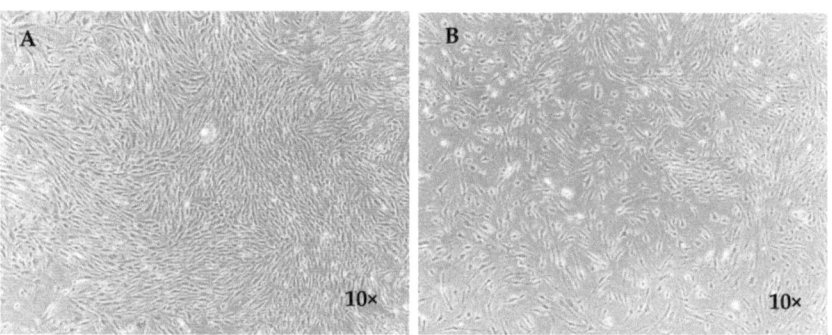

Figure 9. Morphologies of the adherent cells at day 21 taken via a phase-contrast microscope (DS-Fi2-U3, Nikon, Tokyo, Japan) with magnification 10×. (**A**) XSF group and (**B**) FBS group. More spindle-shaped morphologies and higher density of the cells in the XSF group were detected compared with those in the FBS group.

3.6. Flow Cytometry Analysis

Expression of the MSC immunophenotypes of the ADSCs is shown in Table 4 and Figure 10. The percentages of the CD 73, 90 and the hematopoietic markers of both groups were not statistically different. However, the percentage of the CD 105 in the FBS group was significantly higher than that of the XFS group. The cells in both groups expressed CD73 at the highest levels, followed by CD90 and CD105, and they expressed the hematopoietic markers at less than 1%. No statistical difference was detected between the two groups for the expression of CD271 and CD146.

Table 4. The average percentages of immunophenotyping markers of the cells from all subjects. The percentage of CD-105-positive cells in the FBS group was significantly higher than that of the XFS group (* $p < 0.05$).

CD Markers (%)		FBS Group	XSF Group
MSC markers	CD 73	80.12 ± 8.57%	81.26 ± 7.06%
	CD 90	66.26 ± 8.17%	67.28 ± 8.04%
	CD 105	41.58 ± 8.11% *	2.94 ± 1.29%
Hematopoietic markers	CD14, 19, 34, 45	0.16 ± 0.22%	0.25 ± 0.20%
	HLA-DR	0.42 ± 0.14%	0.28 ± 0.08%
	CD271	7.54 ± 7.10%	6.50 ± 3.45%
	CD146	4.08 ± 2.25%	4.14 ± 1.94%

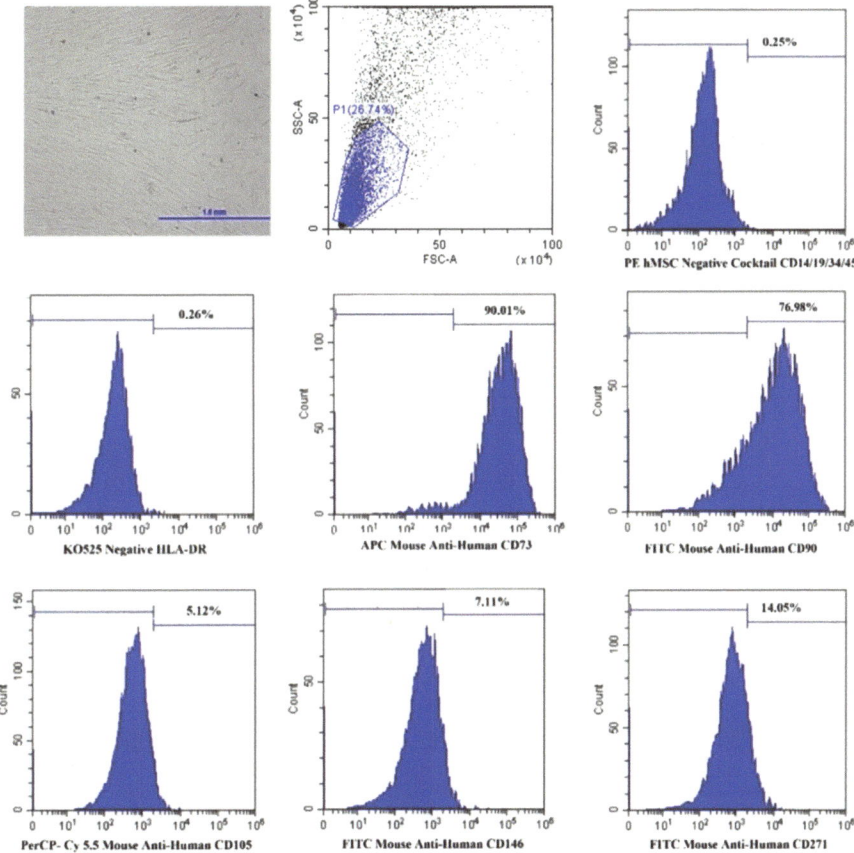

Figure 10. The images of flow cytometry analysis demonstrated the expression profiles of the MSC markers, hematopoietic markers, CD271 and CD146 in the XSF group.

3.7. Cell Proliferation

The proliferation of the ADSCs in the cell–scaffold constructs in the XSF and FBS groups is shown in Figure 11. The cells in both groups proliferated over time until reaching their maximum growth on day 14, and then the growth of the cells decreased on day 21.

The overall growth of the cells in the XSF group was notably higher than it was in the FBS group and this significant difference was detected on day 14 ($p < 0.05$).

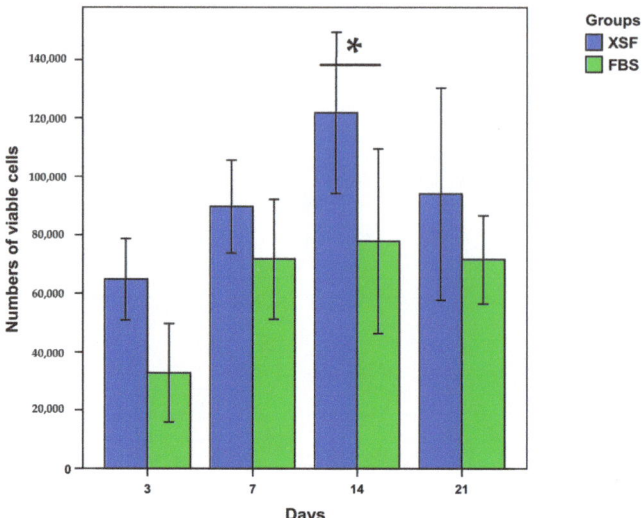

Figure 11. The bar graph shows the numbers of viable cells in the constructs in the XSF and FBS groups over 21 days. The growth of the cells in the XSF group was higher than that of the cells in the FBS group at all time points. The significant difference was detected at day 14 (* $p < 0.05$).

3.8. Cell Differentiation

Expression profiles of the osteoblast-related genes are shown in Figure 12. On day 7, the osteogenic genes in the XSF–OS group, with the exception of Col-1, upregulated to a higher level than those in the FBS–OS and control groups. The genes in both groups downregulated significantly on day 14.

3.9. Morphologies of the Cell–Scaffold Constructs

The SEM pictures demonstrate the behaviors of the ADSCs in the cell–scaffold constructs in the XSF–OS and FBS–OS groups (Figure 13). After seeding, the cells attached and grew well throughout the surfaces of the scaffolds. The cells continued to form multilayer cell sheets covering the entire surfaces over 21 days. By observation, there was no difference in the behaviors of the cells in both groups.

3.10. Experiment In Vivo

All animals tolerated the operation well and were healthy during the observation period. The surgical wounds healed without complications or signs of foreign body reactions.

3.10.1. μ-CT Analysis

The new bone volume fractions are shown in Figure 14. Over the observation period, the new bone volumes in Groups A–C were notably greater than those in Group D ($p > 0.05$). Significant differences were detected between Groups C and D ($p < 0.05$). At week 4 and 8, the new bone volumes in Group C were notably greater than those in Groups A and B. Significant differences were detected between Groups A and C ($p < 0.05$). Interestingly, the new bone formation in Group A was less than that in Group B at all time points ($p > 0.05$). The 3D-constructed images demonstrated that from week 4, the newly formed bone in Groups A–C occurred in the middle portions of the defects, whereas in Group D, it was located at the peripheries. At week 8, the newly formed bone in Groups A–C filled most areas of the roofs of the defects (Figure 15).

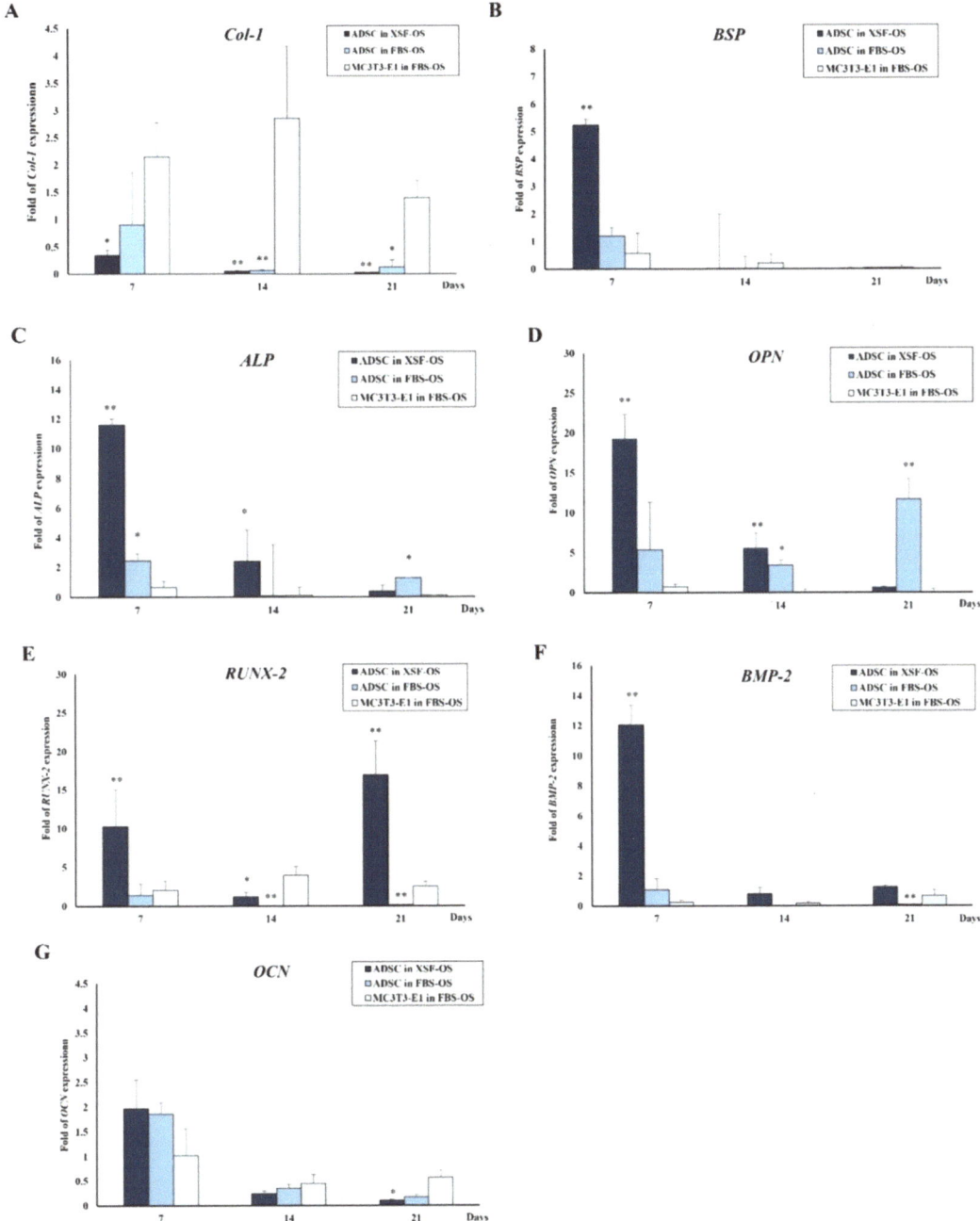

Figure 12. The fold change of gene expression of the cells in the constructs over 21 days; (**A**) Col-1, (**B**) BSP, (**C**) ALP, (**D**) OPN, (**E**) RUNX-2, (**F**) BMP-2 and (**G**) OCN. On day 7, the levels of the osteogenic differentiation genes in the XSF–OS group, with the exception of Col-1, were notably higher than those in the FBS–OS and control groups. The genes significantly downregulated on day 14. The significant differences were at $p < 0.05$ (*) and $p < 0.01$ (**), compared with the control group.

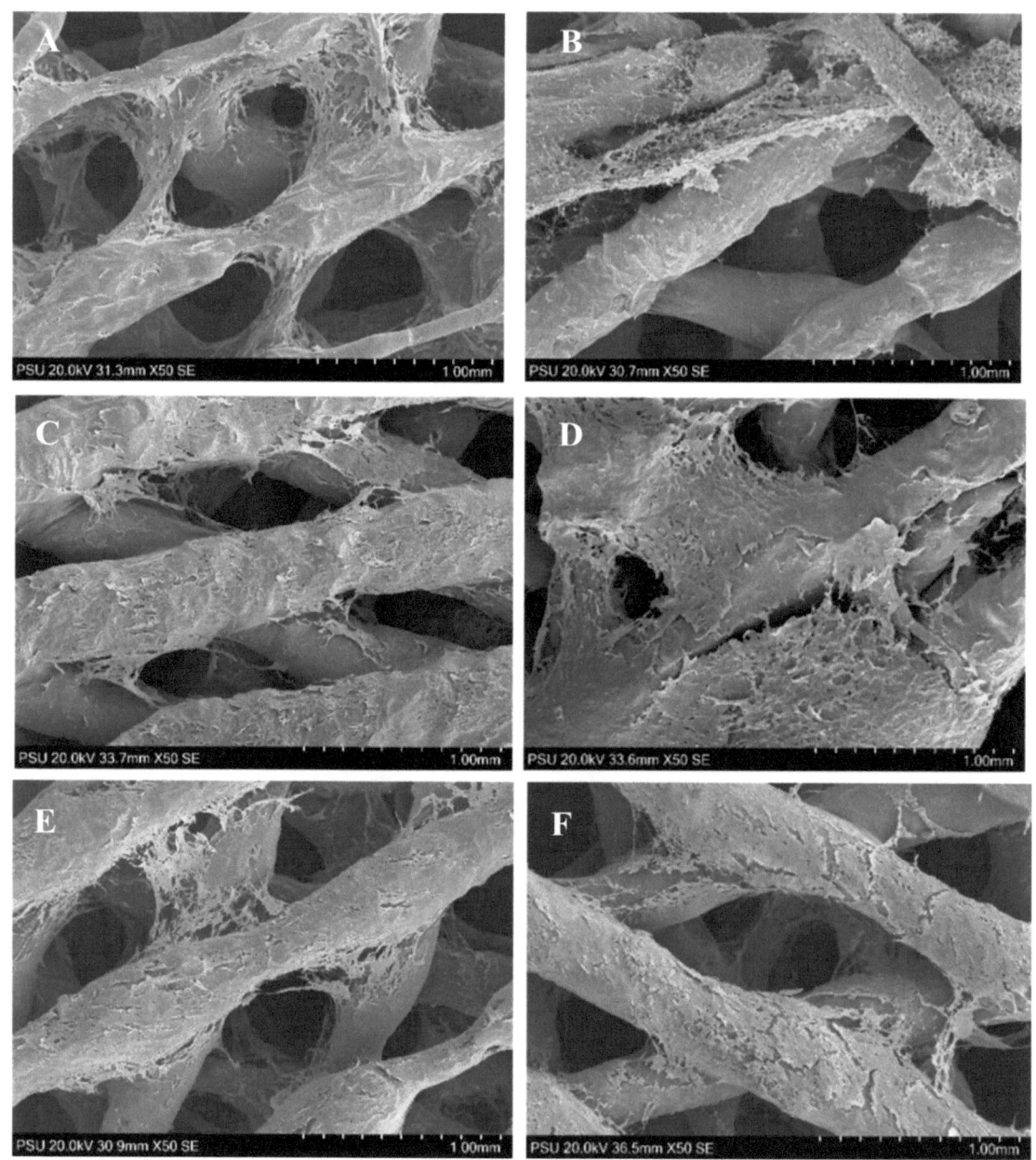

Figure 13. The SEM images of the cell–scaffold constructs in the XSF–OS group (**A,C,E**) and the FBS–OS group (**B,D,F**) at culture days 3 (**A,B**), 14 (**C,D**) and 21 (**E,F**). The cells in both groups attached and grew well on the scaffold surfaces. Dense multilayer cell sheets were observed throughout the scaffolds from day 14 and the morphologies of the cells were difficult to identify.

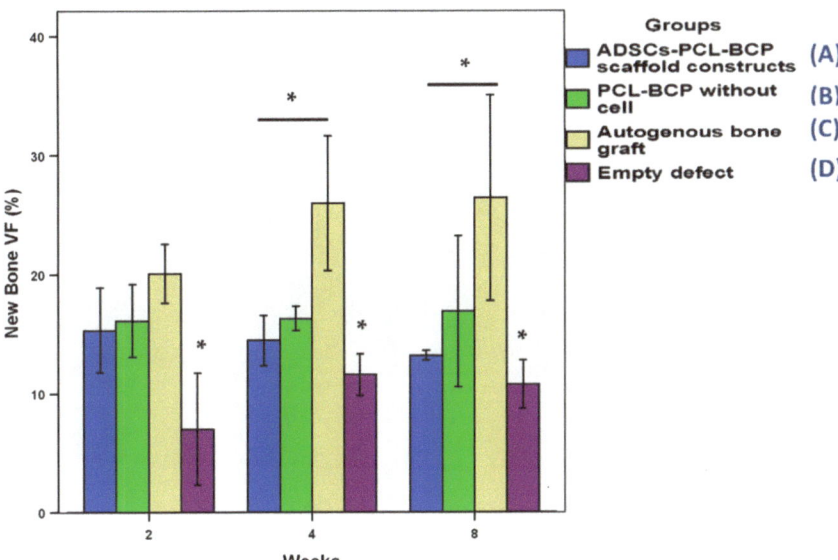

Figure 14. The bar graph demonstrates the new bone VFs for all groups. The new bone VFs in Group C were greater than those in the other groups, whereas those in Group D were less than the other groups over the observation period (*, $p < 0.05$ against group C). From week 4, the new bone volumes in Group C were significantly greater than those in Group A (*, $p < 0.05$). The new bone formation in Group A was less than that in Group B at all time points ($p > 0.05$).

3.10.2. Histological Assessment

Histological features of the implanted areas are shown in Figures 16–18. During histological preparation, the scaffolds in both groups totally dissolved as a result of the histological processes; therefore, the scaffolds were observed as empty spaces within the defects. At week 2 (Figure 16), the scaffolds of Groups A and B were surrounded by dense fibrous tissue. Chronic inflammatory cells were generally found infiltrating around the areas of the collagen membranes rather than the scaffold areas. New bone regeneration was detected extending from the periphery host bone of all groups. At week 4 (Figure 17), the infiltration of the inflammatory cells clearly reduced. Areas of new bone formation in the defects of Groups A–C increased notably compared to week 2. In Groups A and B, the newly formed bone regenerated along the roofs of the defects, whereas in Group C, it was found surrounding the bone graft fragments and in the middle portions of the defects. By observation, the collagen formation within the defects in Group C was denser than that in Groups A and B, which had more adipose tissue in their connective tissue stroma. In Group D, newly formed bone continued to regenerate from the peripheries of the defects. There was less collagen formation in the areas of bone defects compared with the other groups. Remnants of the collagen membranes were still detected in all Groups, but in Group D, the membrane seemed to have collapsed into the defect. At week 8 (Figure 18), in Groups A and B, the larger areas of newly formed bone were detected in some areas within the scaffold frameworks. However, no bone–scaffold integration was detected in either group. New bone bridging of the defects was found only in Group C. In Group D, newly formed bone was detected along the areas of the collagen membrane remnants and there was even less collagen formation within the defect areas.

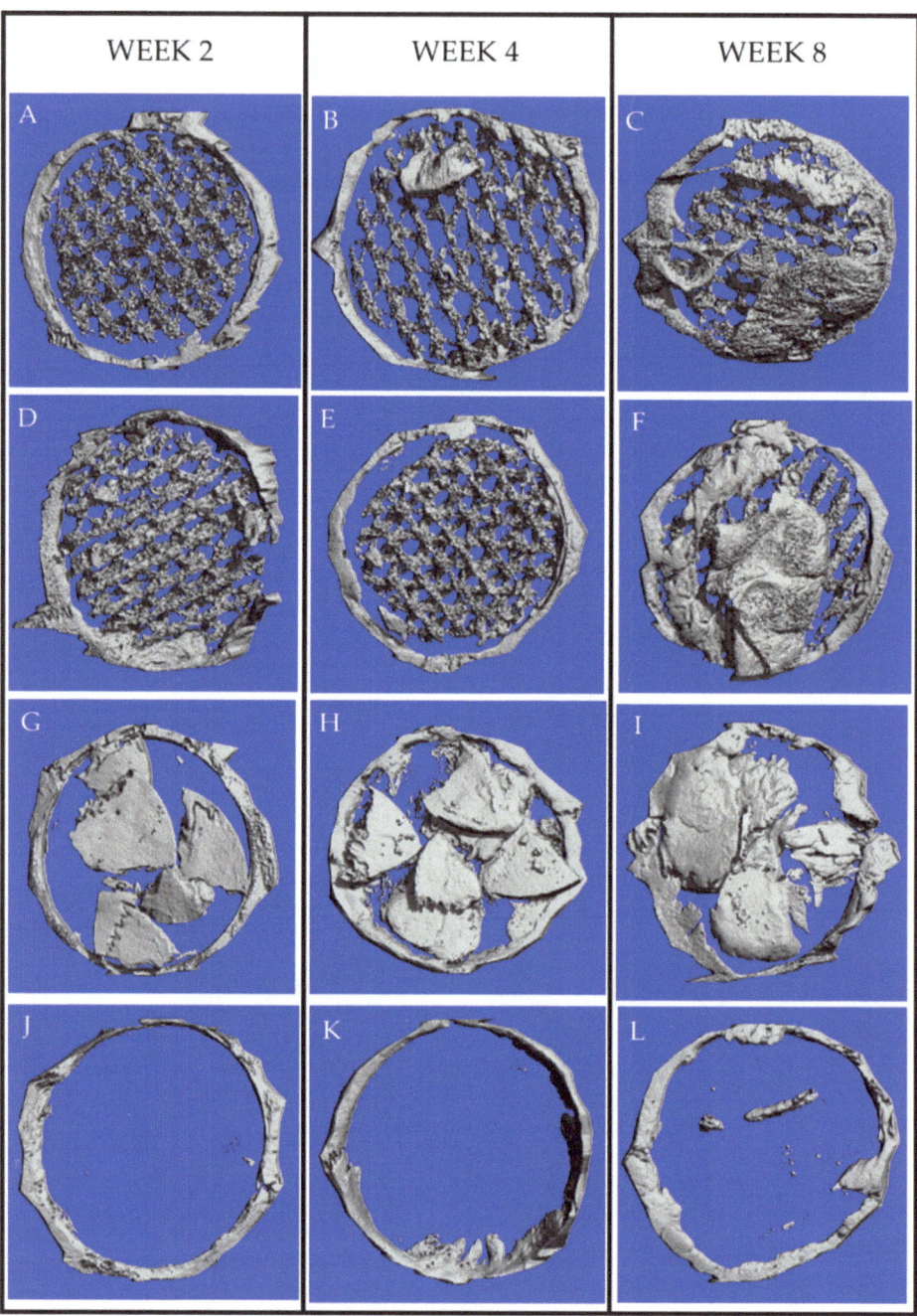

Figure 15. The μ-CT-constructed images demonstrate new bone formation within the defects; (**A**–**C**): Group A, (**D**–**F**): Group B, (**G**–**I**): Group C, and (**J**–**L**): Group D. The new bone formation in Groups A–C was clearly greater than in Group D. At week 8, the newly formed bone in Group A–C almost filled the entire roof of the defects, whereas in Group D, some new bone foci were detected in the middle part of the defects.

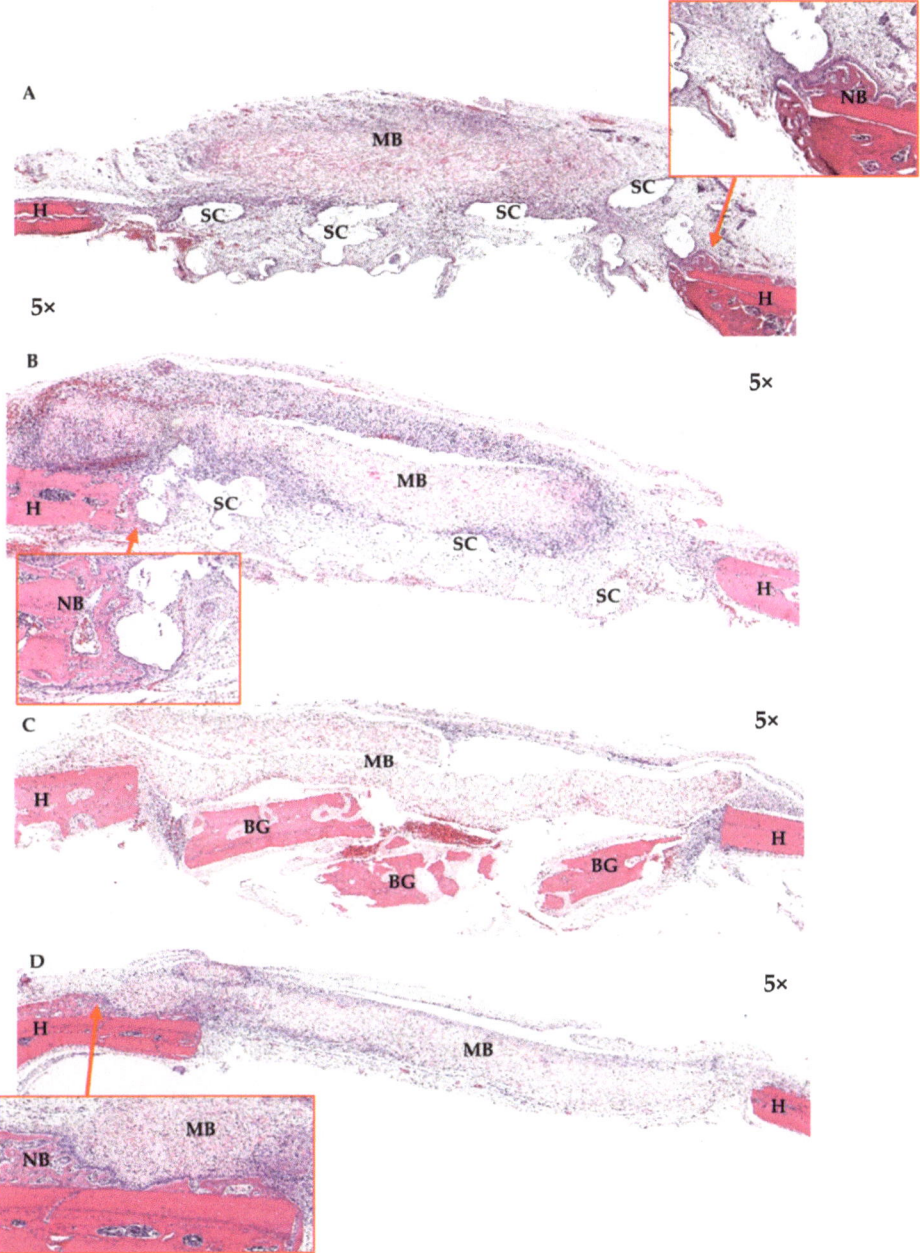

Figure 16. Histological features of the calvarial defects at week 2; (**A**): Group A, (**B**): Group B, (**C**): Group C, and (**D**): Group D. In Groups A and B, the scaffolds and the covering collagen membranes were surrounded by dense fibrous tissue and chronic inflammatory cells. New bone regeneration was detected extending from the periphery host bone (see boxes). In Group C, bone graft fragments were observed along the defect, which had less inflammatory cell infiltration. In Group D, newly formed bone was detected at the margins of the host bone. H = host bone, NB = new bone, SC = scaffold, MB = the collagen membranes, BG = bone graft.

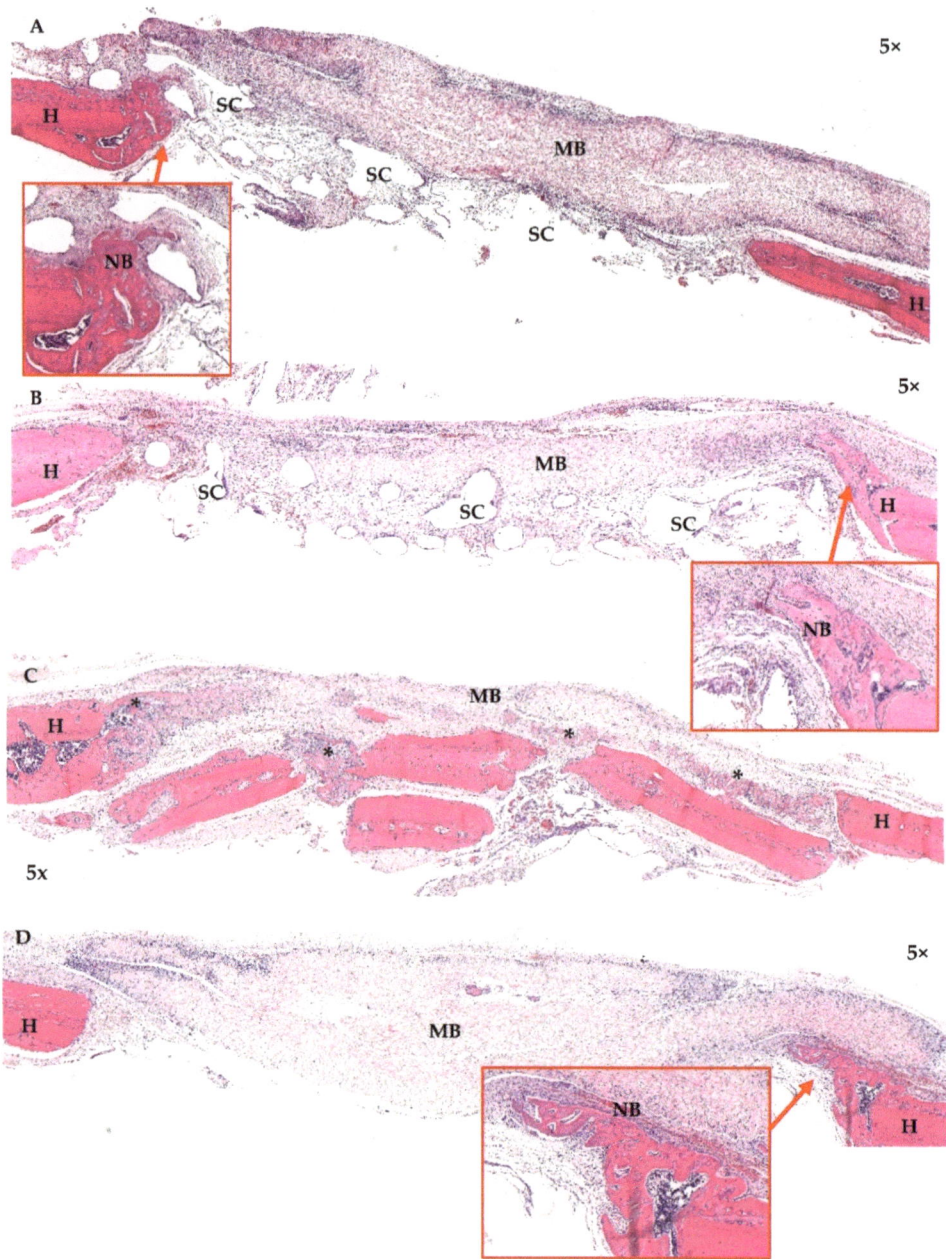

Figure 17. Histological features of the calvarial defects at week 4; (**A**): Group A, (**B**): Group B, (**C**): Group C, and (**D**): Group D. The areas of new bone formation within the defects in Groups A–C were larger than in week 2. In Groups A and B, the newly formed bone came from the peripheries and seemed to regenerate along the roofs of the defects (see boxes), whereas that in Group C was generally found within the middle portions of the defect (*). Remnants of the collagen membranes (MB) were still detected in all Groups. In Group D, newly formed bone regenerated from the peripheries of the defects (see box). H = host bone, NB = new bone, SC = scaffold.

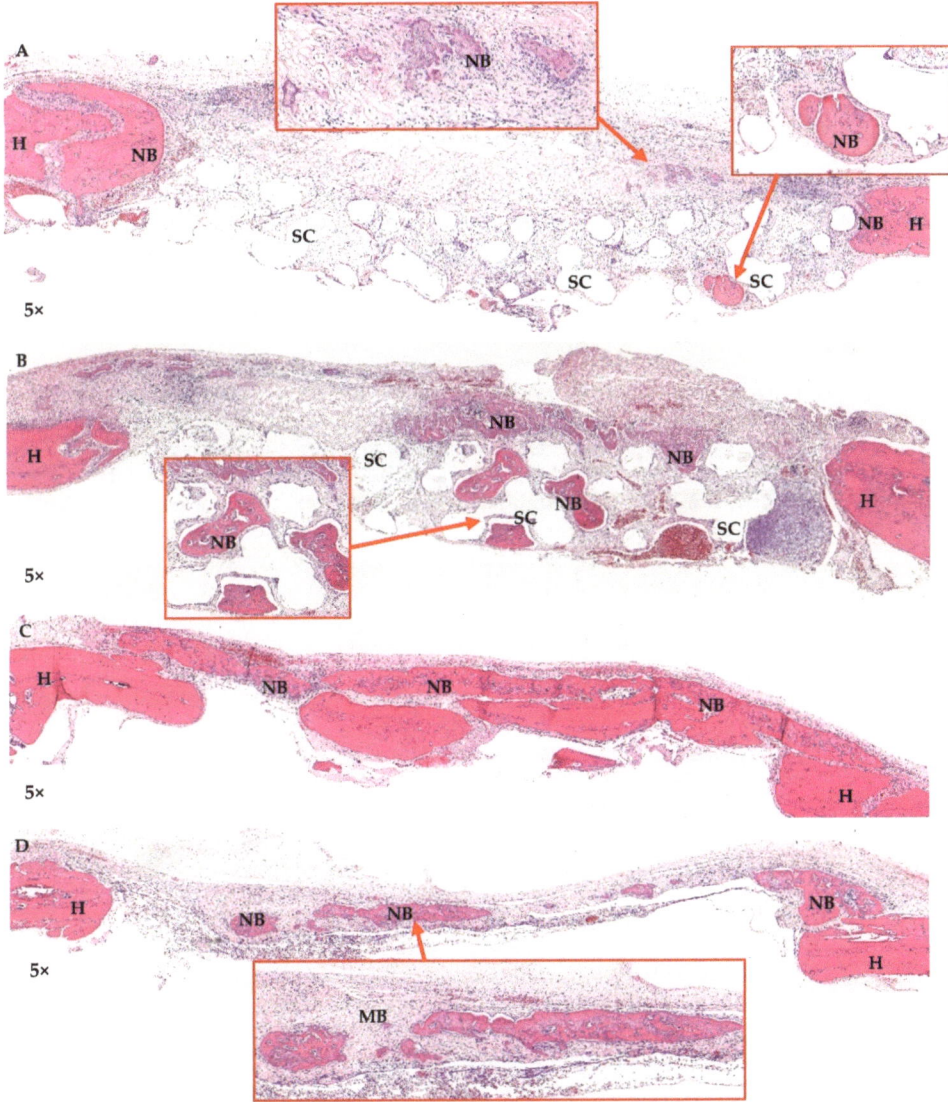

Figure 18. Histological sections of the calvarial defects at week 8; (**A**): Group A, (**B**): Group B, (**C**): Group C, and (**D**): Group D. In Groups A and B, the larger areas of newly formed bone were detected in some areas within the scaffold frameworks (see boxes). New bone bridging of the defects was found only in Group C. In Group D, newly formed bone was detected along the areas of the collagen membrane remnants (MB). H = host bone, NB = new bone, SC = scaffold.

4. Discussion

This study evaluated three major parameters that are clinically relevant to cell-based therapy: the scaffold, the stem cells and the culture medium. This is the first study to combine the PCL–BCP TDP scaffold with buccal fat ADSCs cultured in XSFM and evaluated in vitro and in animal models. For the scaffolds, the PCL and BCP raw materials used for the processing were medical grade and suitable for clinical use. When using the melt-blend processing method, the two components were physically blended without the use of any solvents that might be toxic to the cells. The BCP particles were homogenously dispersed throughout the PCL matrix, as demonstrated by the AR staining. However, the composition of 30% BCP was the maximum amount of filler that could be extruded to become filaments. In the processing, the average diameter of the PCL–BCP filaments was 1.7 ± 0.05 mm, which is in the range of the standard size of 1.74 mm of polymeric filaments used for general 3D printers. In this study, the design of the 3D printed scaffold was still based on the architecture and the interconnecting pore structure of the MSMD scaffold. The scaffolds were designed to have an interconnecting pore system and a pore size of 500 μm to allow vessel and new bone regeneration [38]. The compressive strength of the superior aspects of the scaffolds was 2.98 ± 0.01 MPa, which is comparable to that of human cancellous bone [39], and they could withstand compressive forces of up to 300 N. For the lateral aspects, the scaffolds reached 30% strain at the maximum load of 108.7 ± 4.06 N. Therefore, their mechanical strength would be adequate against wound contraction during the soft tissue healing process. The FTIR analysis indicated separate phases of BCP and PCL without chemical bonding. In addition, the FE-SEM demonstrated poor interfacial adhesion between the materials. An immiscible blend of biodegradable BCP filler and the PCL matrix created several voids inside the filaments that might accelerate degradation of the scaffolds. In animal models, our previous study demonstrated that degradation of the PCL-20% BCP MSCM scaffolds was 30.06 ± 10.48% volume loss over 90 days [3]. The BCP particles distributed throughout the filaments of the scaffolds and exposed on their surfaces would increase the bioactive charges in the calcium–phosphate crystals, confirmed by EDX analysis. Several studies indicated that calcium–phosphate crystals potentially interact with various types of stem cell attached to the materials and can regulate functions of the cells [40–43]. In this study, the PCL–BCP TDP scaffolds induced minimal inflammatory response during the first 2 weeks and new bone could regenerate into the interconnective spaces after 4 weeks. This implied that the scaffolds were biocompatible and suitable for use as an osteoconductive framework.

The adipose tissue from buccal fat pads is easily harvested with minimal tissue site morbidity, and the process is accepted by patients. In this study, the amount of fat tissue harvested from each patient was adequate for the isolation of the stem cells for all of the experiments. The plastic adherence method is cheap, practical and most commonly used for isolating stem cells from several types of tissue. After the processes of enzymatic digestion and centrifugation, the adipose tissue generates a pellet of stromal vascular fraction (SVF). SVF is a heterogeneous mixture of cells that includes ADSCs, vascular endothelial progenitors, pericytes, fibroblasts, smooth muscle cells and various blood cells (44, 45). However, only a small population of ADSCs in SVF, varying from less than 1% to over 15%, can be obtained using this method [44,45]. In the clinical applications of stem cells, a 10^5–10^6 cells/kg/dose is required for therapeutic levels [46,47]. To optimize cell-based therapies, small amounts of the isolated stem cells must be expanded in appropriate culture conditions to obtain sufficient cells. Therefore, the rapid expansion of the cells that can retain their multipotency and reduce exposure to culture reagents is the optimum requirement. In response to the previously mentioned risks of using FBS-containing media, the alternative is to use media containing human serum or xenogeneic serum-free media. Several studies indicated that human serum contains serum proteins, growth factors, growth hormones and essential nutrients for cell metabolism [48–51]. However, volumes of autologous serum are limited and usually inadequate for the entire culture period. In addition, serum from different donors may have different levels of the essential components. XSFM are composed of synthetic and human-derived purified substances without xenogeneic serum supplements. The composition and concentration of the substances are consistent without batch-to-batch variations. Therefore, the media are safe and suitable for clinical applications. Several studies demonstrated that MSCs from different sources, which were expanded in XSFM, potentially retained their phenotypic gene expression, proliferation and multi-differentiation during the multi-passage expansion, similar to those in the FBS-containing medium [30–36]. Interestingly, cells grown in XSFM had more spindle-shaped morphology, which would allow them to be grown in higher densities. Therefore, a greater number of cells can be obtained in a shorter period and can reach their confluence more rapidly. Currently, various commercial XSFM are available for expansion of stem cells. The MesenCult™ medium used in this study is one of the most frequently used media for stem cell culture. Jena et al. assessed the effect of different media, including the MesenCult™ medium, on the population doubling time of bone marrow MSCs [52]. The result showed that the cells in P0, which were cultured in the MesenCult™ medium, had the highest proliferation when compared with the other media ($p < 0.05$). Hoang et al. assessed the efficacy of the MesenCult™ medium for expansion of umbilical cord, bone marrow and adipose-derived MSCs [53]. The results demonstrated that the cells from these different sources retained their biomarker expression from the early to later passages. In addition, they differentiated into osteogenic, adipogenic and chondrogenic lineages. Shahla et al. evaluated the MesenCult™ medium for in vitro expansion of ADSCs as a preliminary protocol for clinical use [30]. The cells were isolated and expanded for five passages in the Mesencult™ medium and FBS-supplemented DMEM. The results demonstrated that the population doubling time of the cells cultured in the Mesencult™ medium was significantly faster than those cultured in the FBS-containing medium ($p < 0.05$). In addition, the cells cultured in the Mesencult™ medium had higher differentiation potential toward osteogenic and adipogenic lineages when compared with those cultured in the FBS-containing medium.

With regard to the consensus between the International Society for Cellular Therapy (ISCT) and the International Federation for Adipose Therapeutics and Science (IFATS) [54], ADSCs should be at least 80% positive to CD13, CD29, CD44, CD73, CD90 and CD105, but less than 2% positive to CD31, CD45 and CD235a. Several studies hypothesize that some subsets of adipose mesenchymal stem cells may arise from the neural crest and are pericytic in origin [55–59]. Therefore, CD271 and 146 are considered to be the specific markers for isolation of MSCs from adipose tissue. Some studies reported that the amount of CD271-positive cells isolated from human subcutaneous adipose tissue was 2.89 to 4.4%, corresponding with the results of our study [60,61]. Therefore, this implies that CD 271- and 146-positive cells were the subpopulation of the majority of ADSCs isolated from the buccal fat tissue. With regard to the result, the cells of both groups expressed the hematopoietic markers at less than 2% and the amount of the CD-73-positive cells of both groups was greater than 80%. However, the CD 90- and 105-positive cells were less than 80% and did not meet the IFATS criteria. This might indicate a heterogeneous population and/or the immune-stimulated status of the cells [62]. Interestingly, the percentage of CD 105 in the XSF group was only $2.94 \pm 1.29\%$, which was significantly less than that in the FBS group. This phenomenon corresponded with the results of previous studies [63–65]. Brohlin et al. compared the MesenCult™ medium and the minimum essential medium-alpha (α-MEM)-containing FBS for culturing ADSCs and bone marrow MSCs (64). The result indicated that the levels of CD73 and CD90 expressed from the cells in both media were not significantly different. However, the levels of CD105 in the cells, which were cultured in the MesenCult™ medium, significantly reduced at passage ten. Decreased expression of CD105 at late passages was also demonstrated in the other studies when MSCs were cultured in the StemPro™ MSC SFM Xeno-free medium (Thermo Fisher Scientific, Waltham, MA, USA) [63,65]. Regarding the results of our study, the growth of the cells in the XSF group was notably higher than that in the FBS group at every observation time point over 21 days. This corresponded to data from several studies, which demonstrated that MSCs cultured in the MesenCult™ medium grew faster than those cultured in FBS-supplemented media [30,36,64,66]. However, Brohlin et al. reported that growth rates of bone-marrow-derived MSCs cultured in the MesenCult™ medium declined after five passages due to cellular senescence during culture [64]. For cell differentiation, most of the osteogenic differentiation genes of the cells in the XSF–OS medium upregulated higher than those in the FBS–OS medium during the first 7 days. The levels of ALP, Runx-2, BSP, OPN and BMP-2 upregulated from the cells in the XSF medium were significantly higher than those of the cells in the FBS medium. It is known that Runx-2 and BMP-2 are the essential transcriptional regulators of osteogenesis, whereas ALP is induced in early calcification during osteoblast development. However, matrix mineralization of the cell–scaffold constructs was not assessed in this study because the calcium and phosphate layers of the BCP filler might confound the result of the positive staining. In vivo, the result demonstrated that the new bone volumes in the defects of Group A that were implanted with the cell–scaffold constructs and Group B that were implanted with the scaffolds alone were notably greater than those of Group D that comprised the empty defects. The newly formed bone in Groups A and B occurred at the middle portions of the defects and filled most areas in the roofs of the defects within 8 weeks. However, the amount of new bone in Group A was lower than that in Group B at all time points. This result contrasted with the findings of several previously mentioned studies. It implies that the osteoconductive property and bioactivity of the PCL–BCP TDP scaffolds were more dominant than the efficacy of the ADSCs. A possible reason might be that there was less homogeneity in the cells that were isolated using the plastic adherence method and rapidly entered the senescence phase of the ADSCs during the culture periods, and this would affect the functions of the cells in the living tissue. These factors must, therefore, be taken into account for stem cell therapy.

5. Conclusions

The 3D printed PCL–BCP scaffolds were biocompatible and suitable for use as the osteoconductive framework. The XSF medium was proved to support the proliferation and differentiation of ADSCs in vitro. Although the cell–scaffold constructs had no benefit in enhancing new bone formation in animal models, the scaffolds and the medium are still practical for further clinical studies and applications.

Author Contributions: Conceptualization, N.T.; methodology, N.T.; investigation, N.T., L.C., P.S., W.S. (Woraporn Supphaprasitt), T.S. and W.S. (Woraluk Srimanok); Surgeons, L.C., N.L., D.S. and S.V.; resources, N.T. and W.S. (Woraporn Supphaprasitt); data curation, N.T. and W.S. (Woraporn Supphaprasitt); writing—original draft preparation, N.T. and W.S. (Woraporn Supphaprasitt); writing—review and editing, N.T. and W.S. (Woraporn Supphaprasitt); visualization, all authors; supervision, N.T.; project administration, N.T. and W.S. (Woraporn Supphaprasitt); All authors have read and agreed to the published version of the manuscript.

Funding: This study was supported by a grant from the Faculty of Dentistry, Prince of Songkla University (DENT 630502).

Institutional Review Board Statement: The human research ethics committee, Faculty of Dentistry, Prince of Songkla University (EC6012-37-P-LR) and the animal experiment ethics committee of the Prince of Songkla University (MHESI 68014/860).

Informed Consent Statement: EC6012-37-P-LR.

Data Availability Statement: The data presented in this study are available on request from the corresponding author.

Acknowledgments: The authors would like to thank The Cranio-Maxillofacial Hard Tissue Engineering Center, Department of Oral and Maxillofacial Surgery and The Research Unit, Faculty of Dentistry, Prince of Songkla University for their support.

Conflicts of Interest: The authors declare no conflict of interest.

References

1. Thuaksuban, N.; Luntheng, T.; Monmaturapoj, N. Physical characteristics and biocompatibility of the polycaprolactone-biphasic calcium phosphate scaffolds fabricated using the modified melt stretching and multilayer deposition. *J. Biomater. Appl.* **2016**, *30*, 1460–1472. [CrossRef] [PubMed]
2. Thuaksuban, N.; Monmaturapoj, N.; Luntheng, T. Effects of polycaprolactone-biphasic calcium phosphate scaffolds on enhancing growth and differentiation of osteoblasts. *Biomed. Mater. Eng.* **2018**, *29*, 159–176. [CrossRef] [PubMed]

3. Thuaksuban, N.; Pannak, R.; Boonyaphiphat, P.; Monmaturapoj, N. In vivo biocompatibility and degradation of novel Polycaprolactone-Biphasic Calcium phosphate scaffolds used as a bone substitute. *Biomed. Mater. Eng.* **2018**, *29*, 253–267. [CrossRef] [PubMed]
4. Wongsupa, N.; Nuntanaranont, T.; Kamolmattayakul, S.; Thuaksuban, N. Assessment of bone regeneration of a tissue-engineered bone complex using human dental pulp stem cells/poly(epsilon-caprolactone)-biphasic calcium phosphate scaffold constructs in rabbit calvarial defects. *J. Mater. Sci. Mater. Med.* **2017**, *28*, 77. [CrossRef]
5. Wongsupa, N.; Nuntanaranont, T.; Kamolmattayakul, S.; Thuaksuban, N. Biological characteristic effects of human dental pulp stem cells on poly-epsilon-caprolactone-biphasic calcium phosphate fabricated scaffolds using modified melt stretching and multilayer deposition. *J. Mater. Sci. Mater. Med.* **2017**, *28*, 25. [CrossRef]
6. Rittipakorn, P.; Thuaksuban, N.; Mai-Ngam, K.; Charoenla, S.; Noppakunmongkolchai, W. Bioactivity of a novel polycaprolactone-hydroxyapatite scaffold used as a carrier of low dose BMP-2: An in vitro study. *Polymers* **2021**, *13*, 466. [CrossRef]
7. Chuenjitkuntaworn, B.; Inrung, W.; Damrongsri, D.; Mekaapiruk, K.; Supaphol, P.; Pavasant, P. Polycaprolactone/hydroxyapatite composite scaffolds: Preparation, characterization, and in vitro and in vivo biological responses of human primary bone cells. *J. Biomed. Mater. Res. A* **2010**, *94*, 241–251. [CrossRef]
8. Darie-Nita, R.N.; Rapa, M.; Frackowiak, S. Special features of polyester-based materials for medical applications. *Polymers* **2022**, *14*, 951. [CrossRef]
9. Elfick, A.P. Poly(epsilon-caprolactone) as a potential material for a temporary joint spacer. *Biomaterials* **2002**, *23*, 4463–4467. [CrossRef]
10. Hutmacher, D.W. Scaffolds in tissue engineering bone and cartilage. *Biomaterials* **2000**, *21*, 2529–2543. [CrossRef]
11. Bose, S.; Tarafder, S. Calcium phosphate ceramic systems in growth factor and drug delivery for bone tissue engineering: A review. *Acta Biomater.* **2012**, *8*, 1401–1421. [CrossRef]
12. De Oliveira Lomelino, R.; Castro-Silva, I.I.; Linhares, A.B.R.; Alves, G.G.; de Albuquerque Santos, S.R.; Gameiro, V.S.; Rossi, A.M.; Granjeiro, J.M. The association of human primary bone cells with biphasic calcium phosphate (betaTCP/HA 70:30) granules increases bone repair. *J. Mater. Sci. Mater. Med.* **2012**, *23*, 781–788. [CrossRef]
13. Chen, W.; Liu, X.; Chen, Q.; Bao, C.; Zhao, L.; Zhu, Z.; Xu, H.H. Angiogenic and osteogenic regeneration in rats via calcium phosphate scaffold and endothelial cell co-culture with human bone marrow mesenchymal stem cells (MSCs), human umbilical cord MSCs, human induced pluripotent stem cell-derived MSCs and human embryonic stem cell-derived MSCs. *J. Tissue Eng. Regen. Med.* **2018**, *12*, 191–203.
14. Diomede, F.; Zini, N.; Gatta, V.; Fulle, S.; Merciaro, I.; D'aurora, M.; La Rovere, R.M.; Traini, T.; Pizzicannella, J.; Ballerini, P.; et al. Human periodontal ligament stem cells cultured onto cortico-cancellous scaffold drive bone regenerative process. *Eur. Cell Mater.* **2016**, *32*, 181–201. [CrossRef]
15. Lendeckel, S.; Jödicke, A.; Christophis, P.; Heidinger, K.; Wolff, J.; Fraser, J.K.; Hedrick, M.H.; Berthold, L.; Howaldt, H.-P. Autologous stem cells (adipose) and fibrin glue used to treat widespread traumatic calvarial defects: Case report. *J. Craniomaxillofac. Surg.* **2004**, *32*, 370–373. [CrossRef]
16. Nuntanaranont, T.; Promboot, T.; Sutapreyasri, S. Effect of expanded bone marrow-derived osteoprogenitor cells seeded into polycaprolactone/tricalcium phosphate scaffolds in new bone regeneration of rabbit mandibular defects. *J. Mater. Sci. Mater. Med.* **2018**, *29*, 24. [CrossRef]
17. Schantz, J.-T.; Hutmacher, D.W.; Lam, C.X.F.; Brinkmann, M.; Wong, K.M.; Lim, T.C.; Chou, N.; Guldberg, R.E.; Teoh, S.H. Repair of calvarial defects with customised tissue-engineered bone grafts II. Evaluation of cellular efficiency and efficacy in vivo. *Tissue Eng.* **2003**, *9* (Suppl. 1), S127–S139. [CrossRef]
18. Shao, X.X.; Hutmacher, D.W.; Ho, S.T.; Goh, J.C.; Lee, E.H. Evaluation of a hybrid scaffold/cell construct in repair of high-load-bearing osteochondral defects in rabbits. *Biomaterials* **2006**, *27*, 1071–1080. [CrossRef]
19. Zhou, Y.; Chen, F.; Ho, S.T.; Woodruff, M.A.; Lim, T.M.; Hutmacher, D.W. Combined marrow stromal cell-sheet techniques and high-strength biodegradable composite scaffolds for engineered functional bone grafts. *Biomaterials* **2007**, *28*, 814–824. [CrossRef]
20. Farre-Guasch, E.; Marti-Page, C.; Hernadez-Alfaro, F.; Klein-Nulend, J.; Casals, N. Buccal fat pad, an oral access source of human adipose stem cells with potential for osteochondral tissue engineering: An in vitro study. *Tissue Eng. Part C Methods* **2010**, *16*, 1083–1094. [CrossRef]
21. Karantalis, V.; Hare, J.M. Use of mesenchymal stem cells for therapy of cardiac disease. *Circ. Res.* **2015**, *116*, 1413–1430. [CrossRef]
22. Kern, S.; Eichler, H.; Stoeve, J.; Kluter, H.; Bieback, K. Comparative analysis of mesenchymal stem cells from bone marrow, umbilical cord blood, or adipose tissue. *Stem Cells* **2006**, *24*, 1294–1301. [CrossRef]
23. Suzuki, E.; Fujita, D.; Takahashi, M.; Oba, S.; Nishimatsu, H. Adipose tissue-derived stem cells as a therapeutic tool for cardiovascular disease. *World J. Cardiol.* **2015**, *7*, 454–465. [CrossRef]
24. Zuk, P.A.; Zhu, M.; Ashjian, P.; De Ugarte, D.A.; Huang, J.I.; Mizuno, H.; Alfonso, Z.C.; Fraser, J.K.; Benhaim, P.; Hedrick, M.H. Human adipose tissue is a source of multipotent stem cells. *Mol. Biol. Cell* **2002**, *13*, 4279–4295. [CrossRef]
25. Broccaioli, E.; Niada, S.; Rasperini, G.; Ferreira, L.M.; Arrigoni, E.; Yenagi, V.; Brini, A.T. Mesenchymal stem cells from Bichat's fat pad: In vitro comparison with adipose-derived stem cells from subcutaneous tissue. *Biores. Open Access* **2013**, *2*, 107–117. [CrossRef]

26. Niada, S.; Ferreira, L.M.; Arrigoni, E.; Addis, A.; Campagnol, M.; Broccaioli, E.; Brini, A.T. Porcine adipose-derived stem cells from buccal fat pad and subcutaneous adipose tissue for future preclinical studies in oral surgery. *Stem Cell Res. Ther.* **2013**, *4*, 148. [CrossRef]
27. Khojasteh, A.; Sadeghi, N. Application of buccal fat pad-derived stem cells in combination with autogenous iliac bone graft in the treatment of maxillomandibular atrophy: A preliminary human study. *Int. J. Oral Maxillofac. Surg.* **2016**, *45*, 864–871. [CrossRef]
28. Gottipamula, S.; Muttigi, M.S.; Kolkundkar, U.; Seetharam, R.N. Serum-free media for the production of human mesenchymal stromal cells: A review. *Cell Prolif.* **2013**, *46*, 608–627. [CrossRef]
29. Cimino, M.; Goncalves, R.M.; Barrias, C.C.; Martins, M.C.L. Xeno-free strategies for safe human mesenchymal stem/stromal cell expansion: Supplements and coatings. *Stem Cells Int.* **2017**, *2017*, 6597815. [CrossRef]
30. Al-Saqi, S.H.; Saliem, M.; Asikainen, S.; Quezada, H.C.; Ekblad, Å.; Hovatta, O.; Le Blanc, K.; Jonasson, A.F.; Götherström, C. Defined serum-free media for in vitro expansion of adipose-derived mesenchymal stem cells. *Cytotherapy* **2014**, *16*, 915–926. [CrossRef]
31. Chase, L.G.; Lakshmipathy, U.; Solchaga, L.A.; Rao, M.S.; Vemuri, M.C. A novel serum-free medium for the expansion of human mesenchymal stem cells. *Stem Cell Res. Ther.* **2010**, *1*, 8. [CrossRef] [PubMed]
32. Chase, L.G.; Yang, S.; Zachar, V.; Yang, Z.; Lakshmipathy, U.; Bradford, J.; Boucher, S.E.; Vemuri, M.C. Development and characterization of a clinically compliant xeno-free culture medium in good manufacturing practice for human multipotent mesenchymal stem cells. *Stem Cells Transl. Med.* **2012**, *1*, 750–758. [CrossRef] [PubMed]
33. Julavijitphong, S.; Wichitwiengrat, S.; Tirawanchai, N.; Ruangvutilert, P.; Vantanasiri, C.; Phermthai, T. A xeno-free culture method that enhances Wharton's jelly mesenchymal stromal cell culture efficiency over traditional animal serum-supplemented cultures. *Cytotherapy* **2014**, *16*, 683–691. [CrossRef]
34. Kandoi, S.; Patra, B.; Vidyasekar, P.; Sivanesan, D.; Verma, R.S. Evaluation of platelet lysate as a substitute for FBS in explant and enzymatic isolation methods of human umbilical cord MSCs. *Sci. Rep.* **2018**, *8*, 12439. [CrossRef] [PubMed]
35. Simões, I.N.; Boura, J.S.; dos Santos, F.; Andrade, P.Z.; Cardoso, C.M.; Gimble, J.M.; da Silva, C.L.; Cabral, J.M. Human mesenchymal stem cells from the umbilical cord matrix: Successful isolation and ex vivo expansion using serum-/xeno-free culture media. *Biotechnol. J.* **2013**, *8*, 448–458. [CrossRef] [PubMed]
36. Swamynathan, P.; Venugopal, P.; Kannan, S.; Thej, C.; Kolkundar, U.; Bhagwat, S.; Ta, M.; Majumdar, A.S.; Balasubramanian, S. Are serum-free and xeno-free culture conditions ideal for large scale clinical grade expansion of Wharton's jelly derived mesenchymal stem cells? A comparative study. *Stem Cell Res. Ther.* **2014**, *5*, 88. [CrossRef] [PubMed]
37. Dominici, M.L.B.K.; Le Blanc, K.; Mueller, I.; Slaper-Cortenbach, I.; Marini, F.C.; Krause, D.S.; Deans, R.J.; Keating, A.; Prockop, D.J.; Horwitz, E.M. Minimal criteria for defining multipotent mesenchymal stromal cells. The international society for cellular therapy position statement. *Cytotherapy* **2006**, *8*, 315–317. [CrossRef] [PubMed]
38. Thuaksuban, N.; Nuntanaranont, T.; Pattanachot, W.; Suttapreyasri, S.; Cheung, L.K. Biodegradable polycaprolactone-chitosan three-dimensional scaffolds fabricated by melt stretching and multilayer deposition for bone tissue engineering: Assessment of the physical properties and cellular response. *Biomed. Mater.* **2011**, *6*, 015009. [CrossRef]
39. Lee, S.H.; Shin, H. Matrices and scaffolds for delivery of bioactive molecules in bone and cartilage tissue engineering. *Adv. Drug Deliv. Rev.* **2007**, *59*, 339–359. [CrossRef]
40. Barradas, A.M.; Yuan, H.; van Blitterswijk, C.A.; Habibovic, P. Osteoinductive biomaterials: Current knowledge of properties, experimental models and biological mechanisms. *Eur. Cell Mater.* **2011**, *21*, 407–429. [CrossRef]
41. Koegler, P.; Clayton, A.; Thissen, H.; Santos, G.N.; Kingshott, P. The influence of nanostructured materials on biointerfacial interactions. *Adv. Drug Deliv. Rev.* **2012**, *64*, 1820–1839. [CrossRef]
42. LeGeros, R.Z. Calcium phosphate-based osteoinductive materials. *Chem. Rev.* **2008**, *108*, 4742–4753. [CrossRef]
43. Wu, X.; Itoh, N.; Taniguchi, T.; Nakanishi, T.; Tanaka, K. Requirement of calcium and phosphate ions in expression of sodium-dependent vitamin C transporter 2 and osteopontin in MC3T3-E1 osteoblastic cells. *Biochim. Biophys. Acta* **2003**, *1641*, 65–70. [CrossRef]
44. Aronowitz, J.A.; Lockhart, R.A.; Hakakian, C.S. Mechanical versus enzymatic isolation of stromal vascular fraction cells from adipose tissue. *Springerplus* **2015**, *4*, 713. [CrossRef]
45. Raposio, E.; Simonacci, F.; Perrotta, R.E. Adipose-derived stem cells: Comparison between two methods of isolation for clinical applications. *Ann. Med. Surg.* **2017**, *20*, 87–91. [CrossRef]
46. Karantalis, V.; Schulman, I.H.; Balkan, W.; Hare, J.M. Allogeneic cell therapy: A new paradigm in therapeutics. *Circ. Res.* **2015**, *116*, 12–15. [CrossRef]
47. Golpanian, S.; Schulman, I.H.; Ebert, R.F.; Heldman, A.W.; DiFede, D.L.; Yang, P.C.; Wu, J.C.; Bolli, R.; Perin, E.C.; Moyé, L.; et al. Concise review: Review and perspective of cell dosage and routes of administration from preclinical and clinical studies of stem cell therapy for heart disease. *Stem Cells Transl. Med.* **2016**, *5*, 186–191. [CrossRef]
48. Aldahmash, A.; Haack-Sorensen, M.; Al-Nbaheen, M.; Harkness, L.; Abdallah, B.M.; Kassem, M. Human serum is as efficient as fetal bovine serum in supporting proliferation and differentiation of human multipotent stromal (mesenchymal) stem cells in vitro and in vivo. *Stem Cell Rev. Rep.* **2011**, *7*, 860–868. [CrossRef]
49. Bieback, K.; Hecker, A.; Kocaömer, A.; Lannert, H.; Schallmoser, K.; Strunk, D.; Klüter, H. Human alternatives to fetal bovine serum for the expansion of mesenchymal stromal cells from bone marrow. *Stem Cells* **2009**, *27*, 2331–2341. [CrossRef]

50. Kocaoemer, A.; Kern, S.; Kluter, H.; Bieback, K. Human AB serum and thrombin-activated platelet-rich plasma are suitable alternatives to fetal calf serum for the expansion of mesenchymal stem cells from adipose tissue. *Stem Cells* **2007**, *25*, 1270–1278. [CrossRef]
51. Tateishi, K.; Ando, W.; Higuchi, C.; Hart, D.A.; Hashimoto, J.; Nakata, K.; Yoshikawa, H.; Nakamura, N. Comparison of human serum with fetal bovine serum for expansion and differentiation of human synovial MSC: Potential feasibility for clinical applications. *Cell Transplant.* **2008**, *17*, 549–557. [CrossRef]
52. Jena, D.; Kharche, S.D.; Singh, S.P.; Rani, S.; Dige, M.S.; Ranjan, R.; Singh, S.K.; Kumar, H. Growth and proliferation of caprine bone marrow mesenchymal stem cells on different culture media. *Tissue Cell* **2020**, *67*, 101446. [CrossRef]
53. Hoang, V.T.; Trinh, Q.-M.; Phuong, D.T.M.; Bui, H.T.H.; Hang, L.M.; Ngan, N.T.H.; Anh, N.T.T.; Nhi, P.Y.; Nhung, T.T.H.; Lien, H.T.; et al. Standardized xeno- and serum-free culture platform enables large-scale expansion of high-quality mesenchymal stem/stromal cells from perinatal and adult tissue sources. *Cytotherapy* **2021**, *23*, 88–99. [CrossRef]
54. Bourin, P.; Bunnell, B.A.; Casteilla, L.; Dominici, M.; Katz, A.J.; March, K.L.; Redl, H.; Rubin, J.P.; Yoshimura, K.; Gimble, J.M. Stromal cells from the adipose tissue-derived stromal vascular fraction and culture expanded adipose tissue-derived stromal/stem cells: A joint statement of the International Federation for Adipose Therapeutics and Science (IFATS) and the International Society for Cellular Therapy (ISCT). *Cytotherapy* **2013**, *15*, 641–648.
55. Billon, N.; Iannarelli, P.; Monteiro, M.C.; Glavieux-Pardanaud, C.; Richardson, W.D.; Kessaris, N.; Dani, C.; Dupin, E. The generation of adipocytes by the neural crest. *Development* **2007**, *134*, 2283–2292. [CrossRef] [PubMed]
56. D'Aquino, R.; De Rosa, A.; Laino, G.; Caruso, F.; Guida, L.; Rullo, R.; Checchi, V.; Laino, L.; Tirino, V.; Papaccio, G. Human dental pulp stem cells: From biology to clinical applications. *J. Exp. Zool. B Mol. Dev. Evol.* **2009**, *312*, 408–415. [CrossRef] [PubMed]
57. Lv, F.J.; Tuan, R.S.; Cheung, K.M.; Leung, V.Y. Concise review: The surface markers and identity of human mesenchymal stem cells. *Stem. Cells* **2014**, *32*, 1408–1419. [CrossRef] [PubMed]
58. Navabazam, A.R.; Sadeghian Nodoshan, F.; Sheikhha, M.H.; Miresmaeili, S.M.; Soleimani, M.; Fesahat, F. Characterization of mesenchymal stem cells from human dental pulp, preapical follicle and periodontal ligament. *Iran. J. Reprod. Med.* **2013**, *11*, 235–242.
59. Stevens, A.; Zuliani, T.; Olejnik, C.; LeRoy, H.; Obriot, H.; Kerr-Conte, J.; Formstecher, P.; Bailliez, Y.; Polakowska, R.R. Human dental pulp stem cells differentiate into neural crest-derived melanocytes and have label-retaining and sphere-forming abilities. *Stem. Cells Dev.* **2008**, *17*, 1175–1184. [CrossRef]
60. Cuevas-Diaz Duran, R.; Gonzalez-Garza, M.T.; Cardenas-Lopez, A.; Chavez-Castilla, L.; Cruz-Vega, D.E.; Moreno-Cuevas, J.E. Age-related yield of adipose-derived stem cells bearing the low-affinity nerve growth factor receptor. *Stem. Cells Int.* **2013**, *2013*, 372164. [CrossRef]
61. Quirici, N.; Scavullo, C.; DE Girolamo, L.; Lopa, S.; Arrigoni, E.; Deliliers, G.L.; Brini, A.T. Anti-L-NGFR and -CD34 monoclonal antibodies identify multipotent mesenchymal stem cells in human adipose tissue. *Stem. Cells Dev.* **2010**, *19*, 915–925. [CrossRef]
62. Bui, H.T.H.; Nguyen, L.T.; Than, U.T.T. Influences of xeno-free media on mesenchymal stem cell expansion for clinical application. *Tissue Eng. Regen. Med.* **2021**, *18*, 15–23. [CrossRef]
63. Bobis-Wozowicz, S.; Kmiotek, K.; Kania, K.; Karnas, E.; Labedz-Maslowska, A.; Sekula, M.; Kedracka-Krok, S.; Kolcz, J.; Boruczkowski, D.; Madeja, Z.; et al. Diverse impact of xeno-free conditions on biological and regenerative properties of hUC-MSCs and their extracellular vesicles. *J. Mol. Med.* **2017**, *95*, 205–220. [CrossRef]
64. Brohlin, M.; Kelk, P.; Wiberg, M.; Kingham, P.J. Effects of a defined xeno-free medium on the growth and neurotrophic and angiogenic properties of human adult stem cells. *Cytotherapy* **2017**, *19*, 629–639. [CrossRef]
65. Cimino, M.; Gonçalves, R.M.; Bauman, E.; Barroso-Vilares, M.; Logarinho, E.; Barrias, C.C.; Martins, M.C.L. Optimization of the use of a pharmaceutical grade xeno-free medium for in vitro expansion of human mesenchymal stem/stromal cells. *J. Tissue Eng. Regen. Med.* **2018**, *12*, e1785–e1795. [CrossRef]
66. Gottipamula, S.; Ashwin, K.M.; Muttigi, M.S.; Kannan, S.; Kolkundkar, U.; Seetharam, R.N. Isolation, expansion and characterization of bone marrow-derived mesenchymal stromal cells in serum-free conditions. *Cell Tissue Res.* **2014**, *356*, 123–135. [CrossRef]

Article

Poly (Butylene Succinate)/Silicon Nitride Nanocomposite with Optimized Physicochemical Properties, Biocompatibility, Degradability, and Osteogenesis for Cranial Bone Repair

Qinghui Zhao and Shaorong Gao *

Institute for Regenerative Medicine, National Stem Cell Translational Resource Center, Shanghai East Hospital, School of Life Sciences and Technology, Tongji University, Shanghai 200092, China
* Correspondence: gaoshaorong@tongji.edu.cn

Abstract: Congenital disease, tumors, infections, and trauma are the main reasons for cranial bone defects. Herein, poly (butylene succinate) (PB)/silicon nitride (Si_3N_4) nanocomposites (PSC) with Si_3N_4 content of 15 w% (PSC15) and 30 w% (PSC30) were fabricated for cranial bone repair. Compared with PB, the compressive strength, hydrophilicity, surface roughness, and protein absorption of nanocomposites were increased with the increase in Si_3N_4 content (from 15 w% to 30 w%). Furthermore, the cell adhesion, multiplication, and osteoblastic differentiation on PSC were significantly enhanced with the Si_3N_4 content increasing in vitro. PSC30 exhibited optimized physicochemical properties (compressive strength, surface roughness, hydrophilicity, and protein adsorption) and cytocompatibility. The m-CT and histological results displayed that the new bone formation for SPC30 obviously increased compared with PB, and PSC30 displayed proper degradability (75.3 w% at 12 weeks) and was gradually replaced by new bone tissue in vivo. The addition of Si_3N_4 into PB not only optimized the surface performances of PSC but also improved the degradability of PSC, which led to the release of Si ions and a weak alkaline environment that significantly promoted cell response and tissue regeneration. In short, the enhancements of cellular responses and bone regeneration of PSC30 were attributed to the synergism of the optimized surface performances and slow release of Si ion, and PSC30 were better than PB. Accordingly, PSC30, with good biocompatibility and degradability, displayed a promising and huge potential for cranial bone construction.

Keywords: nanocomposite; Si_3N_4; cellular response; degradability; bone regeneration

Citation: Zhao, Q.; Gao, S. Poly (Butylene Succinate)/Silicon Nitride Nanocomposite with Optimized Physicochemical Properties, Biocompatibility, Degradability, and Osteogenesis for Cranial Bone Repair. *J. Funct. Biomater.* **2022**, *13*, 231. https://doi.org/10.3390/jfb13040231

Academic Editors: Kunyu Zhang, Qian Feng, Yongsheng Yu, Boguang Yang and Florin Miculescu

Received: 27 September 2022
Accepted: 31 October 2022
Published: 8 November 2022

Publisher's Note: MDPI stays neutral with regard to jurisdictional claims in published maps and institutional affiliations.

Copyright: © 2022 by the authors. Licensee MDPI, Basel, Switzerland. This article is an open access article distributed under the terms and conditions of the Creative Commons Attribution (CC BY) license (https://creativecommons.org/licenses/by/4.0/).

1. Introduction

Cranial bone constructs the neurocranium of the skull that forms a cavity and provides mechanical support to protect the brain [1]. Patients with craniofacial bone defects caused by different disorders (e.g., trauma, infection, tumor resection, and congenital malformation) suffer from problems with chewing, speech, and aesthetics [2]. Large cranial defects (e.g., critical-size defects) lead to a large area of the unprotected brain experiencing remarkable cosmetic deformity [3]. Reconstruction of the cranial defect (Cranioplasty) is commonly carried out to restore the appearance in neurosurgical surgeries, and successful reconstruction of the cranial defect is an integral step to restoring craniofacial function and improving the quality of life [4]. Cranioplasty cosmetically reshapes the cranial defect and provides a physical barrier for the protection of the cerebral structure [4]. Moreover, cranioplasty serves as a treatment measure to control the changes in the brain's blood flow, cerebrospinal fluid, and metabolic requirements [5]. In the development of cranioplasty, some biomaterials (e.g., autograft, allograft, and synthetic biomaterial) have been applied to repair cranial defects [6]. Although autografts are still the standard for bone defect treatment, the high incidence of the donor sites mobility and the limited volume of autografts restrict the large area improvement of bone repair surgery [7]. Accordingly, current biomaterial technology develops advanced functional materials to replace autografts to construct cranial defects. Synthetic materials (metal, ceramic/cement, polymer, and composite) are used for cranial bone construction thanks to the reduced risks of resorption, infection, and reoperation compared with autografts [8].

Degradable polymers are widely applied for bone regeneration owing to good biocompatibility, degradability, mechanical properties, processability, and so on [9]. Poly(butylene succinate) (PB) is a synthetic degradable polymer that exhibits excellent biocompatibility, remarkable toughness, and non-toxicity of degradable products [10]. PB is a semicrystalline polymer that exhibits high fracture energy and a slow degradation rate [11]. These preferable performances of PB make it a promising candidate for bone regeneration applications [10,11]. However, the major shortcoming of PB is the hydrophobic surface property because of very low surface wettability that causes poor interaction with biological fluids, which inhibits cell response [12]. Accordingly, the intrinsic hydrophobic nature and biological inertness of PB may restrict or delay cell adhesion, growth, and bone regeneration [13]. The enhancement of biological properties (e.g., wettability, degradability) of PB for regenerative medicine application is still in development.

Human bone is a natural nanocomposite consisting of organic components (e.g., collagen) and nano-inorganic minerals (e.g., calcium phosphate) that possesses fascinating properties [14]. Inspired by the structure and composition of bone tissue, the design of nano inorganic fillers/polymer composite by integrating the advantages of both organic and inorganic phases can result in the development of high-performance nanocomposites for bone regeneration application [15]. Compared with conventional microparticles, nanoparticles with a large surface area can result in a close combination with a polymer matrix at the interface, offering enhanced mechanical performances while maintaining the favorable biocompatibility and osteoconductivity of the bioactive fillers, thereby improving protein adsorption, cells adhesion, multiplication, and osteoblastic differentiation for bone regeneration [16]. Nanocomposites of bioactive nanomaterials (e.g., bioglass, calcium phosphate, and apatite) and degradable polymers have been increasingly researched and developed for bone repair due to their superior biocompatibility, osteoconductivity, and degradability [17]. The bioactive nanocomposite is a promising class of advanced biomaterial with great potential for bone regeneration thanks to the mimic of the structure/composition and mechanical performances of natural bone tissue [18].

Silicon nitride (Si_3N_4) is a non-oxide ceramic and is regarded as a new biofunctional material with high mechanical properties, good biocompatibility, and bioactivity, which has been applied for bone repair for more than 10 years [19]. Si_3N_4 can be degradable in the biological environment with the slow release of silicon (Si) ions, which boosts the osteoblast response and bone regeneration [20]. In addition, the hydrophilic and negatively charged surface of Si_3N_4 with the bioactive groups of hydroxyl (-OH) and amino (-NH_2) can improve the adsorption of proteins and further facilitate cell adhesion, thereby being applied as a potential biomaterial for bone regeneration application [21]. Si_3N_4 remarkably boosted the adhesion and multiplication of mesenchymal stem cells and improved alkaline phosphatase activity, bone-related gene expression, and bone matrix protein formation [22]. Accordingly, Si_3N_4 is a promising candidate for bone repair thanks to its favourable biocompatibility, osteoconductivity, hydrophilicity, and other bio-properties.

Herein, PB/Si_3N_4 nanocomposites (PSC) with a Si_3N_4 content of 15 w% (PSC15) and 30 w% (PSC30) were fabricated through the solvent casting method, and porous PSC15 and PSC30 were prepared by solvent casting/particle leaching method. The primary goal of this paper was to produce a nanocomposite with good bioactivity and proper degradability for skull defect repair. The effects of Si_3N_4 content on the compressive strength, surface characteristics (e.g., topography, hydrophilicity, and protein adsorption), and degradability of PSC were investigated. The in vitro cell response (e.g., attachment and osteoblastic differentiation) to PSC was assessed, and the in vivo bone regeneration and degradability potential of porous PSC were studied using the skull defect model of rabbits.

2. Materials and Methods

2.1. Materials and Instruments

Poly (butylene succinate) and silicon nitride particles were separately purchased from Anqing Hexing Chemical Co., Ltd., Anqing, Anhui Province, China and Shanghai Xiaohuang Nano Technology Co., Ltd., Shanghai, China. Bicinchoninic acid kit (BCA), Bovine serum albumin (BSA), Fibronectin (Fn), Fluorescein isothiocyanate (FITC), 4,6-diamidino-2-phenylindole dihydrochloride (DAPI), and ALP staining kit (BCIP/NBT) were purchased from Beyotime Biotech Co., Shanghai, China. Sodium dodecyl sulfate (SDS), simulated body fluids (SBF, pH = 7.4), and glycine were purchased from Aladdin Biochemical Technology Co., Ltd., Shanghai, China. α-MEM was purchased from Gibco, Thermo Fisher Scientific, Waltham, MA, USA. Fetal bovine serum was purchased from Hyclone, Australia. Penicillin/streptomycin (P/S), Glutaraldehyde, and Nonidet P-40 (NP-40) were purchased from Sigma, Life Technology, St. Louis, MO, USA. Cell Counting Kit-8 (CCK-8) and p-nitrophenyl phosphate (pNPP) were separately purchased from Sigma-Aldrich and Sangon, Shanghai, China. ARS solution and cetylpyridinium chloride solution were purchased from Servicebio, Wuhan, Hubei Province, China. Trizol reagent was purchased from Life Technologies, Burlington, MA, USA. The samples were characterized with scanning electron microscopy (SEM) with energy dispersive spectrometry (EDS) (S-4800, Hitachi, Tokyo, Japan), X-ray diffraction (XRD; D8, Bruker, Karlsruhe, German), and a Fourier transform infrared spectrometer (FTIR; Nicolet is50, Thermo Fisher Scientific, Waltham, MA, USA). Universal material machine (E44.304, MTS Co., Shenzhen, Guangdong Province, China). Laser confocal 3D microscope (LCM; VK-X 110, Keyence Co., Osaka, Japan) and Contact angle measurement (CAM; JC2000D1, Shanghai Zhongchen Digital Technique Apparatus Co., Shanghai, China).

2.2. Preparation and Characterization of Composites

The dense samples (PB, PSC15 and PSC30) were fabricated by solvent casting. In a few words, PB particles (10 g) were dissolved in Chloroform (10 mL) under stirring to prepare the PB solution. The Si_3N_4 powders with 0 w% (PB), 15 w% (PSC15), and 30 w% (PSC30) in the composites were then added into PB solution with continuous stirring for 6 h at room temperature for uniform dispersion. The mixture was then cast into molds (Φ6 × 6 mm for compressive strength testing and Φ12 × 2 mm for another testing) and dried in a ventilation hood for 24 h to evaporate the solvent.

The dense samples of (PB, PSC15, and PSC30) were characterized with SEM, EDS, XRD, and FTIR. The compressive strength of specimens was performed with a universal material machine. The surface roughness (Ra) and water contact angle of the samples were characterized by LCM and CAM, respectively. For the protein adsorption, the samples were placed into 24-well plates, and then the BSA (10 mg/mL) and Fn (25 μg/mL) solutions were added to the plates, respectively. After incubating at 37 °C for 5 h, the samples were extracted, and the non-absorbed proteins on the samples were removed by washing with phosphate-buffered solution (PBS, TBD, China) twice. After that, the adsorbed proteins were released by adding 1 mL SDS solution, and the protein contents were tested using the BCA assay kit.

2.3. Si ion Release and pH Value Variation after Samples Soaked in SBF

The samples were immersed in simulated body fluids (SBF, pH = 7.4, Shanghai Yuanye Biotechnology Co., Ltd., Shanghai, China). At 1 d, 3 d, 7 d, 14 d, 21 d, and 28 d, the solution was collected, and the concentrations of Si ions in SBF were tested by Inductivity Coupled Plasma (ICP-OES; Agilent IC, Santa Clara, CA, USA). The release of Si ions from the specimens was also determined. Meanwhile, the pH variation of the solution during the whole period was monitored using a pH meter.

2.4. Morphology, Porosity, and Water Absorption and Degradation In Vitro

The porous samples of PB, PSC15, and PSC30 were fabricated with solvent casting/particulate leaching. After different amounts of Si_3N_4 were uniformly dispersed into the PB solution, NaCl particles with sizes of approximately 300 μm were added into the PB solution and stirred for 10 min. Subsequently, the PB solution with NaCl was cast into the molds ($\Phi 6 \times 6$ mm and $\Phi 6 \times 2$ mm) and air-dried overnight. After evaporation, the samples were immersed in water for 2 days to leach NaCl particles, and the water was refreshed every 6 h. The samples were air-dried for 2 days to remove residual water. The morphology of porous samples was observed by SEM. The porosity of samples was determined with the ethanol substitution method according to the following formula:

$$\text{Porosity} = (V - V_e)/(V_0) \times 100\%,$$

where V_0 represents the total volume of samples, and V_e represents the volume of samples immersed in ethanol.

The weight of samples immersed in water for 24 h (M_w) and the weight of dry samples (M_d) were measured. The water absorption was obtained according to the formula:

$$\text{Water absorption} = (M_w - M_d)/M_d \times 100\%.$$

To assess the in vitro degradability of the porous samples, the porous samples (size of $\Phi 12 \times 2$ mm) were weighed (W_d) and then immersed into PBS solution (at 37 °C and pH 7.4) with a constant shaking speed of 60 rpm/min in an orbital shaker for various time. The samples were taken out, rinsed with water, and dried at 37 °C. Finally, the samples were weighed (W_t). The weight loss was obtained according to the formula:

$$\text{Weight loss} = (W_d - W_t)/(W_d) \times 100\%.$$

2.5. Cellular Response to Samples

2.5.1. Cell Culture

The rat bone marrow mesenchymal stem (RBMS) cells were separated from the femur bone marrow of Sprague Dawley rats, the cells at passages 3–5 were cultured in α-MEM supplemented with 10% fetal bovine serum and 1% penicillin/streptomycin in a humidified atmosphere of 5% CO_2 at 37 °C, and the medium was replaced every 2 days.

2.5.2. Cell Morphology

The samples were sterilized with 75% ethanol and UV radiation and then placed in 24-well plates. The cells with a density of 5×10^4 cells/well were cultured on different samples. After incubating for 1 d and 3 d, the medium was removed, and the samples were washed with PBS (3 times) and fixed with glutaraldehyde solution (0.25%) for 2 h. Then, the fixed cells were rinsed with PBS (3 times) and dehydrated by ethanol solution with various concentrations of 10 v%, 30 v%, 50 v%, 70 v%, 85 v%, 90 v%, and 100 v% for 15 min. The cell morphology was observed with SEM. Similarly, after fixation with glutaraldehyde solution (0.25%) for 2 h, the samples were gently rinsed with PBS (3 times). Subsequently, Fluorescein isothiocyanate (FITC, 400 μL) was added to stain the F-actin ring of cells for 40 min under dark conditions and rinsed with PBS 3 times. Afterwards, the nuclei of cells were stained with 4,6-diamidino-2-phenylindole dihydrochloride (DAPI, 400 μL) for 15 min and rinsed with PBS 3 times. In this way, the F-actin rings were stained green, and the nuclei were stained blue. The cell morphology was observed with confocal laser scanning microscopy (CLSM; Nikon A1R, Nikon Co., Tokyo, Japan).

2.5.3. Cell Attachment and Multiplication

The attachment and multiplication of RBMS cells on different samples were investigated with a CCK-8 assay. After culturing for 6 h and 12 h, the specimens were transferred into a 24-well plate. In total, 400 μL of cell medium containing CCK-8 solution (40 μL) were added and incubated for 6 h. Subsequently, the supernatant (100 μL) was transferred into a 96-well plate, and the optical density (OD) value was measured at 450 nm with a microplate reader (MR, 384 SpectraMax, Molecular Devices, Silicon Valley, CA, USA). The OD value of the blank (without samples) was used as a control, and the cell adhesion rate was calculated according to the formula:

$$\text{Cell adhesion ratio} = OD_s/OD_b \times 100\%,$$

where OD_s and OD_b represent the OD values of cells on the samples and blank, respectively. Similarly, at 1 d, 3 d, and 7 d after culturing, the cell multiplication was determined by measuring the OD value of cells on different samples at 450 nm with MR.

2.5.4. ALP/ARS Staining and Quantitative Analysis

The samples were immersed in α-MEM in a humidified atmosphere (at 37 °C) of 5 % CO_2 for 24 h to obtain the extract. ALP/ARS staining was applied to evaluate the effects of the extract on the osteogenic differentiation of the cells. The ALP activity was evaluated by ALP staining and quantification of ALP. At 7 d and 14 d after culturing, the cells were lysed with NP-40 (1%) for 1 h and incubated with pNPP containing $MgCl_2 \cdot 6H_2O$ (1 mmol/L) and glycine (0.1 g/mL) for 2 h. Subsequently, the reaction was terminated by the addition of NaOH solution (0.2 mol/L). The OD value was measured at 405 nm with MR, and the total protein quantity was tested with the BCA kit. The ALP activity was calculated by dividing the measured absorbance by the total protein amount. After being cultured for 14 days, the cells were fixed with 0.25% glutaraldehyde solution for 20 min and stained with BCIP/NBT kit in the dark for 2 h. The reaction was terminated by H_2O, and the stained samples were observed with optical microscopy. The mineralization of the extracellular matrix of cells was evaluated by ARS staining and quantification of calcium nodules. After being cultured for 14 d and 21 d, the cells were immersed in a cetylpyridinium chloride solution for 1 h to extract calcium. Subsequently, the quantitative results of calcium content were obtained by measuring the OD values for different samples at 620 nm with MR. At 21 d after culturing, the cells were fixed with 0.25% glutaraldehyde solution for 20 min and subsequently stained with ARS solution (2%) for 1 h. Then, the stained cells were washed with PBS and observed by optical microscope.

2.5.5. Osteogenic-Related Gene Expressions

After the cells were cultured for 4 d, 7 d, and 14 d, the osteogenic gene expression was tested with RT-PCR. Trizol reagent was applied to extract the total RNA of the cells, and the complementary DNA (cNDA) was obtained by reversely transcribing RNA. Using the cDNA as a template, the expression of osteogenic genes (osteocalcin: OCN, alkaline phosphatase: ALP, Osteopontin: OPN, and runt-associated transcription factor 2: Runx2) was measured with the SYBR® Premix Ex TaqTM system (Takara, Kyoto, Japan). Glyceraldehyde-3-phosphate dehydrogenase (GAPDH) was used as a housekeeping gene for normalization. Table 1 lists the forward and reverse primers.

Table 1. Primer sequences.

Gene	Primers Sequence (F Was Forward, R Was Reverse)
GAPDH	F: CCTGCACCACCAACTGCTTA
	R: GGCCATCCACAGTCTTCTGAG
ALP	F: GGATCAAAGCAGCATCTTACCAG
	R: GCTTTCCCATCTTCCGACACT
OPN	F: GTCCCTTGCCCTGACTACTCT
	R: GACATCTTTTGCAAACCGTGT
OCN	F: CAGACAAGTCCCACACAGCA
	R: CCAGCAGAGTGAGCAGAGAG
Runx2	F: ATCCAGCCACCTTCACTTACACC
	R: GGGACCATTGGGAACTGATAGG

2.6. Implantation of Samples In Vivo

2.6.1. Animal Surgical Procedures

The effects of porous composites on new bone formation in vivo were determined using the rabbit skull defect model. The surgical procedures were permitted by the Animal Experiment Ethics Committee (the project identification code: TJAB03222301) of Shanghai East Hospital, School of Medicine, Tongji University. The 12 New Zealand white rabbits (around 3 kg, 8 months old) were randomly divided into 2 groups (4 w and 12 w). Pentobarbital sodium solution (3%) was used to anesthetize the rabbits by ear vein injection. The skin was sterilized with alcohol, and the cranial bone was exposed by separating the skin and cranial periosteum. Two bone defects (6 mm in diameter) were made on the bilateral sides of the rabbit skull, and PB and PSC30 were implanted into the left and right bone defects, respectively. At 4 w and 12 w after surgery, the rabbits were sacrificed with pentobarbital sodium solution (overdose) and the defective bone of the skulls was harvested and then fixed in phosphate-buffered formalin (10%).

2.6.2. M-CT Images Analysis

The new bone formation for specimens was observed and imaged with microcomputed tomography (m-CT, SkyScan 1272, Bruker, Madison, WI, USA) under 80 KV with a resolution of 5 µm, and the 3D images were reconstructed. Moreover, the bone regeneration:bone volume/total volume (BV/TV), trabecular thickness (Tb.Th), trabecular number (Tb.N), and bone mineral density (BMD) were quantified by CT Analyzer (SkyScan software, CTVOX 2.1.0, Bruker, Madison, WI, USA).

2.6.3. Histological Images Analysis

After decalcifying with 10% EDTA solution for 8 w, the samples were embedded in paraffin, and histological sections with a thickness of 5 µm were obtained. Subsequently, the histological sections of H&E staining were prepared according to the standard protocol. Three microscope images were obtained with microscopy from three random areas for the sample and then evaluated with an Image-Pro Plus. The percentage of the newly formed bone area was determined by testing the number of pixels labeled through histological images. Quantitative analysis of the ratios of new bone and residual material was performed using histological images through Image-Pro Plus, Media Cybernetics, Inc., Rockville, MD, USA.

2.7. Statistical Analysis

Three specimens were utilized in all experiments, and the data were presented as mean ± standard deviation. Statistical significance was performed by applying one-way analysis of variance with Tukey's Post Hoc test; $p < 0.05$ was regarded as statistically significant. The notation "*" denotes $p < 0.05$.

3. Results

3.1. Characterization of Samples

The SEM photos of dense samples are revealed in Figure 1. Under low magnification, PB showed a flat surface, while PSC15 and PSC30 exhibited rough surfaces. Under high magnification, PB also showed a flat surface, while Si_3N_4 particles were observed on the surface of PSC15 and PSC30. The Si_3N_4 particles (size of about 100 nm) were randomly distributed on PSC15 and PSC30, and the Si_3N_4 particles on PSC30 were more abundant than on PSC15.

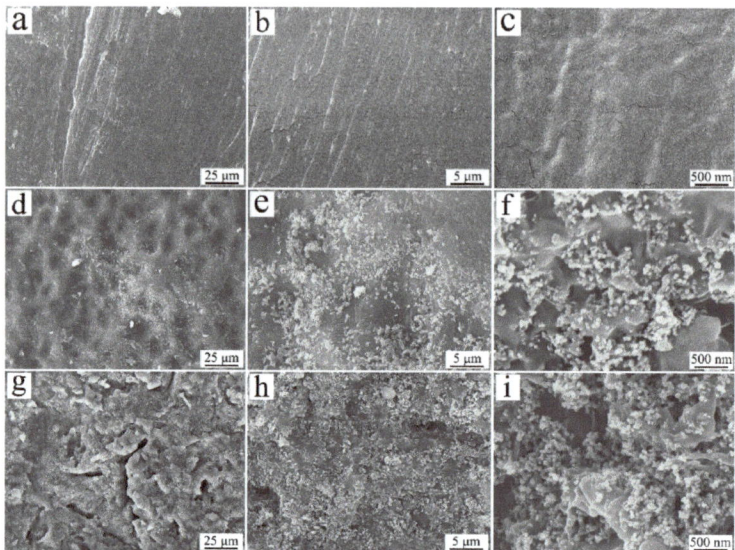

Figure 1. SEM photos of PB (**a–c**), PSC15 (**d–f**), and PSC30 (**g–i**) under various magnifications.

Figure 2a displays the XRD of samples. The diffraction peaks at 19.8°, 23.1°, and 29° were the peaks of PB, which were observed on both PSC15 and PSC30 [23]. No obvious peaks were observed in Si_3N_4, PSC15, and PSC30, indicating that Si_3N_4 exhibited an amorphous phase without crystalline peaks. Figure 2b illustrates the FTIR of the samples. For PB, the peak at 2980 cm^{-1} was the stretching vibration of methylene (-CH$_2$-). The peak at 1718–1731 cm^{-1} was the carbonyl (-C=O), and the peak at 1363–1386 cm^{-1} was the aliphatic group (-C-O-) [24]. For Si_3N_4, the peaks at 3432 cm^{-1} and 1079 cm^{-1} were the amide bond (-N-H), and the peak at 973 cm^{-1} was the stretching vibration of the silicon nitrogen bond (Si-N) [25]. The peaks of PB and Si_3N_4 could be found in PSC15 and PSC30. Figure 2c–e revealed the EDS spectra of the samples. The C and O elements were found in PB, PSC15, and PSC30, while the Si element was seen in both PSC15 and PSC30.

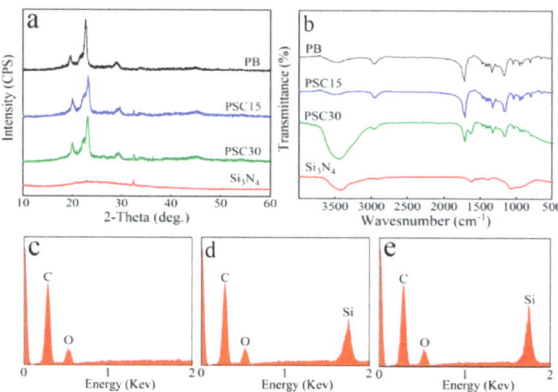

Figure 2. XRD (**a**) and FTIR (**b**) of the samples, and EDS of PB (**c**), PSC15 (**d**), and PSC30 (**e**).

3.2. Physical and Chemical Properties of Samples

Figure 3a–c reveal the specimens' compressive strength, surface roughness, and water contact angle. The compressive strength (Figure 3a) of PB, PSC15, and PSC30 was 31 ± 2.0,

43 ± 2.5, and 52 ± 3.0 MPa. The surface roughness (Figure 3b) of PB, PSC15, and PSC30 was 1.27 ± 0.10, 2.51 ± 0.10, and 3.07 ± 0.15 μm. The water contact angle (Figure 3c) of PB, PSC15, and PSC30 was 84.5 ± 5°, 72.60 ± 5°, and 59.61 ± 4°. Figure 3d reveals the protein adsorption on specimens. The BSA adsorption amount for PB, PSC15, and PSC30 was 7.63 ± 1.5%, 19.38 ± 2.0%, and 34.19 ± 2.5%. The Fn adsorption amount for PB, PSC15, and PSC30 was 5.71 ± 1.5%, 17.64 ± 2.0%, and 27.08 ± 2.5%.

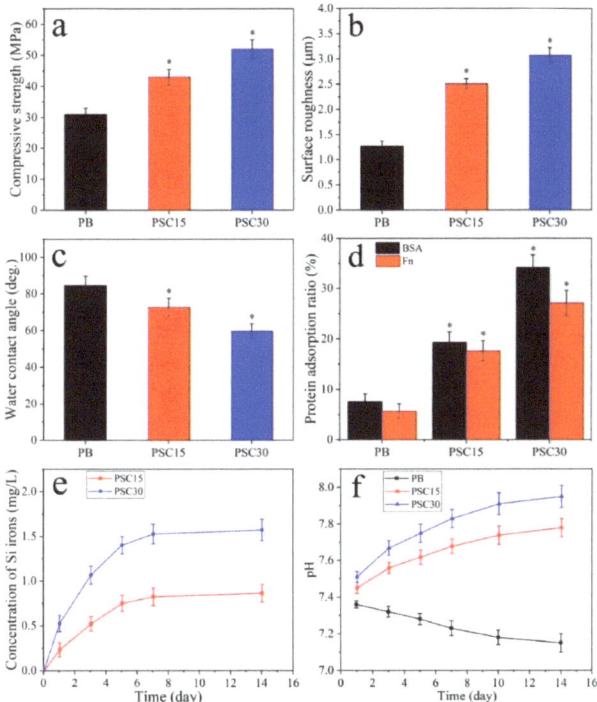

Figure 3. Compressive strength (**a**), surface roughness (**b**), water contact angle (**c**), and protein adsorption (**d**) of specimens, and release of Si ion (**e**) and pH change (**f**) after the samples immersed into SBF for 1 d, 3 d, 5 d, 7 d, 10 d, and 14 d (* $p < 0.05$, vs. PB).

Figure 3e shows the release of Si ions from PSC15 and PSC30 into SBF after immersion for various times. The Si ions exhibited a rapid release at the early stage of immersion (within 5 d) while a slow release at the middle and late stages of immersion (from 6 d to 14 d). At 14 days, the Si ion concentrations for PSC15 and PSC30 were 0.863 mg/L and 1.572 mg/L. Figure 3f shows the pH changes after the specimens were immersed in SBF for various times. The pH values for PSC15 and PSC30 slowly increased with time. At 14 d after soaking, the pH values for PSC15 and PSC30 were 7.78 and 7.95, respectively. However, the pH values for SBF slightly decreased with time. At 14 d after, the pH value for PB was 7.13.

3.3. Characterization of Porous Specimens

Figure 4a–c reveal the SEM photos of the porous specimens. The macropores of all samples showed irregular morphology with pore sizes of approximately 300 μm. The porosity (Figure 4d) of PB, PSC15, and PSC30 was 63.2%, 65.9%, and 68%. Figure 4e displays the water absorption of the samples after they were immersed in water for 6 h. The water absorption for PB, PSC15, and PSC30 was 234.8%, 348.3%, and 379.3%. Figure 4f shows the weight loss of PB, PSC15, and PSC30 after being immersed in PBS for various times. At each time point, the weight loss for PSC30 was higher than PSC15, and PSC15 was higher than PB. At 84 d, the weight loss for PB, PSC15, and PSC30 was 20.58 w%, 47.63 w%, and 67.56 w%.

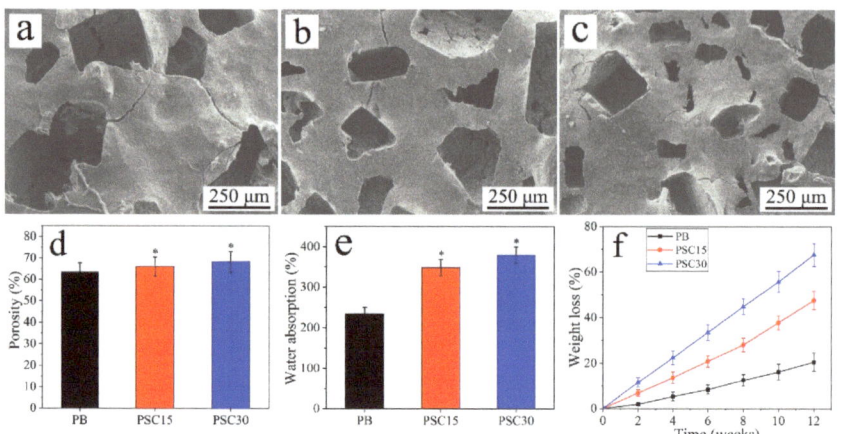

Figure 4. SEM photos of porous specimens of PB (**a**), PSC15 (**b**), and PSC30 (**c**), and porosity (**d**), water absorption (**e**), and weight loss (**f**) of the specimens in PBS after immersion for various times. (* $p < 0.05$, vs. PB).

3.4. Cell Adhesion, Multiplication, and Morphology

Figure 5 demonstrates the CLSM images of the cells on the specimens at different culturing times. The amounts of cells on PSC15 and PSC30 increased with the culturing time, while there was no obvious increase for PB. Further, the number of cells on PSC30 was higher than PSC15, and PSC15 was higher than PB.

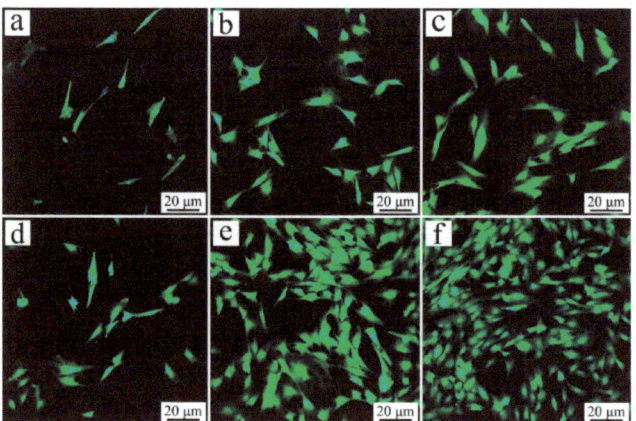

Figure 5. CLSM photos of RBMS cells cultured on PB (**a,d**), PSC15 (**b,e**), and PSC30 (**c,f**) for 1 d (**a–c**) and 3 d (**d–f**).

Figure 6a–f show the SEM photos of cells on the specimens after culturing for various times. On days 1 and 3, only a few cells were observed on PB, while some cells with filopodia spread on the surface of PSC15 and PSC30. More cells with pseudopodia spread better on PSC30 than on PSC15. The number of cells on PSC15 and PSC30 increased with time but there was only a slight increase for PB. Figure 6g reveals the cell adhesion ratio for specimens at various times. At 6 and 12 h, the cell adhesion for PSC15 and PSC30 remarkably increased with time, but there was only a slight increase for PB. Cell adhesion for PSC30 was higher than PSC15, and PSC15 was higher than PB. Figure 6h displays the optical density (OD) value (cell multiplication) of cells on specimens at 1 d, 3 d, and 7 d. The OD values for PB showed a little increase, while the OD values for PSC30 and PSC15 remarkably increased with time, indicating good cytocompatibility. At 3 and 7 days, the OD values for PSC30 were higher than PSC15, and PSC15 were higher than PB.

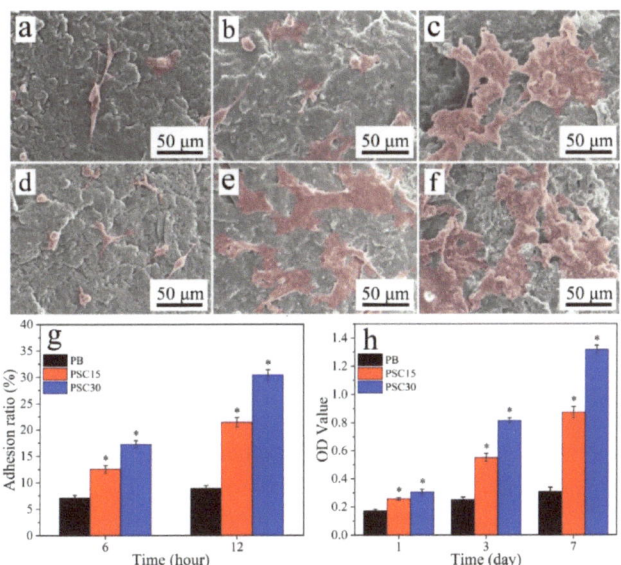

Figure 6. SEM photos of RBMS cells (red areas) cultured on PB (**a**,**d**), PSC15 (**b**,**e**), and PSC30 (**c**,**f**) for 1 d (**a**–**c**) and 3 d (**d**–**f**); adhesion ratio (**g**) and OD values (**h**) of cells on PB, PSC15, and PSC30 for various times (n = 3, * represents $p < 0.05$, compared with PB).

3.5. Osteoblastic Differentiation

3.5.1. ALP Activity and Calcium Nodules

Figure 7a–c display the photos of the ALP staining 14 days after the cells were cultured on the samples. The intensity of ALP staining was the strongest for PSC30, followed by PSC15, and the weakest for PB. Figure 7d–f display the photos of ARS staining at 21 d after culturing. The intensity of ARS staining was the strongest for PSC30, followed by PSC15, and the weakest for PB. Figure 7g displays the quantitative results of the ALP activity of cells after culturing for various times. At 7 d and 14 d, the ALP activity for PSC30 was higher than PSC15 and PB and the lowest for PB. Figure 7 h displays the quantitative results of the calcium content of cells after culturing for various times. At 14 d and 21 d, the calcium content for PSC30 was higher than PSC15 and PB and the lowest for PB.

3.5.2. Expression of Osteoblastic Genes

Figure 8 reveals the expression of osteoblastic differentiation genes (Runx2, ALP, OCN, and OPN) of cells at various times after culturing. The osteoblastic differentiation gene expressions for PSC15 and PSC30 increased with the cultured time but showed a slight change for PB. At 7 d and 14 d, the gene expressions were the highest for PSC30, followed by PSC15, and the lowest for PB.

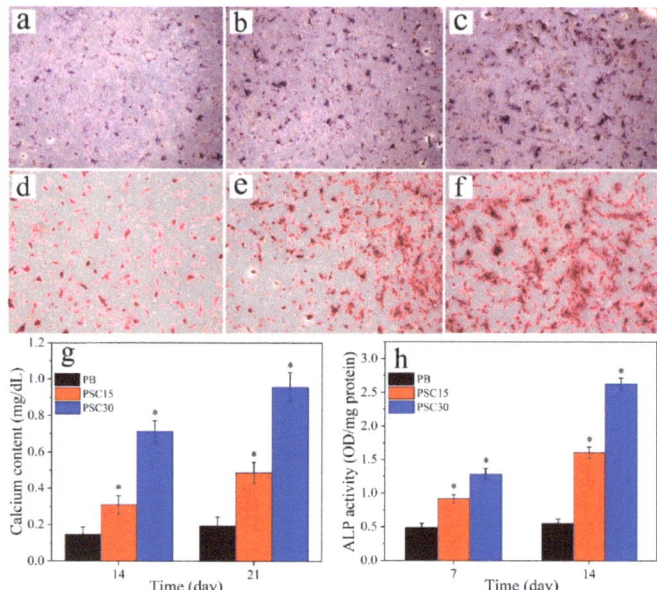

Figure 7. ALP staining (**a**–**c**) at 14 d after culturing and alizarin red staining (**d**–**f**) at 21 d after culturing, and quantitative results of ALP activity (**g**) and calcium content (**h**) (n = 3, * represents $p < 0.05$, compared with PB).

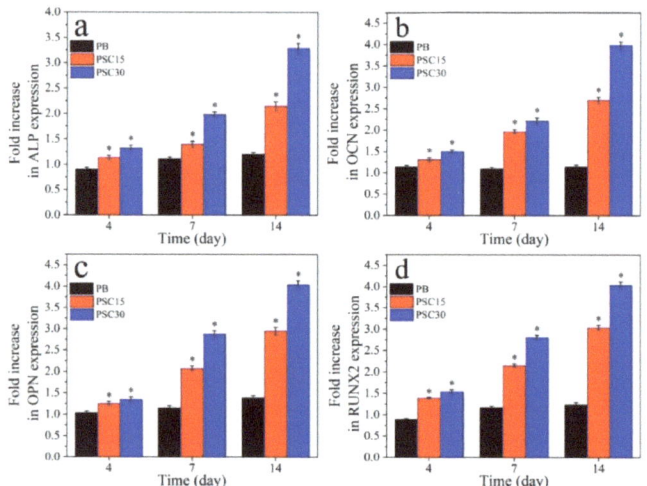

Figure 8. Osteoblastic differentiation genes of ALP (**a**), OCN (**b**), OPN (**c**), and Runx2 (**d**) expression of the cells cultured on PB, PSC15, and PSC30 for various times (n = 3, * represents $p < 0.05$, compared with PB).

3.6. Bone Regeneration and Material Degradation In Vivo

3.6.1. M-CT Valuation

Figure 9a,b shows the macroscopic observation of the samples after being implanted into femur defects of rabbits for 4 w and 12 w. Figure 9c,d displays the m-CT reconstructed images of the samples. At 4 w after surgery, only a small amount of new bone tissue formed along the edge of PB, while some new bone tissues formed for PSC30. At 12 w after surgery, some bone tissues were seen to grow into porous PB while many new bone tissues grew into porous PSC30 and completely repaired the bone defects.

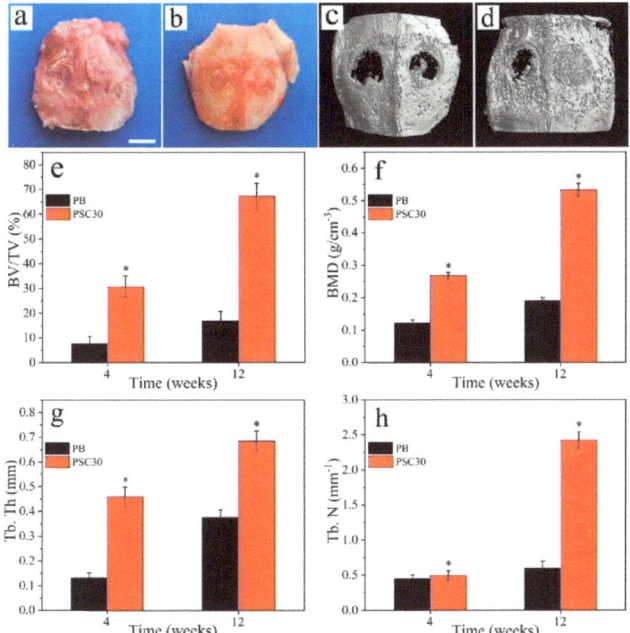

Figure 9. Macroscopic observation ((**a**,**b**), scale = 10 mm) and m-CT images (**c**,**d**) of PB and PSC30 after implanted into bone defects of rabbit femur for 4 w and 12 w, and quantitative results of BV/TV (**e**), BMD (**f**), Tb.Th (**g**), and Tb.N (**h**) (n = 3, * represents $p < 0.05$, compared with PB).

Figure 9e–h shows the quantitative results of new bone formation (BV/TV, BMD, Tb.Th, and Tb.N) for the samples at various times. The new bone formation for both PB and PSC30 increased with time, and that for PSC30 was remarkably higher than PB at both 4 w and 12 w.

3.6.2. Histological Evaluation

Figure 10a–f reveal the H&E staining images of new bone formation for the various samples at 4 w and 12 w after surgery. At 4 w, only a small amount of new bone tissues was seen in the bone defects for PB, while obvious new bone tissues were found for PSC30. At 12 w, the new bone tissues for PB slightly increased compared with 4 w, and the materials in the bone defects slightly reduced accordingly over time. Nevertheless, many new bone tissues were seen for PSC30, and the materials were reduced. Figure 10i,j shows the quantitative results of the new bone area and residual materials at 4 w and 12 w after surgery. The new bone area for PSC30 remarkably increased with time while the residual materials reduced accordingly. In addition, the new bone area for PB slowly increased with time, and the residual materials slowly reduced accordingly. The new bone area for PSC30 was remarkably higher than PB, and the residual materials for PSC30 were remarkably lower than PB.

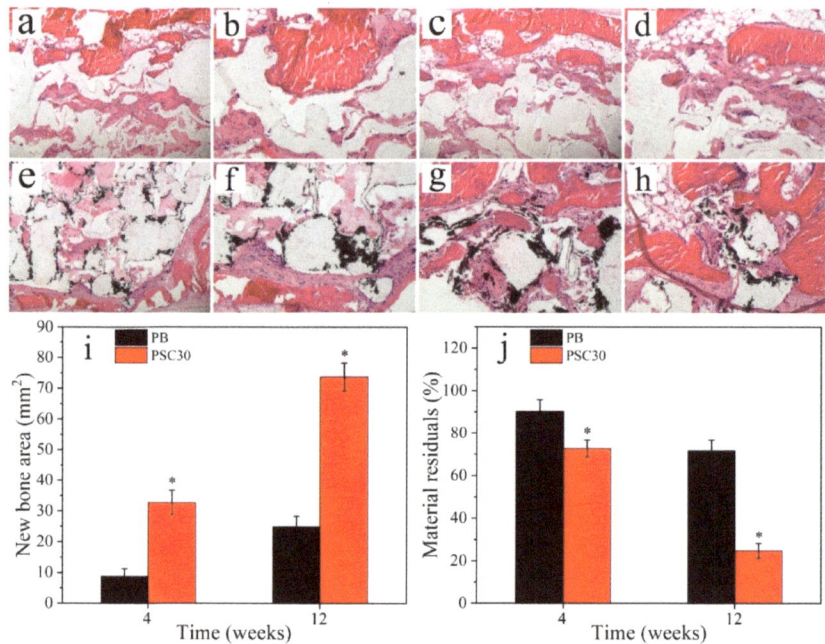

Figure 10. Histological images for PB (**a–d**) and PSC30 (**e–h**) after surgery for 4 w (**a,b,e,f**) and 12 w (**c,d,g,h**); quantitative results of the new bone area (**i**) and residual materials (**j**) (n = 3, * represents $p < 0.05$, compared with PB).

4. Discussion

Interest in the application of nanocomposites with regenerative potential to repair damaged bone tissue has increased because nanocomposites containing degradable polymers and nano bioactive fillers are regarded as a mimic strategy for bone regeneration [26]. Herein, a bioactive nanocomposite was prepared for the construction of cranial bone defects by incorporating Si_3N_4 nanoparticles into the PB matrix. Because the Si_3N_4 nanoparticles reinforced PB, the compressive strength of PB, PSC15, and PSC30 gradually increased, demonstrating that Si_3N_4 content played a critical role in enhancing mechanical properties. Surface character is considered one of the important factors regulating cell behaviors, which exhibits significant effects on the bone–tissue response [27]. Compared to PB with flat surfaces, both PSC15 and PSC30 displayed rough surfaces thanks to the Si_3N_4 nanoparticles exposed on the surface. The surface roughness of PB, PSC15, and PSC30 gradually increased, demonstrating that the Si_3N_4 content played a critical role in the increase in surface roughness. Hydrophilicity is one of the important surface characteristics that affect cellular behaviors. However, the hydrophilic surface tends to improve cell adhesion and spreading on biomaterials compared to a hydrophobic surface [28]. Here, the hydrophilicity of PB, PSC15, and PSC30 gradually increased, revealing that the Si_3N_4 content played a critical role in improving hydrophilicity. Accordingly, compared with PB, the surface properties (hydrophilicity and roughness) of PSC increased, and PSC30 exhibited optimization thanks to the high content of Si_3N_4.

The initial biological response to the biomaterial is protein absorption, which has been demonstrated to be a regulator between the biomaterials and cells [29]. Protein adsorption is generally affected by surface performances, especially the hydrophilicity and topography of biomaterials. Herein, compared with PB, the adsorption of proteins for PSC15 and PSC30 was remarkably enhanced thanks to the presence of Si_3N_4 nanoparticles. The improvement of protein adsorption for PSC15 and PSC30 was ascribed to the hydrophilicity/surface energy and topography/roughness because the hydrophilic groups (-OH, $-NH_2$) of Si_3N_4

and rough surface of PSC with the high surface area could provide more sites for protein binding, leading to an increase in protein adsorption. Apart from the surface properties, the release of Si ions was key to stimulating cell multiplication and osteoblastic differentiation [30]; herein, the gradual release of Si ions from both PSC15 and PSC30 into PBS was ascribed to the degradation of Si_3N_4. The pH value for PB slightly reduced with time due to the production of acid produced by the degradation of PB, while the pH values for PSC15 and PSC30 slowly increased, causing a weak alkaline environment thanks to the production of an alkaline product by degradation of Si_3N_4. A weak alkaline (e.g., pH 7.4~8.0) micro-environment was demonstrated to be useful for osteoblastic differentiation and bone regeneration [31].

Cellular adhesion is the first response in the interaction between cells and biomaterials, which affects subsequent cell multiplication, and further affects osteoblastic differentiation and bone regeneration [32]. Herein, cell adhesion, spreading, and multiplication on PB, PSC15, and PSC30 gradually increased, indicating that the content of Si_3N_4 in the composites played a key role in enhancing cell adhesion and multiplication. The ALP/ARS staining and ALP activity/calcium (nodule) content of the cells cultured on the samples can be used to assess osteogenic differentiation [33]. The staining intensity of ALP/ARS for PB, PSC15, and PSC30 gradually became strong. Moreover, the ALP activity/calcium content for PB, PSC15, and PSC30 gradually increased, indicating that osteogenic differentiation improved with the increase in Si_3N_4 content. Further, the expression of osteogenic-associated genes of cells on the samples could be applied to evaluate osteogenic differentiation [34]. The gene expression of PB, PSC15, and PSC30 gradually increased, indicating that the osteogenic differentiation increased with the increase in the Si_3N_4 content. Accordingly, the content of Si_3N_4 in the composites played a key role in enhancing osteogenic differentiation.

Porous composites for bone regeneration should have the following characteristics: porous structures to promote cell–biomaterial interaction, cell adhesion, and growth; interconnective porous structures to boost transport of nutrients, and mass and regulated factors to allow cell multiplication, survival, and osteoblastic differentiation; pore size is essential for bone regeneration because bone growth requires an optimized pore size of approximately 300 µm [35]. Accordingly, the strategy of a combination of degradable polymer and bioactive material to create porous composites with appropriate porosity is a promising method for bone construction [35]. In this study, porous composites were prepared, and PSC30 exhibited a porous structure with a porosity of approximately 70%. Accordingly, PSC30 was used to construct cranial bone defects in rabbits. A porous composite acts as a temporary template for cell adhesion, multiplication, ensuing osteoblastic differentiation, and eventually resulting in bone regeneration [36]. Consequently, appropriate degradability of the biomaterial is an important factor that affects bone regeneration, and a degradable biomaterial can gradually disappear with time in vivo, thereby producing space for bone tissue ingrowth simultaneously [37]. Herein, the weight loss of PB, PSC15, and PSC30 in PBS increased with soaking time, indicating appropriate degradability. Moreover, the weight loss of PB, PSC15, and PSC30 increased with increasing Si_3N_4 content, indicating that the Si_3N_4 content had obvious effects on degradability. The degradation of Si_3N_4 particles on the macroporous walls created more micropores, which assisted the diffusion of the medium into the porous composites. This diffusion further facilitated the hydrolytic degradation of PB and the dissolution of Si_3N_4. Accordingly, the incorporation of Si_3N_4 particles into the composites increased the degradation rate of the porous composites. The goal of a degradable biomaterial is to boost tissue regeneration at the bone defect and gradually degrade in situ, eventually replacing new bone tissue [38]. The in vivo bone regenerative capability was investigated by implanting the porous composites into rabbit skull defects. The m-CT and histological results displayed that PSC30 gradually degraded and was replaced by newly formed bone tissue, while PB showed slow degradation and thus limited bone regeneration. Compared with PB, PSC30 induced rapid degradation and bone formation. Further, as shown in the histological photos, the porous composite in vivo did not lead to any adverse reactions, indicating good long-term biocompatibility. Collectively, the porous composites exhibited promising bone formation potential.

Cell behaviors and bone tissue regeneration are closely correlative to the surface properties of biomaterials, and the capability to regulate surface characteristics (e.g., composition, topography, roughness, and hydrophilicity) can offer positive effects on responses of cells/tissues [39]. Herein, the surface properties (roughness, hydrophilicity, surface energy, and protein adsorption) of PB, PSC15, and PSC30 gradually increased with the increasing Si_3N_4 content, indicating that the Si_3N_4 content played a pivotal role in the enhancement of surface performances. Moreover, cell adhesion, multiplication, osteoblastic differentiation, and new bone formation for PB and PSC30 remarkably increased. Consequently, the significant improvement of cell response and bone regeneration for PSC30 was ascribed to the enhanced surface performance. Previous studies show that Si ions exhibit significant effects on boosting the multiplication and differentiation of osteoblasts [40]. Moreover, Si ions effectively enhance the gene expression related to the synthesis of the bone matrix, which is essential for bone tissue regeneration [41]. Accordingly, the strong osteogenic outcome induced by PSC30 was attributed to the high content of Si_3N_4, whose chemical composition offered the dissolution products of Si ions for bone regeneration in vivo [42]. Given the special function of Si ions, the degradation of PSC30 led to Si ion release, which resulted in the local pH in a physiological range for cell multiplication, and bone regeneration [43]. Consequently, incorporating Si_3N_4 into PB improved the surface properties of PSC, which remarkably stimulated the cellular responses in vitro and promoted new bone regeneration in vivo. Further, the degradability of PSC30 caused a slow release of Si ions into the local microenvironment that remarkably stimulated the responses of osteoblasts/bone tissues. The incorporation of bioactive nanoparticles into the degradable polymer created a bioactive nanocomposite that has the ability to boost cell attachment, multiplication, and new bone growth along with proper degradability. In conclusion, PSC30 with high content of Si_3N_4 exhibited good biocompatibility and stimulated the responses of cells/bone tissues, which were attributed to the synergism of both optimized surface properties and slow release of Si ions. PSC30 would be a promising candidate and have great potential for constructing cranial bone defects.

5. Conclusions

A bioactive nanocomposite of PSC was created by the addition of Si_3N_4 nanoparticles into PB. In comparison with PB, the incorporation of Si_3N_4 significantly enhanced the compressive strength, surface hydrophilicity, roughness, and protein adsorption of PSC. Furthermore, the addition of Si_3N_4 accelerated the degradation of PSC, which led to the slow release of Si ions. Further, the cell response (adhesion, multiplication, and osteoblastic differentiation) to PSC was remarkably enhanced with the increase in Si_3N_4 content, and PSC30 displayed the highest cell response. Further, PSC30 significantly promoted bone regeneration and gradually degraded in vivo. The high content of Si_3N_4 in PSC led to more positive effects on in vitro cellular response and in vivo bone regeneration. Accordingly, the incorporation of Si_3N_4 into PB created a bioactive nanocomposite that has the ability to boost cell attachment, multiplication, and new bone growth along with proper degradability. The enhancements of cell response/bone regeneration were ascribed to the synergism of the enhanced surface performances and release of Si ions. Subsequently, PSC30 might be a promising candidate and have great potential for the construction of cranial bone defects.

Author Contributions: Conceptualization, methodology, software, validation, formal analysis, investigation, resources, data curation, writing—original draft preparation, Q.Z.; writing—review and editing, visualization, supervision, project administration, funding acquisition, S.G. All authors have read and agreed to the published version of the manuscript.

Funding: This study was supported by funding from the Peak Disciplines (Type IV) of Institutions of Higher Learning in Shanghai, for the research and development of stem cell therapy for major diseases and the establishment of a clinical transformation system (2019CXJQ01).

Institutional Review Board Statement: The Ethics Committee of Shanghai East Hospital, School of Medicine, Tongji University (protocol code TJAB03222301).

Informed Consent Statement: Not applicable.

Data Availability Statement: The data presented in this study are available on request from the corresponding author.

Conflicts of Interest: The authors declare no conflict of interest.

References

1. Frassanito, P.; Bianchi, F.; Pennisi, G.; Massimi, L.; Tamburrini, G.; Caldarelli, M. The growth of the neurocranium: Literature review and implications in cranial repair. *Child Nerv. Syst.* **2019**, *35*, 1459–1465. [CrossRef] [PubMed]
2. Wang, S.; Zhao, Z.J.; Yang, Y.D.; Yang, Y.D.; Qiu, Z.Y.; Song, T.X.; Cui, F.Z.; Wang, X.M.; Zhang, C.Y. A high-strength mineralized collagen bone scaffold for large-sized cranial bone defect repair in sheep. *Regen. Biomater.* **2018**, *5*, 283–292. [CrossRef] [PubMed]
3. Aghali, A. Craniofacial Bone Tissue Engineering: Current Approaches and Potential Therapy. *Cell* **2021**, *10*, 2993. [CrossRef] [PubMed]
4. Ozoner, B. Cranioplasty following severe traumatic brain injury: Role in neurorecovery. *Curr. Neurol. Neurosci.* **2021**, *21*, 62. [CrossRef]
5. Kim, M.J.; Lee, H.B.; Ha, S.K.; Lim, D.J.; Kim, S.D. Predictive factors of surgical site infection following cranioplasty: A study including 3D printed implants. *Front. Neurol.* **2021**, *12*, 745575. [CrossRef]
6. Wu, X.C.; Gauntlett, O.; Zhang, T.F.; Suvarnapathaki, S.; McCarthy, C.; Wu, B.; Camci-Unal, G. Eggshell microparticle reinforced scaffolds for regeneration of critical sized cranial defects. *ACS Appl. Mater. Inter.* **2021**, *13*, 60921–60932. [CrossRef]
7. Zhuang, Y.; Liu, Q.C.; Jia, G.Z.; Li, H.L.; Yuan, G.Y.; Yu, H.B. A biomimetic zinc alloy scaffold coated with brushite for enhanced cranial bone regeneration. *ACS Biomater. Sci. Eng.* **2021**, *7*, 893–903. [CrossRef]
8. Uceda, M.F.I.; Sanchez-Casanova, S.; Escudero-Duch, C.; Vilaboa, N. A Narrative review of cell-based approaches for cranial bone regeneration. *Pharmaceutics* **2022**, *14*, 132. [CrossRef]
9. Zhao, D.Y.; Zhu, T.T.; Li, J.; Cui, L.G.; Zhang, Z.Y.; Zhuang, X.L.; Ding, J.X. Poly(lactic-co-glycolic acid)-based composite bone-substitute materials. *Bioact. Mater.* **2021**, *6*, 346–360. [CrossRef]
10. Cristofaro, F.; Gigli, M.; Bloise, N.; Chen, H.L.; Bruni, G.; Munari, A.; Moroni, L.; Lotti, N.; Visai, L. Influence of the nanofiber chemistry and orientation of biodegradable poly(butylene succinate)-based scaffolds on osteoblast differentiation for bone tissue regeneration. *Nanoscale* **2018**, *10*, 8689–8703. [CrossRef]
11. Penas, M.I.; Perez-Camargo, R.A.; Hernandez, R.; Muller, A.J. A review on current strategies for the modulation of thermomechanical, barrier, and biodegradation properties of poly (butylene succinate) (PBS) and its random copolymers. *Polymers* **2022**, *14*, 1025. [CrossRef] [PubMed]
12. Zarei, M.; El, F.M. Synthesis of hydrophilic poly (butylene succinate-butylene dilinoleate) (PBS-DLS) copolymers containing poly (ethylene glycol) (PEG) of variable molecular weights. *Polymers* **2021**, *13*, 3177. [CrossRef] [PubMed]
13. Wang, H.Y.; Ji, J.H.; Zhang, W.; Zhang, Y.H.; Jiang, J.; Wu, Z.W.; Pu, S.H.; Chu, P.K. Biocompatibility and bioactivity of plasma-treated biodegradable poly(butylene succinate). *Acta Biomater.* **2009**, *5*, 279–287. [CrossRef] [PubMed]
14. Sahoo, N.G.; Pan, Y.Z.; Li, L.; He, C.B. Nanocomposites for bone tissue regeneration. *Nanomedicine* **2013**, *8*, 639–653. [CrossRef] [PubMed]
15. Ding, Z.W.; Li, H.; Wei, J.; Li, R.J.; Yan, Y.G. Developing a novel magnesium glycerophosphate/silicate-based organic-inorganic composite cement for bone repair. *Mater. Sci. Eng. C* **2018**, *87*, 104–111. [CrossRef]
16. Armentano, I.; Dottori, M.; Fortunati, E.; Mattioli, S.; Kenny, J.M. Biodegradable polymer matrix nanocomposites for tissue engineering: A review. *Polym. Degrad. Stabil.* **2010**, *95*, 2126–2146. [CrossRef]
17. Allo, B.A.; Costa, D.O.; Dixon, S.J.; Mequanint, K.; Rizkalla, A.S. Bioactive and biodegradable nanocomposites and hybrid biomaterials for bone regeneration. *J. Funct. Biomater.* **2012**, *3*, 432–463. [CrossRef]
18. Pathmanapan, S.; Periyathambi, P.; Anandasadagopan, S.K. Fibrin hydrogel incorporated with graphene oxide functionalized nanocomposite scaffolds for bone repair—In vitro and in vivo study. *Nanomed.-Nanotechnol.* **2020**, *29*, 102251. [CrossRef]
19. Du, X.Y.; Lee, S.S.; Blugan, G.; Ferguson, S.J. Silicon nitride as a biomedical material: An overview. *Int. J. Mol. Sci.* **2022**, *23*, 6551. [CrossRef]
20. Xu, Z.Y.; Wu, H.; Wang, F.; Kaewmanee, R.; Pan, Y.K.; Wang, D.Q.; Qu, P.Y.; Wang, Z.K.; Hu, G.F.; Zhao, J.; et al. A hierarchical nanostructural coating of amorphous silicon nitride on polyetheretherketone with antibacterial activity and promoting responses of rBMSCs for orthopedic applications. *J. Mater. Chem. B* **2019**, *7*, 6035–6047. [CrossRef]
21. Lee, S.S.; Laganenka, L.; Du, X.Y.; Hardt, W.D.; Ferguson, S.J. Silicon nitride, a bioceramic for bone tissue engineering: A reinforced cryogel system with antibiofilm and osteogenic effects. *Front. Bioeng. Biotech.* **2021**, *9*, 794586. [CrossRef] [PubMed]
22. Dai, Y.; Guo, H.; Chu, L.Y.; He, Z.H.; Wang, M.Q.; Zhang, S.H.; Shang, X.F. Promoting osteoblasts responses in vitro and improving osteointegration in vivo through bioactive coating of nanosilicon nitride on polyetheretherketone. *J. Orthop. Transl.* **2020**, *24*, 198–208. [CrossRef] [PubMed]
23. Veranitisagul, C.; Wattanathana, W.; Wannapaiboon, S.; Hanlumyuang, Y.; Sukthavorn, K.; Nootsuwan, N.; Chotiwan, S.; Phuthong, W.; Jongrungruangchok, S.; Laobuthee, A. Antimicrobial, conductive, and mechanical properties of AgCB/PBS composite system. *J. Chem.* **2019**, *2019*, 3487529. [CrossRef]

24. Neibolts, N.; Platnieks, O.; Gaidukovs, S.; Barkane, A.; Thakur, V.K.; Filipova, I.; Mihai, G.; Zelca, Z.; Yamaguchi, K.; Enachescu, M. Needle-free electrospinning of nanofibrillated cellulose and graphene nanoplatelets based sustainable poly (butylene succinate) nanofibers. *Mater. Today Chem.* **2020**, *17*, 100301. [CrossRef]
25. Gu, D.P.; Liu, S.Y.; Chen, S.W.; Song, K.F.; Yang, B.C.; Pan, D. Tribological Performance of Si3N4-PTFE Composites Prepared by High-Pressure Compression Molding. *Tribol. Trans.* **2020**, *63*, 756–769. [CrossRef]
26. Erol-Taygun, M.; Unalan, I.; Idris, M.I.B.; Mano, J.F.; Boccaccini, A.R. Bioactive glass-polymer nanocomposites for bone tissue regeneration applications: A review. *Adv. Eng. Mater.* **2019**, *21*, 1900287. [CrossRef]
27. Li, M.J.; Fu, X.L.; Gao, H.C.; Ji, Y.R.; Li, J.; Wang, Y.J. Regulation of an osteon-like concentric microgrooved surface on osteogenesis and osteoclastogenesis. *Biomaterials* **2019**, *216*, 119269. [CrossRef]
28. Liu, L.J.; Li, C.J.; Liu, X.X.; Jiao, Y.J.; Wang, F.J.; Jiang, G.S.; Wang, L. Tricalcium phosphate sol-incorporated poly(ε-caprolactone) membrane with improved mechanical and osteoinductive activity as an artificial periosteum. *ACS Biomater. Sci. Eng.* **2020**, *6*, 4631–4643. [CrossRef]
29. Wan, T.; Jiao, Z.X.; Guo, M.; Wang, Z.L.; Wan, Y.Z.; Lin, K.L.; Liu, Q.Y.; Zhang, P.B. Gaseous sulfur trioxide induced controllable sulfonation promoting biomineralization and osseointegration of polyetheretherketone implants. *Bioact. Mater.* **2020**, *5*, 1004–1017. [CrossRef]
30. Akin, S.R.K.; Garcia, C.B.; Webster, T.J. A comparative study of silicon nitride and SiAlON ceramics against *E. coli*. *Ceram. Int.* **2021**, *47*, 1837–1843. [CrossRef]
31. Chen, M.W.; Hu, Y.; Hou, Y.H.; Li, M.H.; Tan, L.; Chen, M.H.; Geng, W.B.; Tao, B.L.; Jiang, H.; Luo, Z.; et al. Magnesium/gallium-layered nanosheets on titanium implants mediate osteogenic differentiation of MSCs and osseointegration under osteoporotic condition. *Chem. Eng. J.* **2022**, *427*, 130982. [CrossRef]
32. Nishida, K.; Anada, T.; Kobayashi, S.; Ueda, T.; Tanaka, M. Effect of bound water content on cell adhesion strength to water-insoluble polymers. *Acta Biomater.* **2021**, *134*, 313–324. [CrossRef] [PubMed]
33. Wu, S.H.; Xiao, Z.L.; Song, J.L.; Li, M.; Li, W.H. Evaluation of BMP-2 Enhances the osteoblast differentiation of human amnion mesenchymal stem cells seeded on nano-hydroxyapatite/collagen/poly(L-Lactide). *Int. J. Mol. Sci.* **2018**, *19*, 2171. [CrossRef] [PubMed]
34. Chen, M.Y.; Liu, Q.Y.; Xu, Y.; Wang, Y.X.; Han, X.W.; Wang, Z.; Liang, J.; Sun, Y.; Fan, Y.J.; Zhang, X.D. The effect of LyPRP/collagen composite hydrogel on osteogenic differentiation of rBMSCs. *Regen. Biomater.* **2021**, *8*, rbaa053. [CrossRef]
35. Hayashi, K.; Munar, M.L.; Ishikawa, K. Effects of macropore size in carbonate apatite honeycomb scaffolds on bone regeneration. *Mat. Sci. Eng. C-Mater.* **2020**, *111*, 110848. [CrossRef]
36. Li, X.; Yin, H.M.; Su, K.; Zheng, G.S.; Mao, C.Y.; Liu, W.; Wang, P.; Zhang, Z.; Xu, J.Z.; Li, Z.M.; et al. Polydopamine-assisted anchor of chitosan onto porous composite scaffolds for accelerating bone regeneration. *ACS Biomater. Sci. Eng.* **2019**, *5*, 2998–3006. [CrossRef]
37. Fu, S.Z.; Ni, P.Y.; Wang, B.Y.; Chu, B.Y.; Peng, J.R.; Zheng, L.; Zhao, X.; Luo, F.; Wei, Y.Q.; Qian, Z.Y. In vivo biocompatibility and osteogenesis of electrospun poly (ε-caprolactone)–poly (ethylene glycol)–poly (ε-caprolactone)/nano-hydroxyapatite composite scaffold. *Biomaterials* **2012**, *33*, 8363–8371. [CrossRef]
38. Bharadwaz, A.; Jayasuriya, A.C. Recent trends in the application of widely used natural and synthetic polymer nanocomposites in bone tissue regeneration. *Mater. Sci. Eng. C* **2020**, *110*, 110698. [CrossRef]
39. Bu, Y.Z.; Ma, J.X.; Bei, J.Z.; Wang, S.G. Surface modification of aliphatic polyester to enhance biocompatibility. *Front. Bioeng. Biotech.* **2019**, *7*, 98. [CrossRef]
40. Pezzotti, G.; Oba, N.; Zhu, W.L.; Marin, E.; Rondinella, A.; Boschetto, F.; McEntire, B.; Yamamoto, K.; Bal, B.S. Human osteoblasts grow transitional Si/N apatite in quickly osteointegrated Si3N4 cervical insert. *Acta Biomater.* **2017**, *64*, 411–420. [CrossRef]
41. Lin, S.X.; Cao, L.Y.; Wang, Q.; Du, J.H.; Jiao, D.L.; Duan, S.Z.; Wu, J.N.; Gan, Q.; Jiang, X.Q. Tailored biomimetic hydrogel based on a photopolymerised DMP1/MCF/gelatin hybrid system for calvarial bone regeneration. *J. Mater. Chem. B* **2018**, *6*, 414–427. [CrossRef] [PubMed]
42. Marin, E.; Boschetto, F.; Zanocco, M.; Adachi, T.; Toyama, N.; Zhu, W.; McEntire, B.J.; Bock, R.M.; Bal, B.S.; Pezzotti, G. KUSA-A1 mesenchymal stem cells response to PEEK-Si3N4 composites. *Mater. Today Chem.* **2020**, *17*, 100316. [CrossRef]
43. Zhao, H.Y.; Xing, H.Y.; Lai, Q.G.; Song, T.X.; Tang, X.P.; Zhu, K.W.; Deng, Y.W.; Zhao, Y.; Liu, W.H.; Xue, R.Q. Mechanical properties, microstructure, and bioactivity of β-Si3N4/HA composite ceramics for bone reconstruction. *Ceram. Int.* **2021**, *47*, 34225–34234. [CrossRef]

Journal of
Functional Biomaterials

Article

Autologous Matrix-Induced Chondrogenesis (AMIC) for Focal Chondral Lesions of the Knee: A 2-Year Follow-Up of Clinical, Proprioceptive, and Isokinetic Evaluation

Paweł Bąkowski [1,*], Kamilla Grzywacz [2], Agnieszka Prusińska [1], Kinga Ciemniewska-Gorzela [1], Justus Gille [3] and Tomasz Piontek [1,4]

1. Department of Orthopedic Surgery, Rehasport Clinic, 60-201 Poznań, Poland
2. Institute of Bioorganic Chemistry Polish Academy of Sciences, 61-704 Poznań, Poland
3. Department of Orthopaedic and Trauma Surgery, University Hospital Schleswig-Holstein, Campus Luebeck, 23562 Luebeck, Germany
4. Department of Spine Disorders and Pediatric Orthopedics, University of Medical Sciences, 60-201 Poznań, Poland
* Correspondence: pawel.bakowski@rehasport.pl

Abstract: (1) Background: The autologous matrix-induced chondrogenesis (AMIC) is a bio-orthopedic treatment for articular cartilage damage. It combines microfracture surgery with the application of a collagen membrane. The aim of the present study was to report a medium-term follow-up of patients treated with AMIC for focal chondral lesions. (2) Methods: Forty-eight patients treated surgically and 21 control participants were enrolled in the study. To evaluate the functional outcomes, the proprioceptive (postural stability, postural priority) and isokinetic (peak value of maximum knee extensor and flexor torque in relation to body mass and the total work) measurements were performed. To evaluate the clinical outcomes, the Lysholm score and the IKDC score were imposed. (3) Results: Compared to the preoperative values, there was significant improvement in the first 2 years after intervention in the functional as well as subjective outcome measures. (4) Conclusions: AMIC showed durable results in aligned knees.

Keywords: AMIC; focal chondral lesions; cartilage chondrogenesis; collagen scaffolds; biomaterials; tissue engineering

1. Introduction

The main function of cartilage is to cover the joint surfaces of the long bones to prevent their wear. Elastic tissue is composed of the chondrocytes which produce prominent amounts of extracellular matrix, rich in proteoglycan, glycosaminoglycans, proteoglycans, and collagen. The cartilage is avascular and aneural, which results in a limited self-healing capability, and therefore the chondral defects are debilitating, often requiring surgical management [1]. For that reason, one of the greatest challenges nowadays for orthopaedic surgeons around the world is the treatment of cartilage injuries. Many recognized surgical procedures have been developed in recent decades, e.g., microfractures, joint surface debridement (which could be further supported with microfracturing) [2], and osteochondral autograft [3] or allograft transplants [4]. It has been observed that the outcomes of the microfracturing cartilage repair have a tendency to regress with time; on the other hand, satisfactory long-term cartilage functionality was observed in patients who underwent the other procedures [5]. Notwithstanding, all of the techniques mentioned could lead to undesirable effects, e.g., donor-site morbidity or auto- and allograft mismatches. Hence, there is growing enthusiasm for cell-based orthobiological techniques to repair chondral defects, which would lead to hyaline cartilage repair.

The autologous matrix-induced chondrogenesis (AMIC) is an innovative, fully arthroscopic orthobiological method of the articular cartilage reconstruction that harnesses the body's natural potential to rebuild the focal cartilage and osteochondral lesions with an area exceeding 1–2 cm^2 [6]. AMIC is combined with microfractures, covering the lesions with a resorbable two-layer collagen membrane to sustain the subsequent blood supply. This creates an appropriate environment for the bone marrow-derived mesenchymal stem cells arising from the subchondral bone to differentiate into chondrocytes.

Functional evaluation is extremely important in making therapeutic decisions. Despite the long-lasting research, there is no single, universal scale that would assess the functional changes in the knee joint after cartilage reconstruction surgery. The proprioception or neuromuscular control is an important part of the knee joint function [7]. It provides the nerve stimulation from the joints, muscles, tendons, and deep tissues, which are processed in the central nervous system and provide the information about the position of the joint, movement, vibration, and pressure. The muscles' strength and endurance are the second most important elements in the function of the knee joint [8]. Rapid muscular atrophy occurs immediately after the surgery as a response to pain, inflammation, and immobilization. Even after a short immobilization, a decrease in muscle strength occurs; in addition, the muscles show greater fatigue, which is translated as a decrease in muscle endurance. Clinical assessments provide a large amount of measurable information about the joint, such as contour disturbances, limb axis deviations, changes in the color and temperature of the skin, as well as joint mobility and stability. This study does not take into account the patient's own feelings, such as joint pain, a sense of fitness or functional limitations, which is why the importance of the patient's subjective assessment is increasingly emphasized [9]. Without such an evaluation and its thorough analysis, it is impossible to obtain a complete picture of the healing of the surgically treated joint.

There are several studies that have evaluated the clinical outcomes of the AMIC for the treatment of focal chondral defects of the knee [10–13]. However, none of them tackled either proprioception or muscle strength and endurance assessments. We have therefore noticed a need to evaluate not only subjective outcomes, but also objective ones. We have compared the results to those for untreated participants.

2. Materials and Methods

2.1. Recruitment Process

All patients treated with an AMIC procedure at the Rehasport Clinic, Poznań (Poland), for full-thickness cartilage tears of the knee between 2011 and 2013 were retrospectively reviewed. All patient-derived data have been fully anonymized and collected according to the clinic's recommendations. The following inclusion criteria were used: (1) magnetic resonance imaging (MRI) evidence; (2) the arthroscopic reconstruction of the knee joint cartilage using the AMIC method; (3) implementation of the RSC rehabilitation protocol; and (4) informed consent to participate in the study. The exclusion criteria included postoperative complications, problems during rehabilitation, and the inability to perform the assessment on any of the proposed dates.

A total of 61 patients were initially selected for the study. Of them, 13 were not eligible to attend the whole assessment. A total of 48 patients were operated on with AMIC, and 26 patients were enrolled in an assessment 24 months after the surgery. Of the 26 patients, 34.6% (n = 9) were women and 65.4% (n = 17) were men. In 18 patients (69%), the lesion was located on the dominant limb. The mean age was 44 y.o. ($\pm$11) and the mean BMI was 26 kg/m^2 ($\pm$4).

A control group of healthy volunteers consisted of 21 participants. The exclusion criteria for the control group were: (1) previous surgery of the knee joint; (2) pain in the knee joint; (3) health contraindications for the biomechanical assessment; and (4) lack of informed consent to participate in the study. The comparison of the research group with the control group is presented in Table 1.

Table 1. Demographic parameters of the research and a control group.

	Research Group	Control Group	p
Age (y.o.)	44.5 ± 11.7 (20–65)	39.1 ± 11.01 (20–65)	n.s.
Body height (cm)	174.0 ± 10.0 (155–194)	181.0 ± 6.2 (173–194)	n.s.
Body mass (kg)	82.0 ± 17.7 (59–110)	83.5 ± 13.3 (65–110)	n.s.
BMI (kg/m^2)	26.9 ± 4.4 (20–37)	25.4 ± 3.7 (19–37)	n.s.

Mean ± standard deviation, and minimal and maximal values are presented in parentheses.

2.2. Surgical Technique

All the surgeries were performed in the same fashion by one experienced surgeon (T.P.) according to the previous report, which describes the AMIC procedure in details and is justified with intraoperative images of the procedure [14]. Summarily, a qualifying diagnostic arthroscopy was performed with classic anteromedial and anterolateral access. Then, the damaged tissue was removed to obtain stable cartilage borders, and the size of each defect was estimated. The chondral defect was cleaned of fibrous tissue and osteophytes with one punch (in cases of smaller cartilage defects, up to 11 mm in diameter) or more (defects larger than 11 mm in diameter). The second part of the procedure was executed under dry arthroscopic conditions. Numerous bores were drilled at 5-mm intervals in the subchondral layer, and Chondro-Gide membrane circles (Figure 1) were placed in the reconstruction area with the porous surface of the membrane facing the bone surface, using a fibrin glue. The stability of the membrane was checked by repeatedly flexing and extending the knee under direct vision. Next, the arthroscopic access wounds were closed.

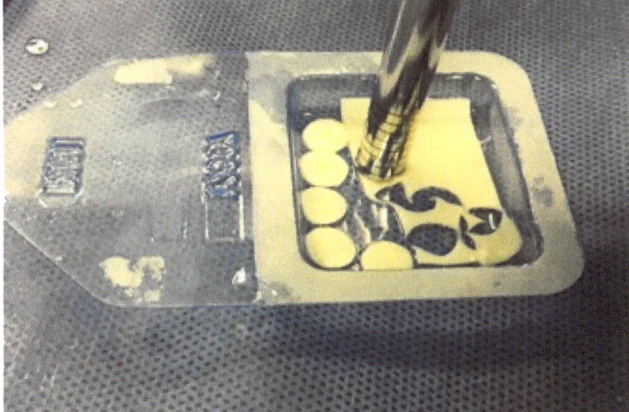

Figure 1. Chondro-Gide membrane circles used in the AMIC surgery.

2.3. Rehabilitation Protocol

Patients were provided with an adjustable-angle orthosis that stabilized the knee at a 15° for the first 24 h after surgery.

Important elements of the rehabilitation protocol of patients after the AMIC cartilage reconstruction were introduced gradually: exercises to improve muscle performance; flexibility exercises—range of motion in the joint and muscle flexibility; exercises for neuromuscular control (proprioception); functional exercises; exercises to correct biomechanical abnormalities; exercises to maintain the efficiency of the cardiovascular system; and psychotherapy. A detailed description of the rehabilitation program is presented in Table 2.

Table 2. The rehabilitation program for patients after the AMIC cartilage reconstruction.

Stage	Rehabilitation Program
I (until the 2nd week)	• anticoagulant exercises • isometric exercises • range of motion-related exercises
II (3rd–8th week)	• anticoagulant exercises • active workout in horizontal position • hip exercises (standing position on one leg) • core stabilization exercises • exercises with the ball: driving the ball into the wall with operated limb in horizontal position • exercises in the swimming pool
III (9th–12th week)	• standing exercises • dynamometric platforms • full-load proprioception exercises • squats up to 90 degrees on an unstable ground • treadmill exercises
IV (13th–24th week)	• external load • warming up on a stationary bike with an increasing load • closed chains exercises • exercises on one leg on a stable and unstable surface • jumping on both legs and on one leg on a trampoline with stops
V (from 24th week)	• jumping exercises aimed at increasing dynamics and power, e.g., double-leg jumps, single-leg jumps and plyometric exercises

2.4. Evaluations

The functional and the clinical outcomes were collected. All patients were assessed postoperatively.

The visual-proprioceptive control was evaluated by measuring the postural strategy (PS) and the postural priority (PP) parameters with the Delos Postural Proprioceptive System. We have implemented and tested dynamic and static Riva tests, as described in [15]. The results for the PS parameter were classified as follows (according to Riva standards [16]): excellent (0.0–1.0°), very good (1.0–2.5°), good (2.5–5.0°), sufficient (5.0–9.0°), or insufficient (>9.0°). The results for the PP parameter were classified as correct (60%), incorrect (40–60%), or bad (<40%).

The isokinetic evaluation of the extensor and flexor muscles was performed using a Biodex 3 dynamometer, where the peak value of maximum knee extensor and flexor torque at 60°/s velocity (Peak Torque, PT), the peak value of maximum knee extensor and flexor torque in relation to body mass at 60°/s velocity (Peak Torque/Body Weight, PT/BW), and the total work of the knee extensors and flexors during the test at 240°/s velocity (W) were measured. The scales for the values of the obtained parameters corresponding to the relation to the body weight and the differences between the limbs have been established according to the published standards [17]. The results for PT/BW for the extensors were classified as: insufficient (<200%), sufficient (200–249%), good (250–299%), very good (300–349%), and excellent (>350%). The results for PT/BW for the flexors were classified as: insufficient (<100%), sufficient (100–149%), good (150–199%), very good (200–249%), and excellent (>250%). The difference of the PT/BW parameter between the limbs were classified as bad (>20%), sufficient (11–20%), or good (<10%). The lowest values required for proper extensor muscle work are 3000 J and for flexors, 1800 J.

The subjective outcomes were assessed with the use of two well-established forms: the International Knee Documentation Committee knee ligament healing standard form (IKDC 2000) and the Lysholm knee scoring scale. The results of the IKDC Knee Index were classified as (according to the standards): very good (90–100 points), good (76–89 points), sufficient (50–75 points), or insufficient (<50 points). The results of the Lysholm scale were classified as (according to the standards): excellent (90–100 points), very good (80–89 points), good (70–79 points), sufficient (60–69 points), or insufficient (<60 points).

2.5. Statistical Analysis

All statistical analyses have been performed in Statistica v. 7.1. The Shapiro-Wilk test was used to assess the normality of the data distribution. Since the data distribution was not normal, all quantitative variables have been shown as median ± standard deviation. The significance of the differences between objective and subjective parameters was determined with the Wilcoxon signed rank test. The correlation between the outcomes was established with the Spearman's rank correlation test. Statistical significance was set at $p < 0.05$.

2.6. Ethics Approval

All subjects gave their informed consent for inclusion before they participated in the study. The study was conducted in accordance with the Declaration of Helsinki, and the protocol was approved by the Bioethics Committee at the Karol Marcinkowski Poznań University of Medical Sciences (no. 830/11).

3. Results

3.1. Objective Knee Evaluation

The numbers of the "correct", "incorrect", and "bad" results of the postural priority did not differ significantly between the patients subjected to the 2-year follow-up and the control group (Figure 2).

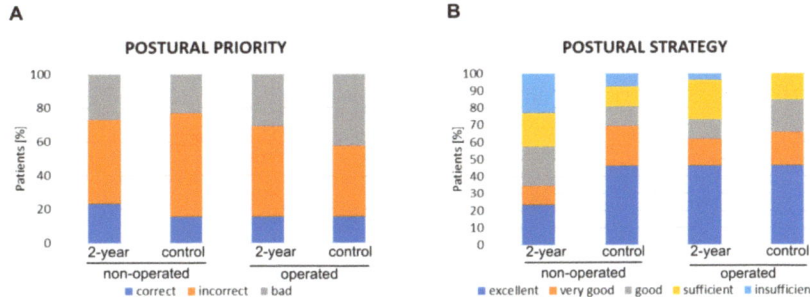

Figure 2. The results of the proprioceptive analysis. The number of patients with correct, incorrect, and bad results in postural priority are shown on the left (**A**). The number of patients with excellent, very good, good, sufficient, and insufficient results in postural strategy are shown on the right (**B**).

A visible difference was noticed in the number of patients with "excellent" results from the postural strategy in the non-operated limb but not in the operated one. "Very good" results of PS were obtained by 25% and 19% in the operated limb, respectively, at 2-year follow-up and in the control group. At 2-year follow-up, there was a lower number of "good" results and a higher number of "sufficient" results in the operated limb when compared to the control group (11% vs. 19% and 23% vs. 15%, respectively).

The median value of the postural priority parameter oscillated around 49%, both for the operated and non-operated limbs in the 2-year follow-up (Table 3 and Figure 3). In the control group, the values of the PP parameter were lower (around 45%), but they were not statistically different in the dominant versus non-dominant limb. No statistically significant differences were found for the PP parameter in the dynamic test.

Table 3. The results of the proprioceptive evaluations.

	2-Year Follow-Up			Control		
	non-operated	operated	p	dominant	non-dominant	p
PP	49.8 ± 9.8	49.3 ± 11.0	n.s.	44.5 ± 11.6	46.3 ± 10.6	n.s.
PS	3.6 ± 7.0	2.7 ± 5.3	n.s.	9.3 ± 9.8	9.6 ± 8.3	n.s.

Data are presented as mean values ± standard deviation. PP: postural priority, PS: postural strategy.

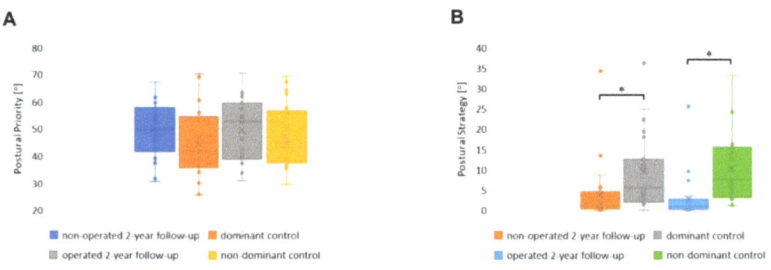

Figure 3. The distribution of the proprioceptive measurements. Box plot diagrams showing the distributions of the proprioceptive parameters. Central lines represent the medians, boxes indicate the range from the 25th to the 75th percentile, whiskers extend 1.5 times the above interquartile range, and outliers are represented as dots. Significance was designated as * $p < 0.05$. (**A**) Postural priority results. (**B**) Postural strategy results.

The median values of the postural strategy parameter reached around 3° ("good") at the 2-year follow-up (Table 3). A significant difference in the value of the postural strategy parameter between operated limbs was noticed when comparing the 2-year follow-up with the control results, both in the operated and non-operated limbs ($p < 0.05$, Figure 2). Significantly lower results (PS of "sufficient") were observed in the control group.

The median of the peak value of maximum knee extensor and flexor torque in relation to body mass at 60°/s velocity (the peak torque/body weight parameter) at 2-year follow-up was significantly lower for the operated limb than for the non-operated limb ($p < 0.001$, Table 4). However, similar differences were also observed in the control group for the dominant versus non-dominant limb ($p < 0.001$). The results for both limbs at the 2-year follow-up were sufficient.

Table 4. The results of the isokinetic evaluation.

	2-Years Follow-Up			Control		
	Extensor Muscles					
	non-operated	operated	p	dominant	non-dominant	p
PT/BW	242 ± 80	203 ± 79	<0.001	263 ± 52	230 ± 66	<0.001
W	2980 ± 876	2814 ± 822	<0.001	3310 ± 702	3153 ± 795	<0.05
	Flexor Muscles					
	non-operated	operated	p	dominant	non-dominant	p
PT/BW	139 ± 46	129 ± 45	<0.05	125 ± 32	123 ± 37	<0.001
W	1916 ± 671	1712 ± 630	n.s.	1784 ± 649	1699 ± 454	n.s.

Data are presented as mean values ± standard deviation. PT/BW: peak torque/body weight [%], W: total work (J).

The median value of the total work at the 2-year follow-up was lower than in the control group. The total work of the extensor muscles reached a critical value of 3000 J only in the control group. The total work of the extensors was significantly lower for the operated limb when compared to the non-operated one at the 2-year follow-up ($p < 0.001$). The difference between the extensors' total work was also noticed for the control group—the dominant limbs were characterized by higher values than non-dominant ones ($p < 0.05$).

The results of the peak torque/body weight of extensors in the operated limb obtained at the 2-year follow-up were statistically significantly lower than for the control group ($p < 0.05$, Figure 4A). The results of the Peak Torque/Body Weight of flexor muscles in the operated limb obtained at 2-year follow-ups were significantly higher than for the control group ($p < 0.05$, Figure 4B), but lower for the non-operated limb ($p < 0.05$).

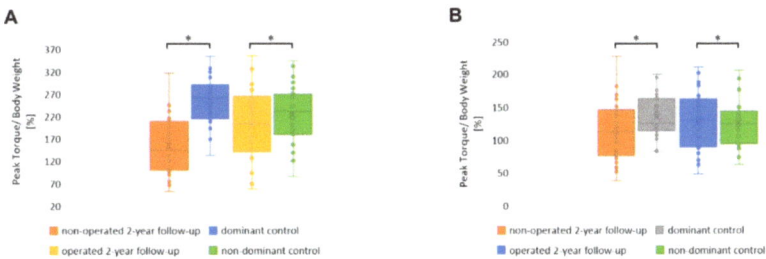

Figure 4. The distribution of the peak value of maximum knee extensor (**A**) and flexor (**B**) torque in relation to body mass at a 60°/s velocity. Box plot diagrams showing the distributions of the parameters. Central lines represent the medians, boxes indicate the range from the 25th to the 75th percentile, whiskers extend 1.5 times the above interquartile range, and outliers are represented as dots. Significance was designated as * with a $p < 0.05$.

Statistically lower results of the total work of knee extensors were observed at the 2-year follow-up when compared to the control group ($p < 0.001$, Figure 5A). The results of the total work of knee extensors in both the operated and non-operated limbs obtained at the 2-year follow-up were statistically significantly lower than for the control group ($p < 0.001$). The results of the total work of the knee flexor muscles were similar (Figure 5B).

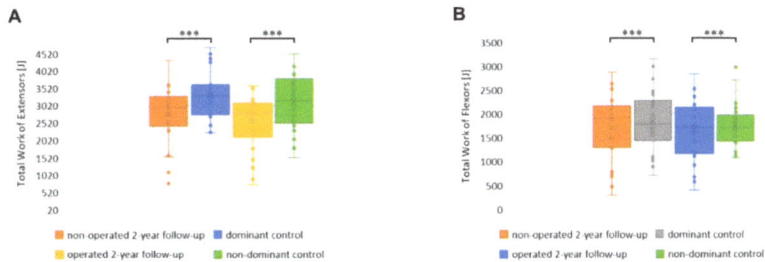

Figure 5. The distribution of the total work of knee extensors (**A**) and flexors (**B**) during the test at 240°/s velocity. Box plot diagrams showing the distributions of the parameters. Central lines represent the medians, boxes indicate the range from the 25th to the 75th percentile, whiskers extend 1.5 times the above interquartile range, and outliers are represented as dots. Significance was designated as *** with a $p < 0.001$.

All objective results were justified by the second-look arthroscopic assessments (Figure 6).

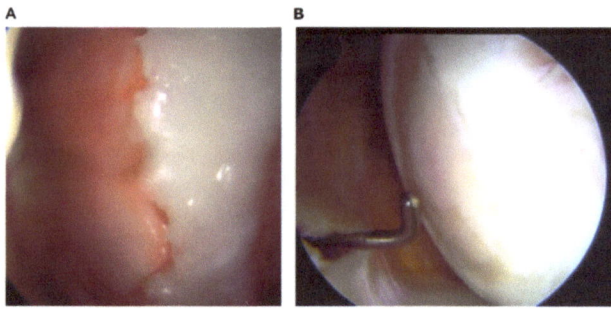

Figure 6. A representative second-look arthroscopic assessment of the AMIC patient before the surgery (**A**) and 2 months postoperatively (**B**).

3.2. Subjective Knee Evaluation

The mean IKDC score increased significantly from insufficient (42 points) at the baseline to good (78 points) at year 2 ($p < 0.001$, Table 5). The mean Lysholm score increased significantly from good (70 points) preoperatively to excellent (90 points) at the 2-year follow-up ($p < 0.001$).

Table 5. IKDC 2000 and Lysholm scores evaluated pre- and post-operatively.

	Baseline	2-Year Follow-Up	p
IKDC	42 ± 14	78 ± 14	<0.001
Lysholm	70 ± 11	90 ± 7	<0.001

Data presented as mean ± standard deviation.

To determine the influence of patient age at the time of operation, patients were divided into 3 subgroups: patients ages ≤ 32 years, 33–46 years, and >46 years. In all groups, a significant improvement was seen for both the IKDC and Lysholm, and there was no significant difference between the age groups.

To determine the body mass index (BMI) influence on the subjective assessments, participants were divided into the following groups: normal (18.5–24.9 BMI), overweight (25.0–29.9 BMI), obesity I (30.0–34.9 BMI), and obesity II (35.0–39.9 BMI). In all groups, a significant improvement was seen for both the IKDC and Lysholm, and there was no significant difference between the BMI groups.

3.3. Correlations between Subjective and Objective Tests

The Spearman Rho coefficient was employed to examine the correlation between the scales. Correlations between subjective and objective tests were investigated for the entire AMIC group. The results of the subjective IKDC assessments for the operated limb correlated well with the results of the isokinetic strength parameters (Table 6).

Table 6. Subjective-objective tests of correlation.

		2-Year Follow-Up	
Subjective Test	Objective Test	r	p
IKDC	PT/BW Ext	0.78	<0.001
IKDC	W Ext	0.59	<0.05
IKDC	PT/BW Flx	0.63	<0.01
IKDC	W Flx	0.53	<0.05
Lysholm	PT/BW Ext	0.32	n.s.
Lysholm	W Ext	0.29	n.s.
Lysholm	PT/BW Flx	0.31	n.s.
Lysholm	W Flx	0.39	n.s.

r—Spearman Rho coefficient. p-values are indicated.

4. Discussion

The results of the present study indicated a significant improvement in all subjective outcome scores and almost all functional scores analyzed up to the 2-year follow-up. The positive effects of the AMIC were seen at the 2-year postoperative visit, with clinically significant improvement in functional outcomes. 30% of the AMIC patients obtained the results of all assessments of the proprioception as well as the muscle strength and endurance within the assumed standards 24 months after the procedure. These data confirm the hypothesis that the clinical benefits may be robust in AMIC-treated patients.

Our results are in accordance with the results of a recent AMIC meta-analysis based on 12 studies (11 level 4 studies and 1 level 1 study) that included a total of 375 patients [10]. Most patients were very satisfied with the result of the index procedure and would choose to undergo the same procedure again if needed.

Moreover, a recent meta-analysis of 2220 surgical procedures of focal chondral defects of the knee, including, except for AMIC, microfractures, osteochondral autograft transplantation, and autologous chondrocyte implantation (ACI), clearly showed that the AMIC procedure performed better overall at approximately three years' follow-up [18]. The authors investigated multiple subjective outcome measures as well as data regarding hypertrophy and the rate of failures and revisions. This analysis added to the knowledge of important comparisons of AMIC versus other procedures. Some prominent examples of these comparisons include: (i) the study by Fossum et al. [19] found no significant differences between AMIC and ACI in terms of the Knee injury and Osteoarthritis Outcome Score (KOOS) and the Lysholm and Visual Analogue Scale (VAS); (ii) Volz et al. [20] reported more effective cartilage repair (clinical outcomes, ICRS Cartilage Injury Standard Evaluation Form-2000, magnetic resonance imaging with the MOCART, and BLOKS and WORMS scores) in AMIC compared to the microfractures; and (iii) Anders et al. [21] presented good, comparable clinical outcomes (modified Cincinnati and ICRS score) of AMIC and microfractures.

Thus far, no studies have examined proprioception or isokinetic assessment in patients subjected to the AMIC procedure. In this report, all AMIC patients had sufficient body control while maintaining the single-legged position with their eyes closed in the static proprioception evaluation test and good body control after 24 months. However, during the dynamic test, the posture control was disturbed according to the norms established by Riva [16]. The muscle strength and endurance of the quadriceps and flexors in the non-operated limb were stronger and more durable than in the operated limb. According to the standards established by Davies [17], the extensor strength of the operated limb was insufficient in relation to the body weight of the examined person, while the value generated by the flexors turned out to be sufficient. However, at the 2-year follow-up, the isokinetic values of both the extensors and flexors were sufficient.

A very important conclusion from this study is the observation that a significant part of the AMIC group obtained positive results in all measured tests when compared to the control group. The comparison of the isokinetic evaluation of the AMIC and the control group revealed no difference in the flexor and extensor muscle endurance. The comparison of proprioceptive results illustrates the positive effect of the balance, stabilization, and proprioception exercises performed by patients after the AMIC cartilage reconstruction following the described rehabilitation protocol. Rozzi and Kusano drew similar conclusions in their research aimed at revealing the influence of proprioceptive training on test results [22,23]. During the evaluation of proprioception in the static test, the control group showed an insufficient degree of control, i.e., too many torso deflections, as shown by the results of the Postural Strategy parameter, and this is the result of the lack of proprioception training. The AMIC group examined in the same test achieved very good PS results.

The IKDC index is a measure of knee joint function, where higher scores indicate higher activity and lower discomfort. Before the AMIC procedure, the function of the knee joint was insufficient. We found it very encouraging that the assessment of the knee joint function improved to a good level, indicating a rapid improvement in activity and only minor ailments in the operated knee joint. Indeed, Chen Chou et al. observed an increase in the IKDC scores 6 months after AMIC surgery, but their results were not as high as in the present study [24]. The IKDC results presented in this study were very similar to those described by Gobbi [25] and higher than the results obtained by AMIC patients in Hoburg's study 12 months after the procedure [26]. The results of IKDC in the present study at 2-year follow-up are lower than those presented by the Victor and Vasso groups [27,28]. However, in all of the cited studies as well as in the present study, the results indicate a good and stable assessment of knee joint function according to the IKDC scale. The results of the IKDC 2000 scale correlated well with the objective isokinetic evaluation.

The Lysholm scores indicated a very good assessment of the knee joint function of the AMIC patients. The obtained results were higher than those presented by other authors. Hoburg showed good results 12 months after the AMIC procedure [26]. In Schagemann [29], Kaiser [13], and Gille [11,30] studies, the results of the Lysholm score revealed good knee function in the 1-year follow-up and very good at the 2-year follow-up.

5. Conclusions

AMIC is an effective and durable treatment, lasting up to 2 years post-surgically, for patients with cartilage defects in the knee. AMIC provides satisfactory results in terms of both subjective esteem of knee function and knee functional rehabilitation, which appear to be sustained in the majority of patients, according to our 2-year follow-up results.

Author Contributions: Conceptualization, P.B., J.G. and T.P.; methodology, P.B. and T.P.; formal analysis, K.G.; investigation, P.B. and K.G.; data curation, A.P. and K.C.-G.; writing—original draft preparation, K.G.; writing—review and editing, P.B., T.P. and K.G. All authors have read and agreed to the published version of the manuscript.

Funding: This research received no external funding.

Data Availability Statement: The data presented in this study are available on request from the corresponding author.

Conflicts of Interest: The authors declare no conflict of interest.

References

1. Dávila Castrodad, I.M.; Mease, S.J.; Werheim, E.; McInerney, V.K.; Scillia, A.J. Arthroscopic Chondral Defect Repair With Extracellular Matrix Scaffold and Bone Marrow Aspirate Concentrate. *Arthrosc. Tech.* **2020**, *9*, e1241–e1247. [CrossRef] [PubMed]
2. Ferkel, R.D.; Zanotti, R.M.; Komenda, G.A.; Sgaglione, N.A.; Cheng, M.S.; Applegate, G.R.; Dopirak, R.M. Arthroscopic treatment of chronic osteochondral lesions of the talus: Long-term results. *Am. J. Sport. Med.* **2008**, *36*, 1750–1762. [CrossRef] [PubMed]
3. Imhoff, A.B.; Paul, J.; Ottinger, B.; Wörtler, K.; Lämmle, L.; Spang, J.; Hinterwimmer, S. Osteochondral transplantation of the talus: Long-term clinical and magnetic resonance imaging evaluation. *Am. J. Sport. Med.* **2011**, *39*, 1487–1493. [CrossRef] [PubMed]
4. Raikin, S.M. Fresh osteochondral allografts for large-volume cystic osteochondral defects of the talus. *J. Bone Jt. Surg. Ser. A* **2009**, *91*, 2818–2826. [CrossRef] [PubMed]
5. Giannini, S.; Battaglia, M.; Buda, R.; Cavallo, M.; Ruffilli, A.; Vannin, F. Surgical Treatment of Osteochondral Lesions of the Talus by Open-Field Autoliguous Chondrocyte Implantation: A 10-Year Follow-up Clinical and Magnetic Resonance Imaging T2-Mapping Evaluation. *Am. J. Sport. Med.* **2009**, *37*, 112–118. [CrossRef]
6. Mikrofrakturierung, B.P.M. Ein neues Konzept zur Knorpeldefektbehandlung. *Arthroskopie* **2005**, *18*, 193–197.
7. Lattanzio, P.J.; Petrella, R.J. Knee proprioception: A review of mechanisms measurements, and implications of muscular fatigue. *Orthopedics* **1998**, *21*, 463–471. [CrossRef]
8. Kannus, P. Isokinetic evaluation of muscular performance: Implications for muscle testing and rehabilitation. *Int. J. Sport. Med.* **1994**, *15*, S11–S18. [CrossRef]
9. Collins, N.J.; Misra, D.; Felson, D.T.; Crossley, K.M.; Roos, E.M. Measures of knee function: International Knee Documentation Committee (IKDC) Subjective Knee Evaluation Form, Knee Injury and Osteoarthritis Outcome Score (KOOS), Knee Injury and Osteoarthritis Outcome Score Physical Function Short Form (KOOS-PS), Knee Outcome Survey Activities of Daily Living Scale (KOS-ADL), Lysholm Knee Scoring Scale, Oxford Knee Score (OKS), Western Ontario and McMaster. *Arthritis Care Res.* **2011**, *63*, S208–S228.
10. Kim, J.H.; Heo, J.W.; Lee, D.H. Clinical and Radiological Outcomes After Autologous Matrix-Induced Chondrogenesis Versus Microfracture of the Knee: A Systematic Review and Meta-analysis With a Minimum 2-Year Follow-up. *Orthop. J. Sport Med.* **2020**, *8*, 2325967120959280. [CrossRef]

11. Gille, J.; Reiss, E.; Freitag, M.; Schagemann, J.; Steinwachs, M.; Piontek, T.; Reiss, E. Autologous Matrix-Induced Chondrogenesis for Treatment of Focal Cartilage Defects in the Knee: A Follow-up Study. *Orthop. J. Sport Med.* **2021**, *9*, 2325967120981872. [CrossRef]
12. Migliorini, F.; Eschweiler, J.; Maffulli, N.; Schenker, H.; Baroncini, A.; Tingart, M.; Rath, B. Autologous matrix-induced chondrogenesis (AMIC) and microfractures for focal chondral defects of the knee: A medium-term comparative study. *Life* **2021**, *11*, 183. [CrossRef]
13. Kaiser, N.; Jakob, R.P.; Pagenstert, G.; Tannast, M.; Petek, D. Stable clinical long term results after AMIC in the aligned knee. *Arch. Orthop. Trauma Surg.* **2020**, *141*, 1845–1854. [CrossRef]
14. Piontek, T.; Ciemniewska-Gorzela, K.; Szulc, A.; Naczk, J.; Słomczykowski, M. All-arthroscopic AMIC procedure for repair of cartilage defects of the knee. *Knee Surg. Sport. Traumatol. Arthrosc.* **2012**, *20*, 922–925. [CrossRef]
15. Bąkowski, P.; Ciemniewska-Gorzela, K.; Bąkowska-Żywicka, K.; Stołowski, Ł.; Piontek, T. Similar outcomes and satisfaction of the proprioceptive versus standard training on the knee function and proprioception, following the anterior cruciate ligament reconstruction. *Appl. Sci.* **2021**, *11*, 3494. [CrossRef]
16. Riva, D.S.G.P. Per ritrovare l'equilibrio. *Sport Med.* **1999**, *5*, 55–58.
17. Davies, G.J. *Compendium of Isokinetics in Clinical Usage and Rehabilitation Techniques*, 4th ed.; S&S Publishers: New York, NY, USA, 1992.
18. Migliorini, F.; Eschweiler, J.; Schenker, H.; Baroncini, A.; Tingart, M.; Maffulli, N. Surgical management of focal chondral defects of the knee: A Bayesian network meta-analysis. *J. Orthop. Surg. Res.* **2021**, *16*, 543. [CrossRef]
19. Fossum, V.; Hansen, A.K.; Wilsgaard, T.; Knutsen, G. Collagen-covered autologous chondrocyte implantation versus autologous matrix-induced chondrogenesis: A randomized trial comparing 2 methods for repair of cartilage defects of the knee. *Orthop. J. Sport. Med.* **2019**, *7*, 2325967119868212. [CrossRef]
20. Volz, M.; Schaumburger, J.; Frick, H.; Grifka, J.; Anders, S. A randomized controlled trial demonstrating sustained benefit of autologous matrix-induced chondrogenesis over microfracture at five years. *Int. Orthop.* **2017**, *41*, 797–804. [CrossRef]
21. Anders, S.; Volz, M.; Frick, H.; Gellissen, J. A randomized, controlled trial comparing autologous matrix-induced chondrogenesis (AMIC(R)) to microfracture: Analysis of 1- and 2-year follow-Up data of 2 Centers. *Open Orthop. J.* **2013**, *7*, 133–143. [CrossRef]
22. Kusano, T.; Jakob, R.P.; Gautier, E.; Magnussen, R.A.; Hoogewoud, H.; Jacobi, M. Treatment of isolated chondral and osteochondral defects in the knee by autologous matrix-induced chondrogenesis (AMIC). *Knee Surg. Sport. Traumatol. Arthrosc.* **2012**, *20*, 2109–2115. [CrossRef] [PubMed]
23. Rozzi, S.L.; Lephart, S.M.; Sterner, R.; Kuligowski, L. Balance training for persons with functionally unstable ankles. *J. Orthop. Sport. Phys. Ther.* **1999**, *29*, 478–486. [CrossRef] [PubMed]
24. Chou, A.C.C.; Lie, D.T.T. Clinical Outcomes of an All-Arthroscopic Technique for Single-Stage Autologous Matrix-Induced Chondrogenesis in the Treatment of Articular Cartilage Lesions of the Knee. *Arthrosc. Sport Med. Rehabil.* **2020**, *2*, e353–e359. [CrossRef] [PubMed]
25. Gobbi, A.; Karnatzikos, G.; Sankineani, S.R. One-step surgery with multipotent stem cells for the treatment of large full-thickness chondral defects of the knee. *Am. J. Sport. Med.* **2014**, *42*, 648–657. [CrossRef]
26. Hoburg, A.; Leitsch, J.M.; Diederichs, G.; Lehnigk, R.; Perka, C.; Becker, R.; Scheffler, S. Treatment of osteochondral defects with a combination of bone grafting and AMIC technique. *Arch. Orthop. Trauma Surg.* **2018**, *138*, 1117–1126. [CrossRef]
27. Dhollander, A.; Moens, K.; Van Der Maas, J.; Verdonk, P.; Almqvist, K.F.; Victor, J. Treatment of patellofemoral cartilage defects in the knee by autologous matrix-induced chondrogenesis (AMIC). *Acta Orthop. Belg.* **2014**, *80*, 251–259.
28. Schiavone Panni, A.; Del Regno, C.; Mazzitelli, G.; D'Apolito, R.; Corona, K.; Vasso, M. Good clinical results with autologous matrix-induced chondrogenesis (Amic) technique in large knee chondral defects. *Knee Surg. Sport Traumatol. Arthrosc.* **2018**, *26*, 1130–1136. [CrossRef]
29. Schagemann, J.; Behrens, P.; Paech, A.; Riepenhof, H.; Kienast, B.; Mittelstädt, H.; Gille, J. Mid-term outcome of arthroscopic AMIC for the treatment of articular cartilage defects in the knee joint is equivalent to mini-open procedures. *Arch. Orthop. Trauma Surg.* **2018**, *138*, 819–825. [CrossRef]
30. Gille, J.; Behrens, P.; Volpi, P.; de Girolamo, L.; Reiss, E.; Zoch, W.; Anders, S. Outcome of Autologous Matrix Induced Chondrogenesis (AMIC) in cartilage knee surgery: Data of the AMIC Registry. *Arch. Orthop. Trauma Surg.* **2013**, *133*, 87–93. [CrossRef]

Article

Increased UHMWPE Particle-Induced Osteolysis in Fetuin-A-Deficient Mice

Christina Polan [1,*], Christina Brenner [1], Monika Herten [1], Gero Hilken [2], Florian Grabellus [3], Heinz-Lothar Meyer [1], Manuel Burggraf [1], Marcel Dudda [1], Willi Jahnen-Dechent [4], Christian Wedemeyer [5] and Max Daniel Kauther [6]

1. Department of Trauma, Hand and Reconstructive Surgery, University Hospital Essen, University Duisburg-Essen, 45147 Essen, Germany
2. Central Animal Laboratory, University Hospital Essen, University Duisburg-Essen, 45147 Essen, Germany
3. Institute of Pathology and Neuropathology, University Hospital Essen, University Duisburg-Essen, 45147 Essen, Germany
4. Helmholtz-Institute for Biomedical Engineering, RWTH Aachen University Hospital, 52074 Aachen, Germany
5. Department of Orthopaedic Surgery, St. Barbara Hospital Gladbeck, 45964 Gladbeck, Germany
6. Department of Trauma Surgery and Orthopedics, Pediatric Orthopedics, Agaplesion Diakonieklinikum Rotenburg (Wümme), 27356 Rotenburg, Germany
* Correspondence: christina.polan@uk-essen.de; Tel.: +49-201-723-1301

Citation: Polan, C.; Brenner, C.; Herten, M.; Hilken, G.; Grabellus, F.; Meyer, H.-L.; Burggraf, M.; Dudda, M.; Jahnen-Dechent, W.; Wedemeyer, C.; et al. Increased UHMWPE Particle-Induced Osteolysis in Fetuin-A-Deficient Mice. J. Funct. Biomater. 2023, 14, 30. https://doi.org/10.3390/jfb14010030

Academic Editors: Kunyu Zhang, Qian Feng, Yongsheng Yu and Boguang Yang

Received: 8 December 2022
Revised: 30 December 2022
Accepted: 31 December 2022
Published: 4 January 2023

Copyright: © 2023 by the authors. Licensee MDPI, Basel, Switzerland. This article is an open access article distributed under the terms and conditions of the Creative Commons Attribution (CC BY) license (https://creativecommons.org/licenses/by/4.0/).

Abstract: Particle-induced osteolysis is a major cause of aseptic prosthetic loosening. Implant wear particles stimulate tissue macrophages inducing an aseptic inflammatory reaction, which ultimately results in bone loss. Fetuin-A is a key regulator of calcified matrix metabolism and an acute phase protein. We studied the influence of fetuin-A on particle-induced osteolysis in an established mouse model using fetuin-A-deficient mice. Ten fetuin-A-deficient ($Ahsg^{-/-}$) mice and ten wild-type animals ($Ahsg^{+/+}$) were assigned to test group receiving ultra-high molecular weight polyethylene (UHMWPE) particle implantation or to control group (sham surgery). After 14 days, bone metabolism parameters RANKL, osteoprotegerin (OPG), osteocalcin (OC), alkaline phosphatase (ALP), calcium, phosphate, and desoxypyridinoline (DPD) were examined. Bone volume was determined by microcomputed tomography (µCT); osteolytic regions and osteoclasts were histomorphometrically analyzed. After particle treatment, bone resorption was significantly increased in $Ahsg^{-/-}$ mice compared with corresponding $Ahsg^{+/+}$ wild-type mice ($p = 0.007$). Eroded surface areas in $Ahsg^{-/-}$ mice were significantly increased ($p = 0.002$) compared with $Ahsg^{+/+}$ mice, as well as the number of osteoclasts compared with control ($p = 0.039$). Fetuin-A deficiency revealed increased OPG ($p = 0.002$), and decreased levels of DPD ($p = 0.038$), OC ($p = 0.036$), ALP ($p < 0.001$), and Ca ($p = 0.001$) compared with wild-type animals. Under osteolytic conditions in $Ahsg^{-/-}$ mice, OPG was increased ($p = 0.013$), ALP ($p = 0.015$) and DPD ($p = 0.012$) were decreased compared with the $Ahsg^{+/+}$ group. Osteolytic conditions lead to greater bone loss in fetuin-A-deficient mice compared with wild-type mice. Reduced fetuin-A serum levels may be a risk factor for particle-induced osteolysis while the protective effect of fetuin-A might be a future pathway for prophylaxis and treatment.

Keywords: implants; arthroplasty; aseptic prosthetic loosening; osteolysis; particle-induced osteolysis; mouse calvaria osteolysis model; polyethylene particles; wear particles

1. Introduction

The aseptic osteolysis, or particle disease, is the major cause of long-term failure of arthroplasty [1] and is evoked by wear debris particles generated by shear and friction forces. Wear particles from prosthesis components such as ultra-high molecular weight polyethylene (UHMWPE), titanium, ceramic, or cement particles are phagocytosed by macrophages, which subsequently trigger an inflammatory cascade via the release of proinflammatory cytokines, leading to loss of bone substance and loosening of endoprosthesis [2,3]. The inflammatory reaction depends on the number, size, material, and shape of the abrasive material. The inflammatory reaction increases with the particle concentration [4,5].

Bone tissue, which consists of minerals such as calcium and phosphate in the form of hydroxyapatite organic material, undergoes constant modification. An imbalance of the bone formation and degradation in favor of bone loss leads to various pathologies such as, among other things, the aseptic osteolysis in prostheses. In addition to numerous other cells, such as stem cells or immune cells, bone tissue consists of bone-degrading osteoclasts, bone-building osteoblasts, and osteocytes [6]. Bone resorption is regulated by the release of messenger substances, hormones, and cytokines, which regulate the differentiation and activation of osteoclasts.

Key regulator of bone metabolism is the RANK-RANKL-OPG system, which consists of the messenger substances osteoprotegerin (OPG), the ligand of the receptor activator of the nuclear factor κB (RANKL) and the associated receptor (RANK) [7–9]. The ratio of bone-protective OPG to bone-destructive RANKL determines the balance between bone formation and bone resorption that is imbalanced in aseptic loosening [3,10]. Macrophages can increase RANKL release through the secretion of proinflammatory cytokines such as IL-1β and TNF-α and promote bone resorption by stimulating osteoclast function [11–13].

Fetuin-A (alpha2-Heremans-Schmid glycoprotein (Ahsg)) is highly enriched in the mineralized bone matrix [14,15]. Fetuin-A plays an essential role in the solubility of serum calcium, inhibits the precipitation of calcium through the formation of calcium-protein complexes and influences the formation of mineralized bone [16–19]. A recent study suggested that fetuin-A is a multifaceted protective factor that locally counteracts calcification, modulates macrophage polarization, and attenuates inflammation and fibrosis, thus preserving tissue function [20].

The majority of publications support that fetuin-A acts as an indirect anti-inflammatory agent [20,21]. Fetuin-A is known as negative acute phase protein in injury and infection [22–24]. A recent review concluded that fetuin-A is down-regulated in degenerative joint disease and can be used as a biomarker in the diagnosis and treatment of arthritis [25]. In a murine calvaria model of particle-induced osteolysis, a single dose of fetuin-A minimized bone resorption under osteolytic conditions demonstrating a protective effect of Fetuin-A [26]. This calvaria model is an established animal model for the investigation of particle-induced osteolysis [27,28]. Following peri-implant osteolysis in human joint endoprostheses, polyethylene particles are applied in a very high concentration to the cranial calotte of the calvaria model. This results in an aseptic inflammatory reaction, the formation of granulation tissue rich in macrophages, and a change in local bone metabolism with the activation of osteoclasts.

Given the pleiotropic effects of fetuin-A in bone metabolism, the question of whether fetuin-A deficiency might also play a role in particle-induced osteolysis prompted us to further investigate the influence of fetuin-A on particle-induced osteolysis in an established mouse model of aseptic loosening using homozygot wild-type ($Ahsg^{+/+}$) and homozygot fetuin-A-deficient ($Ahsg^{-/-}$) mice. Our hypothesis was that in absence of fetuin-A the particle-induced inflammatory response and osteolysis is increased.

2. Materials and Methods

2.1. Animals and Ethics

The animal experiment was carried out in accordance with the international regulations for use and care of laboratory animals and the reporting of in vivo experiments (ARRIVE) guidelines and was approved by the national district government (State Office for Nature, Environment and Consumer Protection of North Rhine-Westphalia, LANUV AZ 9.93.2.10.34.07.138). Ten 12-week-old male fetuin-A-deficient mice ($Ahsg^{-/-}$, ILAR strain BL6, Ahsgtm1, wja, originated from backcrossing of the original C57BL/6-129/Sv hybrid mice) as well as ten age- and gender-matching wild-type $Ahsg^{+/+}$ (wt) mice of the

C57 Bl 6/N strain (Charles River Wiga GmbH, Sulzfeld, Germany) were used [29,30]. The specific pathogen-free animals were kept in cages in a climate-controlled room (22 °C; 45–54% relative humidity) with a 12 h light/dark cycle in groups until surgery and as single animals after surgery. The cages were equipped with bedding, nesting material, and shelter (spacious plastic house). Food and water were allowed ad libitum. The animals were monitored on a daily interval. In case of any abnormality, the individual stress of the animal concerned was evaluated with an experiment-specific score sheet. On the first and last experimental days, the animals were housed in metabolic cages for urine collection.

2.2. Ultra-High Molecular Weight Polyethylene UHMWPE Particles

The Clariant Company (Gersthofen, Germany) supplied the UHMWPE polyethylene particles (Ceridust VP 3610). More than 35% of the particles were smaller than 1 μm showing a mean particle size (provided as equivalent circle diameter) of 1.75 ± 1.43 μm (range 0.05–11.6 μm). The surface of the particles was rougher, and the particles were shorter in length, displaying a more uniform size distribution in relation to particles obtained in hip joint simulators. To remove putative endotoxins and for disinfection, the particles were washed twice in 70% ethanol at room temperature for 24 h, followed by washing in sterile phosphate buffered saline and drying [31]. The endotoxin content of the particles was controlled (Limulus Amebocyte Lysate (LAL) Assay) to be lower than the detection level of <0.25 EU/mL.

2.3. Animal Surgery

Ten $Ahsg^{-/-}$ mice and ten wt mice were randomly assigned to either particle implantation or a sham operation ($n = 5$ per group and type) (see graphical abstract Figure 1). An anesthetic mixture of ketamine and xylazine (CEVA Tiergesundheit, Düsseldorf, Germany) was administered intraperitoneally before the animals were anesthetized with isoflurane followed by an intraperitoneal injection of 0.1 g/kg bodyweight (BW) ketamin/0.01 g/kg BW xylazine in NaCl. Before surgery, 300 μL blood was drawn by orbital puncture. A 10 mm incision was made along the calvarial sagittal midline suture exposing an area of 10 mm × 10 mm in the periosteum. The animals in the particle groups received a total volume of 30 μL dried UHMWPE particles (corresponding to 6×10^6 particles), which were distributed over the periosteum using a sterile sharp surgical spoon as previously described [32]. In the sham controls, the incision was closed without any further intervention.

Subsequently, each animal received buprenorphine (0.1 mg/kg BW, Temgesic®, Reckitt Benckiser, Slough, UK) for postoperative analgesia.

Fourteen days after operation, the animals were sacrificed using CO_2 aspiration, and blood samples were taken immediately by cardiac puncture. The skulls were removed and fixed in 10% formalin. Apart from the conduction of the surgery, further analyses (outcome assessment and data analysis) were performed in a blinded condition.

2.4. Microcomputed Tomography

The radiological examination of the skulls was performed by microcomputer tomograph (μCT) (X-ray microcomputer tomograph 1072, Skyscan, Aartselaar, Belgium). Two-dimensional X-ray images were realized using the TomoNT program (Skyscan) with a 19.23× magnification for the images. The X-ray source was set to a voltage of 80 kV at an amperage of 100 μA. At an exposure time of 4.9 s and a rotation angle of 0.9° per image, a total angle of 180° was scanned. A field correction was also activated. The greyscales were selected so that the tissue was radiographically displayed mainly around the bone window. The cone-beam reconstruction program generated a total of about 700 sectional images. A quantitative analysis of bone volumes and bone surfaces was performed using a CT-analysis program (CTAn, Skyscan). Five defined cuboids (8 mm³) along the sagittal suture were selected for investigation of the following parameters: bone volume (BV), bone surface (BS), and bone volume per tissue volume (BV/TV). Subsequently, a three-dimensional model of the calvaria was created (CTVol Surface Rendering, Skyscan).

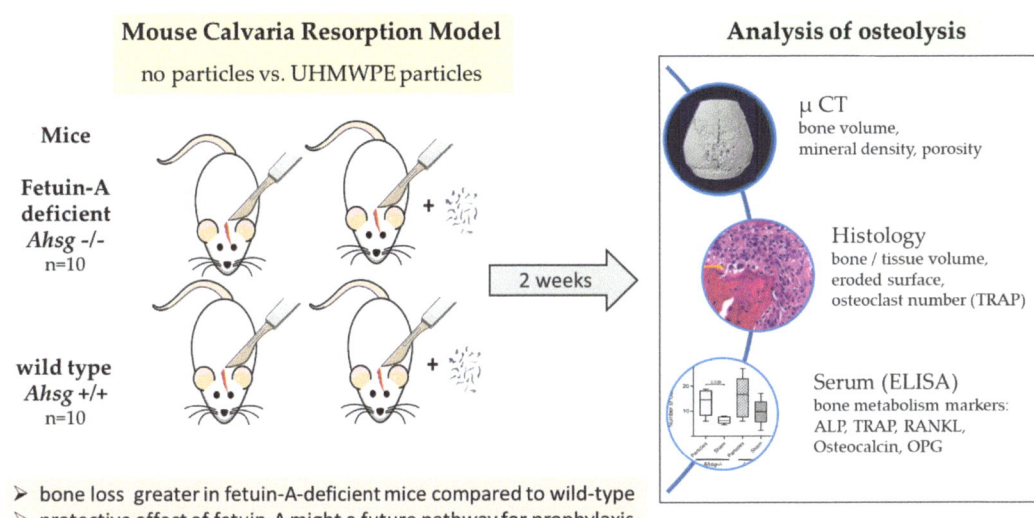

Figure 1. Graphical abstract illustrating the different group distribution as well as the applied methods.

2.5. Bone Histomorphometry

The calvaria were prepared from the skulls and decalcified for seven days (Osteosoft®, Merck, Darmstadt, Germany) and embedded in paraffin. Sectional images of 4 µm thickness were prepared and stained with hematoxylin and eosin (HE) to determine the area of bone resorption in the midline suture. Additionally, the osteoclasts were identified as large multinucleated tartrate resistant alkaline phosphatase (TRAP) positive cells (Sigma Aldrich, Deisenhofen, Germany). The area of bone resorption containing the osteolytic areas was defined as eroded surface (ES) and measured with special software (Image Tool 3.0, University of Texas, San Antonio, TX, USA) [33].

2.6. Analyses of the Parameters of Bone Metabolism

Serum was stored at −70 °C until analysis. All assays were performed according to the instructions of the manufacturers. Calcium, phosphate, and ALP were determined using an RX Monza Analyzer (Randox, Antrim, UK). For OPG and RANKL, immunoassays were used (Quantikine assays, R&D Systems, Minneapolis, MN, USA), while OC was analyzed with an immunoradiometric assay (Immutopics, San Clemente, CA, USA). Deoxypyridinoline (DPD), a crosslink product of collagen molecules, was determined in urine samples (Quidel, San Diego, CA, USA). Creatinine was measured in urine samples within two hours after collection using a Siemens ADVIA 2400 analyzer (Siemens Medical Solutions, Fernwald, Germany). Variations in urine concentration were eliminated using the ratio of DPD [nmol/L] and the urine creatinine concentration [mmol/L], resulting in a value without units.

2.7. Statistical Analysis

Estimation of the number of cases for changes in bone volume resulted in a case number of 5 with a power of 80% and a probability of error of <0.05%.

The data were analyzed using the statistics program SPSS 27 (IBM, New York, NY, USA). For all continuous parameters, the mean value, standard deviation, minimum, maximum, and median were determined based on descriptive statistics. The individual groups were examined for normal distribution using the Shapiro–Wilk test for small samples. Parametric values were tested for significance using the two-way analysis of variance (ANOVA) to analyze an influence of particles (main effect), group (main effect), or the combination of particles and group (interaction effect). Normally distributed data sets were examined with a two-sided t-test for independent samples to study the difference in the means between pairs of groups. For nonparametric values, the Wilcoxon rank-sum test and the Mann–Whitney U test were performed. A p-value < 0.05 was considered as significant.

3. Results

No surgical or postoperative complications occurred. All animals were euthanized at the end of the follow-up period.

3.1. Micro Computed Tomography

Three-dimensional reconstructions of the calvaria showed the qualitative extent of particle-induced osteolysis (Figure 2). Particle treatment led to bone loss and circular osteolysis, especially close to the sagittal suture where most of the particles were implanted (Figure 2A,B), whereas the sham-operated mice did not have any osteolytic lesions (Figure 2C,D). The osteolytic regions were largest in the fetuin-A-deficient particle-treated group (Figure 2A).

The analysis of bone volume in the target region showed an influence of UHMWPE particle implantation on the radiologically quantifiable bone tissue of both genotypes (Figure 3). Osteolytic conditions resulted in a significant decrease in bone volume in the fetuin-A-deficient group ($p = 0.007$) as well as in the wt group ($p = 0.032$) compared with the respective sham-operated groups (Table 1) (Figure 3A). Fetuin-A-deficient animals displayed a significant lower bone volume after particle treatment than wild-type animals ($p = 0.007$). Osteolytic conditions led to a significant reduction in the ratio BV/TV in fetuin-A-deficient mice ($p = 0.001$) as well as in wild-type mice ($p = 0.02$) compared with the respective sham-operated groups. After particle treatment, fetuin-A-deficient mice had a significantly lower BV/TV ratio than wt mice ($p = 0.001$) (Table 2) (Figure 3B). No significant differences were found between the groups regarding the bone surface in µCT (data not shown).

Table 1. Bone volume of Fetuin-A-deficient mice ($Ahsg^{-/-}$) vs. wild-type ($Ahsg^{+/+}$) mice under osteolytic conditions (particles) vs. sham-operated animals. Data are presented in mm^3 as median and interquartiles range (IOR) and as mean ± standard deviation (SD) with $n = 5$ per group.

Bone Volume	$Ahsg^{-/-}$		$Ahsg^{+/+}$	
[mm^3]	Particles	Sham	Particles	Sham
Median (IQR)	0.50 (0.02)	0.53 (0.02)	0.52 (0.01)	0.54 (0.03)
Mean ± SD	0.50 ± 0.01	0.53 ± 0.01	0.52 ± 0.01	0.54 ± 0.02

Table 2. Quotient of bone volume/tissue volume of Fetuin-A-deficient ($Ahsg^{-/-}$) mice vs. wild-type ($Ahsg^{+/+}$) mice under osteolytic conditions (particles) vs. sham-operated animals × 10^{-2}. Data are presented in mm^3 as median and interquartiles range (IOR) and as mean ± standard deviation (SD) with $n = 5$ per group.

Bone Volume/ Tissue Volume	$Ahsg^{-/-}$		$Ahsg^{+/+}$	
[×10^{-2}]	Particles	Sham	Particles	Sham
Median (IQR)	5.69 (0.18)	6.02 (0.05)	6.01 (0.15)	6.19 (0.31)
Mean ± SD	5.71 ± 0.09	6.03 ± 0.03	6.00 ± 0.09	6.25 ± 0.17

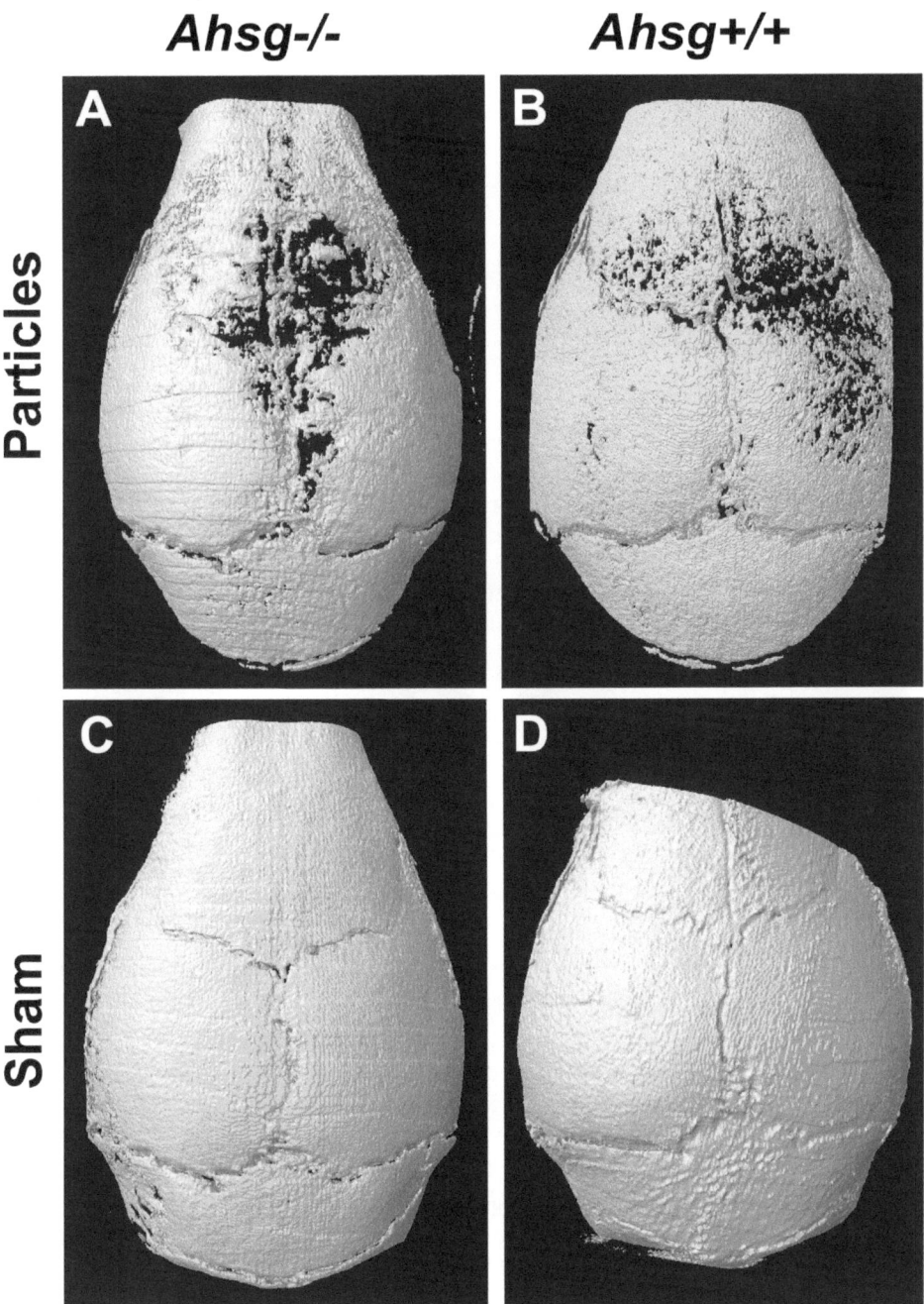

Figure 2. Three-dimensional reconstruction of the calvaria. After 14 days, the calvarial bone was analyzed by µCT. (**A**) Particle treatment in mice with the $Ahsg^{-/-}$ background led to severe osteolysis. (**B**) In wild-type mice, less osteolysis was seen after particle treatment. (**C**) Sham-operated fetuin-A-deficient mice, and (**D**) sham-operated wild-type mice did not show any signs of osteolysis.

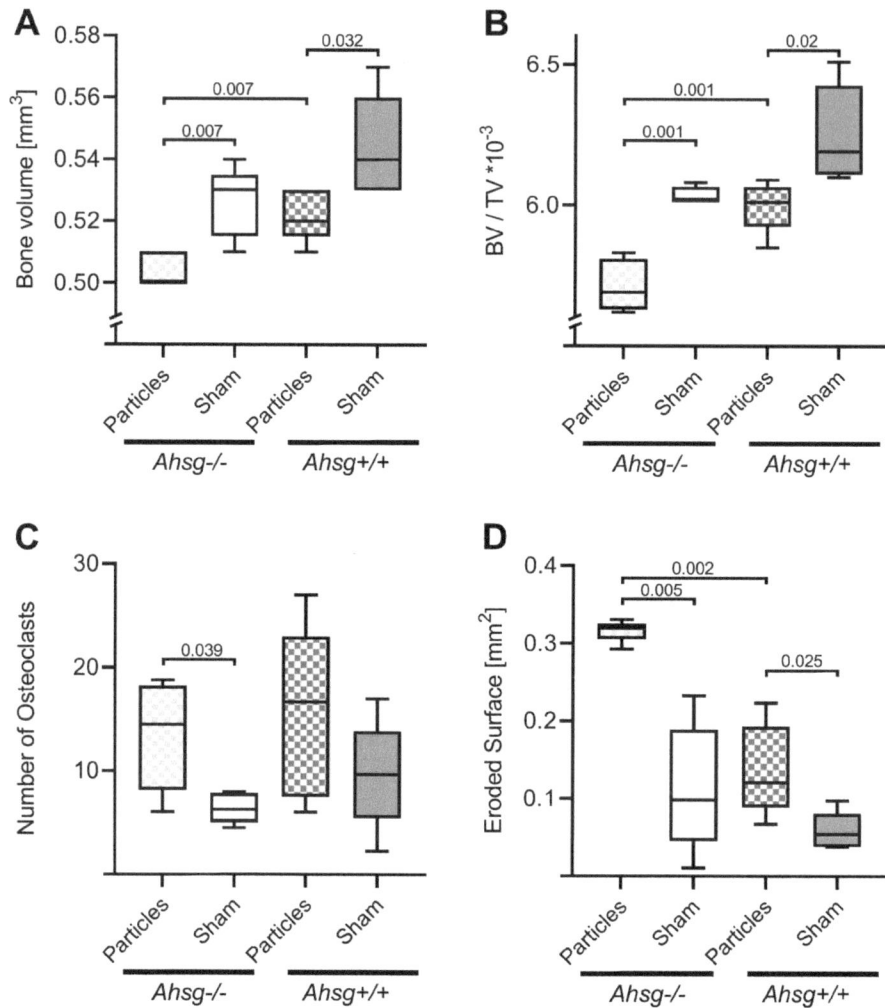

Figure 3. Bone volume and ratio of bone volume/tissue volume analyzed by μCT analysis and number of osteoclasts and eroded surface area at two weeks after surgery as box plots measured in in the histological sections. (**A**) Bone volume. (**B**) Ratio of bone volume/tissue volume. (**C**) Number of osteoclasts. (**D**) Eroded surface area.

3.2. Bone Histomorphometry

Under osteolytic conditions, the granulation tissue of both groups was infiltrated with macrophages and polynucleated giant cells in the area adjacent to the implanted UHMWPE particles (Figure 4A,B). At the same time, macrophages were found in typical resorption lacunae in the calvarial bone resulting in osteolytic lesions. In the sham groups, granulation tissue was also present but displayed only few inflammatory reactions. TRAP staining revealed an increased number of osteoclasts in both particle-treated groups compared with the sham groups, which was statistically significant for fetuin-A-deficient mice ($p = 0.039$) (Figures 3C and 5), (Table 3).

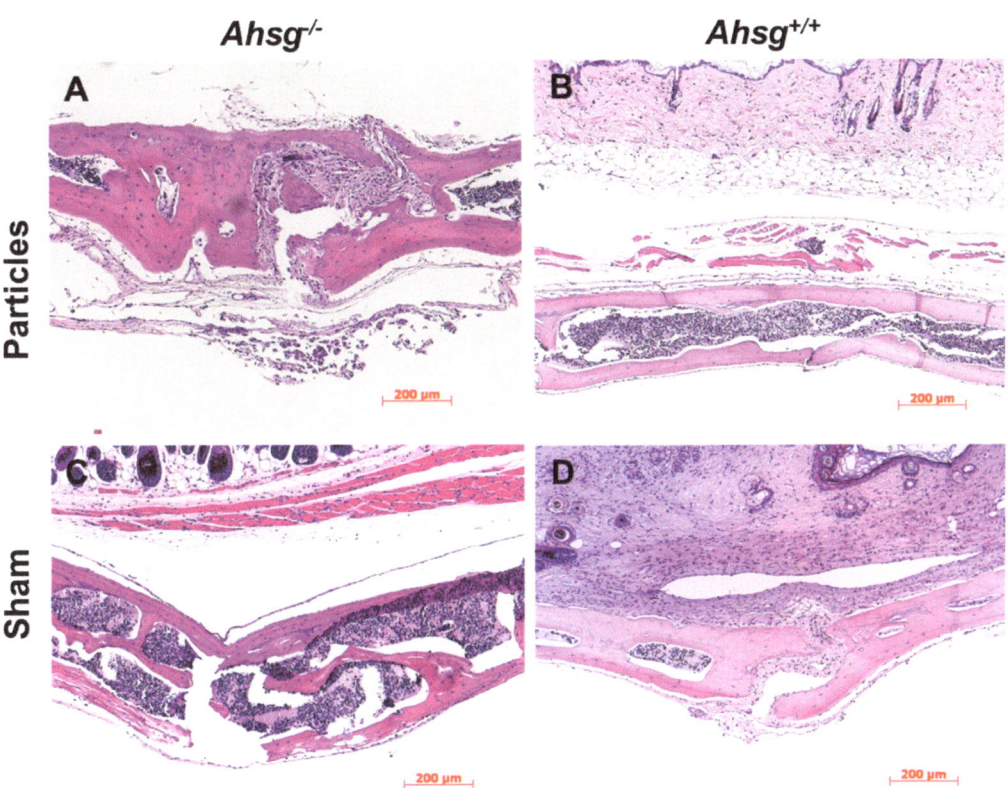

Figure 4. HE staining of calvarial sections. After 14 days, the calvarial bone was fixed, embedded and histologically stained. (**A**) Particle treatment in mice with the $Ahsg^{-/-}$ background led to a highly eroded surface. (**B**) In wild-type mice, less eroded surface was seen after particle treatment. (**C**) Sham-operated fetuin-A-deficient mice and (**D**) sham-operated wild-type mice did not show any signs of osteolysis. Scale bar indicates 200 µm.

Table 3. Number of osteoclast in fetuin-A-deficient ($Ahsg^{-/-}$) mice vs. wild-type ($Ahsg^{+/+}$) mice under osteolytic conditions (particles) vs. sham-operated animals. Data are presented as median and interquartiles range (IQR) and as mean ± standard deviation (SD) with n = 5 per group.

Osteoclast Number	$Ahsg^{-/-}$		$Ahsg^{+/+}$	
	Particles	Sham	Particles	Sham
Median (IQR)	14.5 (10.1)	6.3 (2.9)	16.7 (15.5)	9.7 (8.4)
Mean ± SD	13.5 ± 5.3	6.4 ± 1.5	15.5 ± 8.3	9.7 ± 5.3

Histomorphometric investigation of the eroded surface (ES) in the osteolytic areas showed that particle treatment led to a significant increase in ES in both genetic backgrounds, $Ahsg^{+/+}$ (p = 0.025) as well as $Ahsg^{-/-}$ (p = 0.005) (Figure 3D) (Table 4). Fetuin-A-deficient mice displayed the largest ES compared with wild-type mice under osteolytic conditions (p = 0.002). In the sham-operated animals, no difference was detected between the groups.

Table 4. Eroded surface area of fetuin-A-deficient ($Ahsg^{-/-}$) mice vs. wild-type ($Ahsg^{+/+}$) mice under osteolytic conditions (particles) vs. sham-operated animals. Data are presented in mm² as median and interquartiles range (IQR) and as mean ± standard deviation (SD) with n = 5 per group.

Eroded Surface Area [mm²]	$Ahsg^{-/-}$		$Ahsg^{+/+}$	
	Particles	Sham	Particles	Sham
Median (IQR)	0.32 (0.02)	0.10 (0.14)	0.12 (0.10)	0.05 (0.04)
Mean ± SD	0.32 ± 0.01	0.11 ± 0.08	0.14 ± 0.06	0.06 ± 0.02

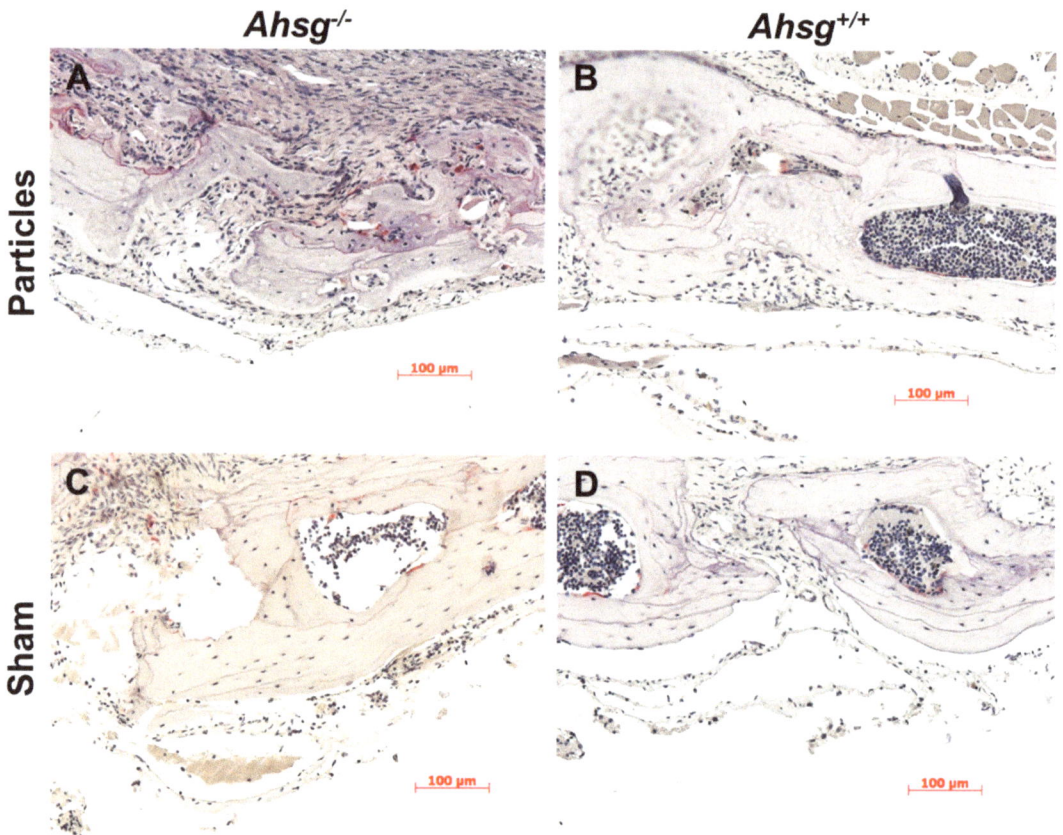

Figure 5. TRAP staining of calvarial sections. After 14 days, the calvarial bone was fixed, embedded and the osteoclast stained by tartrate resistant alkaline phosphatase. Particle treatment in the mice with the $Ahsg^{-/-}$ background (**A**) and in the wild-type mice (**B**) led to a visible increase in osteoclasts compared with the sham-operated animals (**C**,**D**). Both sham-operated fetuin-A-deficient mice (**C**) and sham-operated wild-type mice (**D**) had fewer osteoclasts. Scale bar indicates 100 μm.

3.3. Parameters of Bone Metabolism

We determined bone- and mineral-related serum chemistry to monitor bone metabolism in $Ahsg^{-/-}$ and $Ahsg^{+/+}$ mice before and after particle-induced osteolysis (Figure 6).

Preoperatively, the $Ahsg^{-/-}$ mice had significantly lower DPD values than the $Ahsg^{+/+}$ animals (p = 0.038) (Table 5) (Figure 6A).

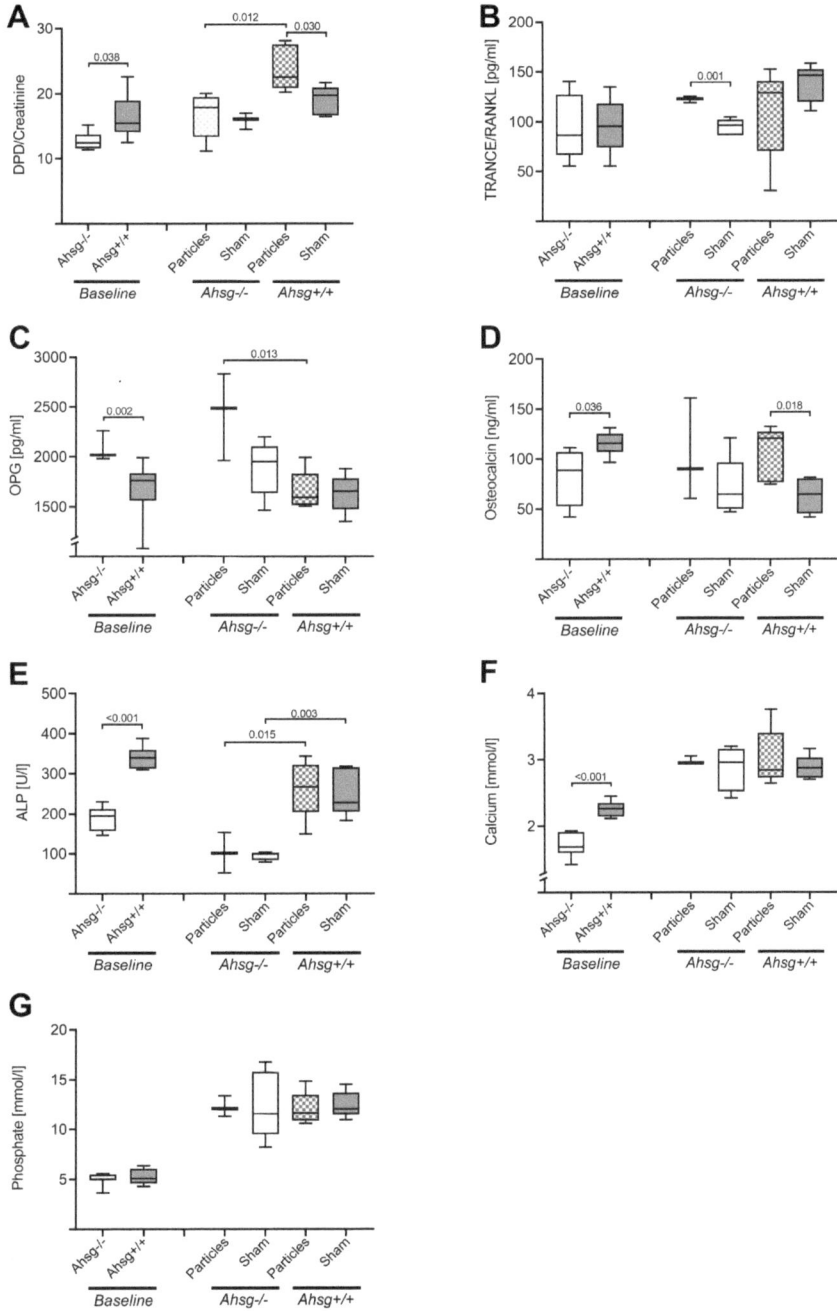

Figure 6. Bone metabolism parameters as box plots. The figure shows the single values of $n = 5$ animals per group with the median as line. $Ahsg^{-/-}$ = fetuin-A-deficient animals, $Ahsg^{+/+}$ = wild-type mice. Significant differences between the groups are indicated with *p*-values. (**A**) DPD/Creatinine. (**B**) RANKL. (**C**) Osteoprotegerin (OPG). (**D**) Osteocalcin. (**E**) Alkaline phosphatase (ALP). (**F**) Serum calcium. (**G**) Serum phosphate.

Table 5. Bone metabolism parameter in urine and serum at day 0 and at day 14 in Fetuin-A deficient (Ahsg$^{-/-}$) mice vs. wild type (Ahsg$^{+/+}$) mice a day 0 and at day 14 under osteolytic conditions (particles) vs. sham-operated animals. Data are presented as median and interquartils range (IOR) in the upper row and as mean ± standard deviation (SD) in the lower row with $n = 10$ per group at day 10 and $n = 5$ per group at day 14. The respective units were [nmol/mmol] for DPD/Kreatinin, [pg/mL] for RANKL and Osteoprotegerin, [ng/mL] for Osteocalcin and [mmol/L] for Calcium and Phosphate.

Parameter	Day 0 (Baseline) ($n = 10$ per Group)		Day 14 ($n = 5$ per Group)			
	Ahsg$^{-/-}$	Ahsg$^{+/+}$	Ahsg$^{-/-}$		Ahsg$^{+/+}$	
			Particles	Sham	Particles	Sham
DPD/Creatinin	12.4 (2.1) 12.7 ± 1.5	15.5 (4.8) 16.4 ± 3.5	17.9 (6.1) 16.7 ± 3.5	16.1 15.9 ± 1.2	22.5 (6.7) 23.8 ± 3.5	19.71 (4.21) 19.0 ± 2.2
RANKL	86.5 (60.3) 96.1 ± 31.0	95.6 (43.9) 95.4 ± 27.5	122.6 122.2 ± 3.2	96.2 (15.1) 94.51 ± 7.9	128.6 (69.9) 110.1 ± 47.0	146.4 (32.6) 138.0 ± 18.6
Osteo-protegerin	2021 (29) 2046 ± 89	1760 (272) 1685 ± 259	2485 2426 ± 438	1950 (467) 1884 ± 275	1593 (316) 1654 ± 199	1653 (309) 1632 ± 189
Osteocalcin	88.7 (54.1) 82.8 ± 29.3	116.1 (18.8) 115.7 ± 12.0	90.0 104.0 ± 51.5	64.7 (46.4) 71.5 ± 29.5	120.9 (50.9) 105.7 ± 27.1	64.6 (35.0) 63.1 ± 17.7
Alkaline Phosphatase	195.0 (56.1) 186.9 ± 30.2	339.6 (46.3) 340.4 ± 25.4	100.8 101.5 ± 50.8	98.3 (18.5) 94.6 ± 10.7	266.8 (118.0) 263.5 ± 72.1	227.6 (109.8) 253.3 ± 58.9
Calcium	1.69 (0.31) 1.72 ± 0.17	2.27 (0.20) 2.26 ± 0.12	2.95 2.98 ± 0.6	3.0 (0.6) 2.9 ± 0.3	2.8 (0.7) 3.0 ± 0.4	2.9 (0.3) 2.9 ± 0.2
Phosphate	5.0 (0.6) 5.0 ± 0.6	5.1 (1.5) 5.2 ± 0.8	12.1 12.2 ± 1.1	11.6 (6.3) 12.4 ± 3.4	11.6 (2.6) 12.1 ± 1.6	12.0 (2.2) 12.5 ± 1.3

After particle treatment, the Ahsg$^{-/-}$ mice showed significantly lower DPD concentrations compared with the wild-type animals ($p = 0.012$). Particle addition led to significant higher DPD values for the wt animals ($p = 0.030$) while there were no differences in Ahsg$^{-/-}$ animals compared with sham control.

The baseline RANKL values of both genetic backgrounds were comparable (Table 5) (Figure 6B). In fetuin-A-deficient mice, particle treatment led to increased RANKL values compared with the sham-operated mice ($p = 0.001$). In osteolytic conditions, the Ahsg$^{-/-}$ mice showed similar RANKL values compared with the Ahsg$^{+/+}$ mice. In the wt animals, there was no difference in the RANKL values between the particle-treated and the sham-operated mice.

Preoperative OPG levels were significantly increased in Ahsg$^{-/-}$ mice ($p = 0.002$) compared with wt animals (Table 5) (Figure 6C). In the postoperative examination under osteolytic conditions, a significantly increased OPG value was detected in the Ahsg$^{-/-}$ mice in comparison to the Ahsg$^{+/+}$ mice ($p = 0.013$). There was no statistically significant difference between the particle-treated and the sham-operated mice of either genetic background.

Baseline levels of osteocalcin revealed significantly lower values in the fetuin-A-knockout animals compared with the wild-type mice ($p = 0.036$) (Table 5) (Figure 6D). Particle treatment resulted in significantly higher OC concentrations in the wild-type mice compared with sham-operated animals ($p = 0.018$). The other groups did not differ significantly.

The fetuin-A-deficient mice had significantly lower systemic alkaline phosphatase levels compared with the wild-type mice ($p < 0.001$) (Table 5) (Figure 6E). Ahsg$^{-/-}$ mice had significantly decreased ALP activity compared with wt animals in both treatment groups, after particle treatment ($p = 0.015$) and in the sham animals ($p = 0.003$).

Before surgery, the fetuin-A-deficient animals had a lower serum calcium level than the wild-type mice ($p < 0.001$) (Table 5) (Figure 6F). Postoperatively, no significant differences in calcium level were detected between the subgroups (genetic background and treatment).

The baseline phosphate levels did not differ between the genetic background groups (Table 5) (Figure 6G). Likewise, postoperative scores also did not differ between the different genetic background groups or between the different surgical groups.

4. Discussion

4.1. In Absence of Fetuin-A, the Particle-Induced Inflammatory Response and Osteolysis Is Increased

A previous study suggested a possible attenuating role of the hepatic serum protein fetuin-A a key regulator of calcified matrix metabolism, and particle-induced osteolysis [26].

Our hypothesis, that in absence of fetuin-A the particle-induced inflammatory response and osteolysis is increased, was confirmed. After particle treatment and comparison with wild-type mice, in fetuin-A-deficient mice the bone volume and the ratio of bone volume to tissue volume was significantly decreased while the eroded surface area was significantly increased. Additionally, the number of osteoclasts was significantly increased compared with sham animals. These results provide novel insights demonstrating that bone resorption was significantly increased in fetuin-A-deficient mice compared with wild-type mice under osteolytic conditions. This was confirmed by significant changes in bone metabolism markers compared with wild-type animals in fetuin-A-deficient animal: OPG was increased, DPD, Ca, ALP, and OC were decreased.

4.2. Comparing the Phenotypes before Particle Treatment: Fetuin-A-Deficient Mice Presented Differences Regarding Bone Micro-Structure in µ-CT and Serum Analytes

In our study, the phenotype of wild-type mice and fetuin-A-deficient mice maintained against the genetic background C57BL/6 presented differences regarding bone microstructure in µ-CT and serum analytes. The µCT analysis presented reduced BV/TV in the $Ahsg^{-/-}$ mice, which is in line with the findings of Schäfer et al. [16]. The calcium levels in the $Ahsg^{-/-}$ mice were lower than those in the wild-type and the $Ahsg^{-/-}$ mice with a DBA/2 genetic background described by Schäfer et al. [16]. This emphasizes the importance of using mice with matching genetic backgrounds, as was performed in this study. The phenotype of $Ahsg^{-/-}$ under osteolytic conditions was accentuated by increased OPG levels in the present study. Lower levels of alkaline phosphatase activity in $Ahsg^{-/-}$ mice might relate to the genetic background and were not influenced by particle stimulation.

4.3. Potential Mechanisms of Fetuin-A in Particle Treatment: Fetuin-A Attenuates UHMWPE Particle-Induced Local Inflammation by Aiding the Removal of Calcified Cellular and Tissue Debris

We used a standardized model of UHMWPE particle treatment to simulate the effect of particles generated by wear and corrosion of joint replacement prostheses in wild-type and $Ahsg^{-/-}$ mice. The wild-type mice developed significantly increased osteolysis after UHMWPE particle treatment indicating the validity of the calvarial model as used in previous studies [28]. The fetuin-A-deficient mice showed increased particle-induced osteolysis compared with the wild-type mice, a hitherto unknown influence of the glycoprotein.

Established facts as the role of fetuin-A in mineralized matrix metabolism explains the increase in particle-induced osteolysis in $Ahsg^{-/-}$ mice. Fetuin-A attenuates UHMWPE particle-induced local inflammation by aiding the removal of calcified cellular and tissue debris. In the absence of fetuin-A such debris triggers a vicious cycle of defective debris clearance and inflammation, and therefore of increased cell death and decreased wound healing and osteogenesis [17]. In addition to aiding calcified debris removal, fetuin-A can exercise an anti-inflammatory action through associated anti-inflammatory polyanions and TGF-ß, as well as though neutralization of HMGB1, a late mediator of cell-damage associated systemic inflammation [17,24,34,35]. Our histological findings of pronounced inflammatory granulomatous tissue on the calvariae of fetuin-A-deficient mice after particle implantation underscore this hypothesis. We propose that the anti-inflammatory action of fetuin-A is due to the mobilization and removal of potentially pro-inflammatory debris, and may be further enhanced by fetuin-A-associated anti-inflammatory agents collectively neutralizing the pro-inflammatory influence of danger associated intracellular molecules such as HMGB1 and the overt pro-inflammatory cytokines TNF-α and IL-1 [20,35–37]. Failure to clear cell debris is known to cause cytokine-mediated induction of osteoclasts [38,39]. The capsules and interface membranes of patients with particle-induced osteolysis showed various strongly expressed apoptosis-related pathways, such as the Fas receptor, BAK, and caspase-3 cleaved [39]. Jersmann et al. showed that fetuin-A leads to increased opsonization of apoptotic cells and macropinocytosis by human macrophages [40]. The authors of this study concluded that fetuin-A is an attractive candidate for future therapeutic intervention in inflammatory diseases [40]. Fetuin-A deficiency might lead to a decrease in the opsonization of apoptotic cells leading to pronounced locally restricted inflammation followed by increased calvarial osteolysis.

The critical role of fetuin-A in mineralized tissue remodeling was initially demonstrated in unchallenged homozygous $Ahsg^{-/-}$ mice. Up to 70% of these mice suffer epiphysiolysis of the distal femur, causing deformed growth plates and foreshortened proximal limb bones [41,42]. The elevated pre-operative OPG levels in the fetuin-A-deficient mice in our study may reflect an overall osteoanabolic state associated with bone healing following the femoral epiphysiolysis required to heal the slipped growth plates. The enhanced osteolysis found in fetuin-A-deficient mice following induced bone damage suggests that the elevated OPG level cannot compensate for the strong pro-inflammatory and osteocatabolic particle-mediated stimulus.

In summary, we found increased osteoclast activity and decreased osteoblast activity following the local UHMWPE particle challenge of fetuin-A-deficient mice.

4.4. Correlation between Fetuin-A and Bone Metabolism in Humans: Fetuin-A Is Slightly Possitivly Associated with Areal Bone Mineral Density in Older Adults

While an osteoprotective effect was demonstrated in the present animal study of fetuin-A deficiency as well as in a prior study on fetuin-A substitution, the effects of the fetuin-A background in humans with aseptic loosening has been recently investigated. In a large cohort of 4714 patients, Fink et al. [43] reported a small positive association between fetuin-A and areal bone mineral density in older adults, but they did not find an association between fetuin-A and the risk of a clinical fracture. Further, Steffen et al. demonstrated that a more favorable outcome after proximal femur fracture was associated with higher fetuin-A serum levels and age during hospitalization [44]. We suggest studying the fetuin-A levels in patients with aseptic loosening as well as the possible association of existing functional fetuin-A polymorphisms on aseptic loosening. In case of confirmed low levels of serum fetuin-A in these patients, a fetuin-A substitution to normal serum levels can be considered as a new treatment option in joint arthroplasty. Any side-effects of fetuin-A intake/application should be investigated beforehand.

We report enhanced particle-induced osteolysis in fetuin-A-deficient mice, but our study has limitations. Despite the irrefutable evidence of disturbed mineral metabolism and bone anomalies in fetuin-A-deficient mice [16,17,20,29,41,42,45,46], these results may not apply to humans. Particle-induced osteolysis was analyzed after fourteen days in the calvaria model, while aseptic prosthetic loosening in humans is a process that occurs over many years. As another limitation, the calvaria model is not analogous to the clinical situation since there is no load-bearing implant [47]. Furthermore, we did not analyze the substitution or over-expression of fetuin-A to prove a possible protective effect on particle-induced osteolysis.

5. Conclusions

In conclusion, this study further links clinically relevant particle disease to the glycoprotein fetuin-A. We believe that the interaction of fetuin-A with osteoclast function, inflammation, and apoptosis ultimately leads to in increased particle-induced osteolysis in fetuin-A-deficient mice. Thus, reduced fetuin-A serum levels can be a risk factor for particle-induced osteolysis. Further clinical studies are needed to implement these findings. The protective effect of fetuin-A can be an approach for prophylaxis of loosening of orthopedic endoprostheses and treatment of incipient osteolysis to minimize the number of revision surgeries.

Author Contributions: Conceptualization, C.P., C.B., W.J.-D., C.W. and M.D.K.; methodology, C.P., C.B., G.H., F.G., C.W. and M.D.K.; validation, M.B., H.-L.M., C.W.; M.D.K.; formal analysis, C.P.; C.B.; M.H.; F.G.; M.B. and H.-L.M.; investigation, C.P., G.H., F.G. and M.D.K.; resources, C.P.; C.B., G.H., F.G., M.D., W.J.-D. and M.D.K.; writing—original draft preparation, C.P., M.H., W.J.-D. and M.D. K.; writing—review and editing, C.P., C.B., M.H., G.H., F.G., M.B., H.-L.M., M.D., W.J.-D., C.W. and M.D.K.; visualization, C.P., M.H., M.B., H.-L.M. and M.D.K.; supervision, C.P., C.B. and M.D.K.; project administration, C.P. and M.D.K.; funding acquisition, M.D., W.J.-D., C.W. and M.D.K. All authors have read and agreed to the published version of the manuscript.

Funding: The study was supported by an IFORES grant from the University Hospital Essen, Germany and by grants from the Deutsche Forschungsgemeinschaft WE 3634/1-1 (CW), project ID 322900939 and 403041552 (W.J.-D.). We acknowledge support by the Open Access Publication Fund of the University of Duisburg-Essen, Germany.

Institutional Review Board Statement: The animal study protocol was approved by the National District Government (LANUV AZ 9.93.2.10.34.07.138).

Informed Consent Statement: Not applicable.

Data Availability Statement: Data available on request from the authors.

Acknowledgments: The authors thank the Clariant Company for supplying the particles. We thank PD Bröcker-Preuß of the Central Laboratory of the University Hospital Essen, Germany, for coordination of the clinical chemistry. We thank Bodo Levkau for permission to use the μCT facility in the Department of Pathophysiology at the University Hospital Essen, Germany. The authors thank the WTZ Research Support Service (supported in part by the Deutsche Krebs-hilfe/Comprehensive Cancer Center financing). We also thank Kaye Schreyer for editing the manuscript.

Conflicts of Interest: The authors declare no conflict of interest.

References

1. Sadoghi, P.; Liebensteiner, M.; Agreiter, M.; Leithner, A.; Bohler, N.; Labek, G. Revision surgery after total joint arthroplasty: A complication-based analysis using worldwide arthroplasty registers. *J. Arthroplast.* **2013**, *28*, 1329–1332. [CrossRef] [PubMed]
2. Gallo, J.; Goodman, S.B.; Konttinen, Y.T.; Raska, M. Particle disease: Biologic mechanisms of periprosthetic osteolysis in total hip arthroplasty. *Innate. Immun.* **2013**, *19*, 213–224. [CrossRef] [PubMed]
3. Goodman, S.B.; Gallo, J. Periprosthetic Osteolysis: Mechanisms, Prevention and Treatment. *J. Clin. Med.* **2019**, *8*, 2091. [CrossRef] [PubMed]
4. Kobayashi, A.; Bonfield, W.; Kadoya, Y.; Yamac, T.; Freeman, M.A.; Scott, G.; Revell, P.A. The size and shape of particulate polyethylene wear debris in total joint replacements. *Proc. Inst. Mech. Eng. H* **1997**, *211*, 11–15. [CrossRef]
5. Matthews, J.B.; Besong, A.A.; Green, T.R.; Stone, M.H.; Wroblewski, B.M.; Fisher, J.; Ingham, E. Evaluation of the response of primary human peripheral blood mononuclear phagocytes to challenge with in vitro generated clinically relevant UHMWPE particles of known size and dose. *J. Biomed. Mater. Res.* **2000**, *52*, 296–307. [CrossRef] [PubMed]
6. Datta, H.K.; Ng, W.F.; Walker, J.A.; Tuck, S.P.; Varanasi, S.S. The cell biology of bone metabolism. *J. Clin. Pathol.* **2008**, *61*, 577–587. [CrossRef] [PubMed]
7. Drees, P.; Eckardt, A.; Gay, R.E.; Gay, S.; Huber, L.C. Mechanisms of disease: Molecular insights into aseptic loosening of orthopedic implants. *Nat. Clin. Pract. Rheumatol.* **2007**, *3*, 165–171. [CrossRef] [PubMed]
8. Revell, P.A. The combined role of wear particles, macrophages and lymphocytes in the loosening of total joint prostheses. *J. R. Soc. Interface* **2008**, *5*, 1263–1278. [CrossRef]
9. Khosla, S. Minireview: The OPG/RANKL/RANK system. *Endocrinology* **2001**, *142*, 5050–5055. [CrossRef]
10. Trouvin, A.P.; Goeb, V. Receptor activator of nuclear factor-kappaB ligand and osteoprotegerin: Maintaining the balance to prevent bone loss. *Clin. Interv. Aging* **2010**, *5*, 345–354. [CrossRef]
11. Ingham, E.; Fisher, J. Biological reactions to wear debris in total joint replacement. *Proc. Inst. Mech. Eng. H* **2000**, *214*, 21–37. [CrossRef] [PubMed]
12. Baumann, B.; Seufert, J.; Jakob, F.; Noth, U.; Rolf, O.; Eulert, J.; Rader, C.P. Activation of NF-kappaB signalling and TNFalpha-expression in THP-1 macrophages by TiAlV- and polyethylene-wear particles. *J. Orthop. Res.* **2005**, *23*, 1241–1248. [CrossRef] [PubMed]
13. Wei, S.; Kitaura, H.; Zhou, P.; Ross, F.P.; Teitelbaum, S.L. IL-1 mediates TNF-induced osteoclastogenesis. *J. Clin. Investig.* **2005**, *115*, 282–290. [CrossRef] [PubMed]

14. Dickson, I.R.; Poole, A.R.; Veis, A. Localisation of plasma alpha2HS glycoprotein in mineralising human bone. *Nature* **1975**, *256*, 430–432. [CrossRef] [PubMed]
15. Triffitt, J.T.; Gebauer, U.; Ashton, B.A.; Owen, M.E.; Reynolds, J.J. Origin of plasma alpha2HS-glycoprotein and its accumulation in bone. *Nature* **1976**, *262*, 226–227. [CrossRef]
16. Schäfer, C.; Heiss, A.; Schwarz, A.; Westenfeld, R.; Ketteler, M.; Floege, J.; Müller-Esterl, W.; Schinke, T.; Jahnen-Dechent, W. The serum protein α2–Heremans-Schmid glycoprotein/fetuin-A is a systemically acting inhibitor of ectopic calcification. *J. Clin. Investig.* **2003**, *112*, 357–366. [CrossRef]
17. Jahnen-Dechent, W.; Heiss, A.; Schäfer, C.; Ketteler, M. Fetuin-A regulation of calcified matrix metabolism. *Circ. Res.* **2011**, *108*, 1494–1509. [CrossRef]
18. Schinke, T.; Amendt, C.; Trindl, A.; Poschke, O.; Muller-Esterl, W.; Jahnen-Dechent, W. The serum protein alpha2-HS glycoprotein/fetuin inhibits apatite formation in vitro and in mineralizing calvaria cells. A possible role in mineralization and calcium homeostasis. *J. Biol. Chem.* **1996**, *271*, 20789–20796. [CrossRef]
19. Heiss, A.; Eckert, T.; Aretz, A.; Richtering, W.; Van Dorp, W.; Schäfer, C.; Jahnen-Dechent, W. Hierarchical role of fetuin-A and acidic serum proteins in the formation and stabilization of calcium phosphate particles. *J. Biol. Chem.* **2008**, *283*, 14815–14825. [CrossRef]
20. Rudloff, S.; Janot, M.; Rodriguez, S.; Dessalle, K.; Jahnen-Dechent, W.; Huynh-Do, U. Fetuin-A is a HIF target that safeguards tissue integrity during hypoxic stress. *Nat. Commun.* **2021**, *12*, 549. [CrossRef]
21. Jirak, P.; Stechemesser, L.; More, E.; Franzen, M.; Topf, A.; Mirna, M.; Paar, V.; Pistulli, R.; Kretzschmar, D.; Wernly, B.; et al. Clinical implications of fetuin-A. *Adv. Clin. Chem.* **2019**, *89*, 79–130. [CrossRef]
22. Wang, H.; Sama, A.E. Anti-inflammatory role of fetuin-A in injury and infection. *Curr. Mol. Med.* **2012**, *12*, 625–633. [CrossRef]
23. Dziegielewska, K.M.; Andersen, N.A.; Saunders, N.R. Modification of macrophage response to lipopolysaccharide by fetuin. *Immunol. Lett.* **1998**, *60*, 31–35. [CrossRef]
24. Ombrellino, M.; Wang, H.; Yang, H.; Zhang, M.; Vishnubhakat, J.; Frazier, A.; Scher, L.A.; Friedman, S.G.; Tracey, K.J. Fetuin, a negative acute phase protein, attenuates TNF synthesis and the innate inflammatory response to carrageenan. *Shock* **2001**, *15*, 181–185. [CrossRef]
25. Pappa, E.; Perrea, D.S.; Pneumaticos, S.; Nikolaou, V.S. Role of fetuin A in the diagnosis and treatment of joint arthritis. *World J. Orthop.* **2017**, *8*, 461–464. [CrossRef]
26. Jablonski, H.; Polan, C.; Wedemeyer, C.; Hilken, G.; Schlepper, R.; Bachmann, H.S.; Grabellus, F.; Dudda, M.; Jäger, M.; Kauther, M.D. A single intraperitoneal injection of bovine fetuin-A attenuates bone resorption in a murine calvarial model of particle-induced osteolysis. *Bone* **2017**, *105*, 262–268. [CrossRef]
27. Merkel, K.D.; Erdmann, J.M.; McHugh, K.P.; Abu-Amer, Y.; Ross, F.P.; Teitelbaum, S.L. Tumor necrosis factor-alpha mediates orthopedic implant osteolysis. *Am. J. Pathol.* **1999**, *154*, 203–210. [CrossRef]
28. Wedemeyer, C.; Neuerburg, C.; Pfeiffer, A.; Heckelei, A.; Bylski, D.; von Knoch, F.; Schinke, T.; Hilken, G.; Gosheger, G.; von Knoch, M.; et al. Polyethylene particle-induced bone resorption in alpha-calcitonin gene-related peptide-deficient mice. *J. Bone Miner. Res.* **2007**, *22*, 1011–1019. [CrossRef]
29. Jahnen-Dechent, W.; Schinke, T.; Trindl, A.; Müller-Esterl, W.; Sablitzky, F.; Kaiser, S.; Blessing, M. Cloning and targeted deletion of the mouse fetuin gene. *J. Biol. Chem.* **1997**, *272*, 31496–31503. [CrossRef]
30. Westenfeld, R.; Schäfer, C.; Smeets, R.; Brandenburg, V.M.; Floege, J.; Ketteler, M.; Jahnen-Dechent, W. Fetuin-A (AHSG) prevents extraosseous calcification induced by uraemia and phosphate challenge in mice. *Nephrol. Dial. Transpl.* **2007**, *22*, 1537–1546. [CrossRef]
31. Wedemeyer, C.; von Knoch, F.; Pingsmann, A.; Hilken, G.; Sprecher, C.; Saxler, G.; Henschke, F.; Loer, F.; von Knoch, M. Stimulation of bone formation by zoledronic acid in particle-induced osteolysis. *Biomaterials* **2005**, *26*, 3719–3725. [CrossRef]
32. Kauther, M.D.; Bachmann, H.S.; Neuerburg, L.; Broecker-Preuss, M.; Hilken, G.; Grabellus, F.; Koehler, G.; von Knoch, M.; Wedemeyer, C. Calcitonin substitution in calcitonin deficiency reduces particle-induced osteolysis. *BMC Musculoskelet Disord* **2011**, *12*, 186. [CrossRef]
33. Parfitt, A.M. Bone histomorphometry: Proposed system for standardization of nomenclature, symbols, and units. *Calcif. Tissue Int.* **1988**, *42*, 284–286. [CrossRef]
34. Wang, H.C.; Zhang, M.H.; Soda, K.; Sama, A.; Tracey, K.J. Fetuin protects the fetus from TNF. *Lancet* **1997**, *350*, 861–862. [CrossRef]
35. Li, W.; Zhu, S.; Li, J.H.; Huang, Y.; Zhou, R.R.; Fan, X.G.; Yang, H.A.; Gong, X.; Eissa, N.T.; Jahnen-Dechent, W.; et al. A Hepatic Protein, Fetuin-A, Occupies a Protective Role in Lethal Systemic Inflammation. *PLoS ONE* **2011**, *6*, e16945. [CrossRef]
36. Köppert, S.; Büscher, A.; Babler, A.; Ghallab, A.; Buhl, E.M.; Latz, E.; Hengstler, J.G.; Smith, E.R.; Jahnen-Dechent, W. Cellular Clearance and Biological Activity of Calciprotein Particles Depend on Their Maturation State and Crystallinity. *Front. Immunol.* **2018**, *9*, 1991. [CrossRef]
37. Köppert, S.; Ghallab, A.; Peglow, S.; Winkler, C.F.; Graeber, S.; Buscher, A.; Hengstler, J.G.; Jahnen-Dechent, W. Live Imaging of Calciprotein Particle Clearance and Receptor Mediated Uptake: Role of Calciprotein Monomers. *Front. Cell Dev. Biol.* **2021**, *9*, 633925. [CrossRef]
38. Landgraeber, S.; Toetsch, M.; Wedemeyer, C.; Saxler, G.; Tsokos, M.; von Knoch, F.; Neuhauser, M.; Loer, F.; von Knoch, M. Over-expression of p53/BAK in aseptic loosening after total hip replacement. *Biomaterials* **2006**, *27*, 3010–3020. [CrossRef]

39. Landgraeber, S.; von Knoch, M.; Loer, F.; Wegner, A.; Tsokos, M.; Hussmann, B.; Totsch, M. Extrinsic and intrinsic pathways of apoptosis in aseptic loosening after total hip replacement. *Biomaterials* **2008**, *29*, 3444–3450. [CrossRef]
40. Jersmann, H.P.A.; Dransfield, I.; Hart, S.P. Fetuin/alpha(2)-HS glycoprotein enhances phagocytosis of apoptotic cells and macropinocytosis by human macrophages. *Clin. Sci.* **2003**, *105*, 273–278. [CrossRef]
41. Brylka, L.J.; Köppert, S.; Babler, A.; Kratz, B.; Denecke, B.; Yorgan, T.A.; Etich, J.; Costa, I.G.; Brachvogel, B.; Boor, P.; et al. Post-weaning epiphysiolysis causes distal femur dysplasia and foreshortened hindlimbs in fetuin-A-deficient mice. *PLoS ONE* **2017**, *12*, e0187030. [CrossRef] [PubMed]
42. Seto, J.; Busse, B.; Gupta, H.S.; Schäfer, C.; Krauss, S.; Dunlop, J.W.C.; Masic, A.; Kerschnitzki, M.; Zaslansky, P.; Boesecke, P.; et al. Accelerated Growth Plate Mineralization and Foreshortened Proximal Limb Bones in Fetuin-A Knockout Mice. *PLoS ONE* **2012**, *7*, e47338. [CrossRef]
43. Fink, H.A.; Bůžková, P.; Garimella, P.S.; Mukamal, K.J.; Cauley, J.A.; Kizer, J.R.; Barzilay, J.I.; Jalal, D.I.; Ix, J.H. Association of Fetuin-A With Incident Fractures in Community-Dwelling Older Adults: The Cardiovascular Health Study. *J. Bone Min. Res.* **2015**, *30*, 1394–1402. [CrossRef] [PubMed]
44. Steffen, C.J.; Herlyn, P.K.E.; Cornelius, N.; Mittlmeier, T.; Fischer, D.C. Evaluation of fetuin-A as a predictor of outcome after surgery for osteoporotic fracture of the proximal femur. *Arch. Orthop. Trauma. Surg.* **2020**, *140*, 1359–1366. [CrossRef] [PubMed]
45. Babler, A.; Schmitz, C.; Buescher, A.; Herrmann, M.; Gremse, F.; Gorgels, T.; Floege, J.; Jahnen-Dechent, W. Microvasculopathy and soft tissue calcification in mice are governed by fetuin-A, magnesium and pyrophosphate. *PLoS ONE* **2020**, *15*, e0228938. [CrossRef] [PubMed]
46. Herrmann, M.; Babler, A.; Moshkova, I.; Gremse, F.; Kiessling, F.; Kusebauch, U.; Nelea, V.; Kramann, R.; Moritz, R.L.; McKee, M.D.; et al. Lumenal calcification and microvasculopathy in fetuin-A-deficient mice lead to multiple organ morbidity. *PLoS ONE* **2020**, *15*, e0228503. [CrossRef]
47. Langlois, J.; Hamadouche, M. New animal models of wear-particle osteolysis. *Int. Orthop.* **2011**, *35*, 245–251. [CrossRef] [PubMed]

Disclaimer/Publisher's Note: The statements, opinions and data contained in all publications are solely those of the individual author(s) and contributor(s) and not of MDPI and/or the editor(s). MDPI and/or the editor(s) disclaim responsibility for any injury to people or property resulting from any ideas, methods, instructions or products referred to in the content.

Article

Surface Modified β-Ti-18Mo-6Nb-5Ta (wt%) Alloy for Bone Implant Applications: Composite Characterization and Cytocompatibility Assessment

Michael Escobar [1], Oriol Careta [2], Nora Fernández Navas [3], Aleksandra Bartkowska [1], Ludovico Andrea Alberta [3], Jordina Fornell [1], Pau Solsona [1], Thomas Gemming [3], Annett Gebert [3], Elena Ibáñez [2], Andreu Blanquer [2,*], Carme Nogués [2], Jordi Sort [1,4] and Eva Pellicer [1,*]

[1] Departament de Física, Universitat Autònoma de Barcelona, 08193 Bellaterra (Cerdanyola del Vallès), Spain
[2] Departament de Biologia Cel lular, Fisiologia i Immunologia, Universitat Autònoma de Barcelona, 08193 Bellaterra (Cerdanyola del Vallès), Spain
[3] Institute for Complex Materials, Leibniz Institut für Festkörper- und Werkstoffforschung Dresden e.V., 01069 Dresden, Germany
[4] Institució Catalana de Recerca i Estudis Avancats (ICREA), Pg. Lluís Companys 23, 08010 Barcelona, Spain
* Correspondence: andreu.blanquer@uab.cat (A.B.); eva.pellicer@uab.cat (E.P.)

Abstract: Commercially available titanium alloys such as Ti-6Al-4V are established in clinical use as load-bearing bone implant materials. However, concerns about the toxic effects of vanadium and aluminum have prompted the development of Al- and V-free β-Ti alloys. Herein, a new alloy composed of non-toxic elements, namely Ti-18Mo-6Nb-5Ta (wt%), has been fabricated by arc melting. The resulting single β-phase alloy shows improved mechanical properties (Young's modulus and hardness) and similar corrosion behavior in simulated body fluid when compared with commercial Ti-6Al-4V. To increase the cell proliferation capability of the new biomaterial, the surface of Ti-18Mo-6Nb-5Ta was modified by electrodepositing calcium phosphate (CaP) ceramic layers. Coatings with a Ca/P ratio of 1.47 were obtained at pulse current densities, $-j_c$, of 1.8–8.2 mA/cm^2, followed by 48 h of NaOH post-treatment. The thickness of the coatings has been measured by scanning electron microscopy from an ion beam cut, resulting in an average thickness of about 5 µm. Finally, cytocompatibility and cell adhesion have been evaluated using the osteosarcoma cell line Saos-2, demonstrating good biocompatibility and enhanced cell proliferation on the CaP-modified Ti-18Mo-6Nb-5Ta material compared with the bare alloy, even outperforming their CaP-modified Ti-6-Al-4V counterparts.

Keywords: titanium alloy; electrodeposition; ceramic coating; indentation; corrosion resistance; cell proliferation

1. Introduction

For many years, commercially available pure α-titanium (cp-Ti) and its alloys, such as Ti-6Al-4V, have been widely used as bone implant materials. However, Ti-6Al-4V, being a dual-phase α + β alloy, presents a higher modulus of elasticity (125 ± 2 GPa [1]) compared with that of bone (11.4–21.2 GPa [2]). This hinders the transfer of load to the bone, and it produces a stress-shielding effect that can cause implant loosening and may result in premature bone failure [3–5]. Additionally, it has been pointed out that the release of V and Al ions can cause hematological and biochemical alterations, osteomalacia, peripheral neuropathy, and Alzheimer's disease [4,6]. One way to overcome these problems is with the use of β-phase Ti alloys free of V and/or Al. These alloys possess high specific strength, good fatigue resistance, sufficient toughness, excellent corrosion resistance, good formability, and, most importantly, a low elastic modulus in the range of 33–85 GPa [7,8]. Moreover, β-phase stabilizing elements usually confer these alloys excellent biocompatibility [5,9].

Such is the case for Mo, which shows an increased ability to stabilize the β phase [9,10] and low cytotoxicity, being an essential element for the human body [11]. In a similar way, besides not having toxic effects on cells, Nb and Ta are both regarded as biocompatible elements that are able to promote the generation of vascularized, loose connective tissue when in contact with the organism [5]. Nb is also used to further ensure the stabilization of the β phase and improve the alloy's processability during fabrication, while Ta enhances the mechanical properties and corrosion resistance of the resulting alloy [12]. Thus, it has been demonstrated that Ti alloys containing Mo, Nb, and/or Ta exhibit good biocompatibility and corrosion resistance in human body fluids without causing allergic reactions; at the same time, they have a low elastic modulus and overall superior mechanical properties, which make them suitable for applications as environmentally neutral construction materials and biomedical devices, more specifically orthopedic implants [8,10,13,14].

Unfortunately, the metallic implant surface in its as-cast or standard finished state, e.g., ground or polished with a natural passive layer, usually does not provide sufficient bioactivity. Namely, it is not very efficient in stimulating bone tissue formation and healing, hindering in this way the bone recovery process [15]. To overcome this drawback, the deposition of calcium phosphates (often denoted as CaP for simplicity) onto the implant has been proposed as a strategy to increase the osseointegration [15–17]. Different techniques to produce this type of coating are available. Alternatively, dry techniques directly deposit particles onto the substrate without the need of a solvent [15]. Among them, thermal spraying (plasma, flame, and high-velocity oxygen fuel (HVOF) spraying) and physical vapor deposition (PVD) (e.g., magnetron sputtering) can be mentioned. However, these methods either suffer from high production costs or yield coatings with high internal residual stress, low crystallinity, and poor control over the structure of the coating [18]. In addition, some of them operate at temperatures higher than 1000 °C, which is above the β transus temperature of Ti alloys, thus affecting the alloy microstructure underneath the coating. Alternatively, wet techniques include sol-gel and electrodeposition, which, thanks to their low production costs and high flexibility, are a promising alternative to their dry counterparts [15]. More specifically, electrodeposition can be regarded as a very convenient method to deposit CaP since it provides better control over the coating thickness, the grain size, and the microstructure to a great extent [18,19].

Electrodeposited CaP coatings can be categorized into four main compositions, namely dicalcium phosphate dihydrate (DCPD) or brushite, octacalciumphosphate (OCP), hydroxyapatite (HA), and tricalcium phosphate (TCP) [15]. Of them, HA ($Ca_{10}(PO_4)_6(OH)_2$), with a Ca/P atomic ratio of 1.67, is chemically and structurally similar to the mineral of human hard tissue. Indeed, it exhibits good stability when surrounded by the variable pH and corrosive environment of the body fluid [20,21]. Closely related to it, calcium-deficient hydroxyapatite ($Ca_{10-x}(PO_4)_{6-x}(HPO_4)_x(OH)_{2-x}$, CDHA) shows higher solubility and better bioresorbable properties, and it can induce the precipitation of bone-like apatite more efficiently than HA [20–23]. Therefore, by surface-modifying β-phase Ti alloys with CDHA, it should be possible to produce an implant material with improved mechanical properties and biocompatibility.

Most electrolyte formulations for CaP electrodeposition consist of calcium nitrate tetrahydrate ($Ca(NO_3)_2 \cdot 4H_2O$) and ammonium dihydrogen phosphate ($NH_4H_2PO_4$) at pH = 4–5 [24]. At sufficiently large current densities, corresponding to more negative potentials than the water decomposition, the electrodeposition process involves water/proton reduction at the cathode (Equations (1) and (2)), which cause pH value variations at its near-surface regions and, ultimately, the dissociation of dihydrogen phosphate ions via an acid-base reaction (Equations (3) and (4)):

$$2H_2O + 2e^- \rightarrow H_2 \uparrow + 2OH^- \qquad (1)$$

$$2H^+ + 2e^- \rightarrow H_2 \uparrow \qquad (2)$$

Depending on the pH value rise, then, the resulting hydrogen phosphate ions (HPO_4^{2-}) for pH values between 7.2 and 12.3 (Equation (3)) or the phosphate ions (PO_4^{3-}) for pH > 12.3 (Equation (4)), precipitate together with Ca^{2+} and OH^- ions on the cathode, yielding the above-mentioned CaP phases [24]. The eventual precipitation of HA is given as an example in Equation (5):

$$H_2PO_4^- \rightarrow HPO_4^{2-} + H^+ \quad (3)$$

$$HPO_4^{2-} \rightarrow PO_4^{3-} + H^+ \quad (4)$$

$$5Ca^{2+} + 3PO_4^{3-} + OH^- \rightarrow Ca_5(PO_4)_3(OH) \quad (5)$$

If the current densities are lower, then other species different from water (NO_3^- or O_2) contribute to creating the required scenario for CaP precipitation [19]. Note that although the term "electrodeposition" is frequently utilized in this field, the formation of a coating does not involve the reduction of ions to their zero-valent state as would be the case with metal electrodeposition. Instead, the electroreduction of electrolyte species generates the agents and/or required pH value for subsequent chemical reactions and the precipitation of CaP phases on the electrode.

Although direct current (DC) electrodeposition has been used to coat metallic substrates with CaP layers, pulse current (PC) electrodeposition shows several advantages. Local concentrations of calcium and phosphate ions change under DC electrodeposition, especially for long deposition times, which makes the control of the stoichiometry of the CaP layer (particularly when HA is targeted) difficult. The introduced resting time enables the ions to diffuse and restore the initial concentration of the electrolyte in the whole solution. In addition, the evolved hydrogen gas bubbles under DC become adsorbed onto the surface, giving rise to porosity and worsening the mechanical properties of the coatings. During the resting time between the pulses, hydrogen bubbles can migrate from the working electrode, allowing for a more homogeneous and dense deposition of the CaP [15,18].

In the present work, with the objective of following the research and experimentation on the fabrication of materials for their possible application as bone implants, a new β-phase Ti alloy (Ti-18Mo-6Nb-5Ta (wt%)) was synthesized by arc melting, and the surface of disk-shaped specimens was further coated with CaP-based films by PC electrodeposition. The mechanical properties of both the base alloy and the resulting calcium-deficient HA (CDHA) coatings grown on top were studied by nanoindentation and scratch tests. For the sake of comparison, commercially available Ti-6Al-4V alloy was also coated with CDHA. Finally, cytocompatibility and cell adhesion of the composite biomaterials were evaluated using the Saos-2 cell line, and the results were compared with those for the bare alloys (β-Ti-18Mo-6Nb-5Ta and Ti-6Al-4V).

2. Materials and Methods

2.1. Materials Preparation

Commercial Ti-6Al-4V rod (grade 5-ASTM B348) of 3 mm in diameter was purchased from Goodfellow Cambridge Limited. The new β-phase Ti alloy was fabricated by adding the corresponding amounts of titanium (Ti), molybdenum (Mo), niobium (Nb), and tantalum (Ta) to the proposed composition. The pieces of each metal were then placed in a mini arc-melting system (MAM-1) with an arc-melting generator type Lorch Handy TIG 180 DC basic plus pumping system HVT 51/G under high vacuum. The material was melted several times in order to obtain a homogeneous blend, and subsequently, the melted alloy was forced to enter a cold mold by suction so that 4 mm diameter rods could be obtained by suction casting.

Both commercial Ti-6Al-4V and crafted Ti-18Mo-6Nb-5Ta (referred also as TiAlV and TiMoNbTa for simplicity) rods were cut into disks with a thickness of 2.5 mm and ultrasonically rinsed in ethanol and ultrapure water for 5 min each. Then, the disks were embedded in a thermoplastic resin for polishing with SiC papers up to 4000 grit. Once the disks were extracted from the resin, they were again rinsed in ethanol and ultrapure water, as mentioned before.

For the deposition of CaP, electrical contacts were welded to the metallic disks with Sn-Cu wire to be used as working electrodes. The area of the working electrode was limited by wrapping the disks with Teflon tape, so only their polished faces were available for electrodeposition. Then, the disks were immersed in an electrolytic vessel containing 100 mL of a solution consisting of 0.042 M $Ca(NO_3)_2 \cdot 4H_2O$ and 0.025 M $NH_4H_2PO_4$ (pH = 4.17 at 24 °C). The electrolytic vessel was equipped with a thermostat jacket that maintained the temperature of the electrolytic bath at 65 °C with a heating circulator, Julabo F12. In the electrolytic cell, the reference electrode employed consisted of a double junction Ag/AgCl with 0.5 M KNO_3 outer solution and 3 M KCl inner solution, and a platinum wire was used as the counter electrode. For the electrodeposition, a PC method was implemented with varying pulse current densities (j_c) of −0.3, −1.8, −5, and −8.2 mA/cm^2, and pulse on and off times of 1 and 2 s, respectively. Although t_{on} and t_{off} values of the order of minutes are frequently applied in the literature [24], shorter times of 0.5–2 s have also been used [25–27]. The j_c values were selected considering prior literature in the field of CaP electrosynthesis [15,27]. The number of pulses was adjusted accordingly to have similar overall charges passing through the system, namely 506 cycles (−0.3 mA/cm^2), 167 cycles (−1.8 mA/cm^2), 34 cycles (−5 mA/cm^2), and 20 cycles (−8.2 mA/cm^2). Note that here the concept of charge is different from the 'deposition charge' used in metal electrodeposition. Electrodeposition was performed with a potentiostat/galvanostat, Autolab PGSTAT302N. After electrodeposition, some of the samples were subjected to a NaOH post-treatment for 48 h or 72 h at room temperature and dipped several times in ultrapure water to remove the NaOH excess afterwards.

2.2. Characterization of Structural, Physical, and Physicochemical Properties

The composition of the TiMoNbTa alloy, as well as the morphology and composition of the CaP coatings, were studied by field-emission scanning electron microscopy (FESEM) on a Zeiss Merlin, which was equipped with energy-dispersive X-ray (EDX) spectroscopy detector. EDX was performed five times throughout the surface of the samples at an applied voltage of 15 kV, and the mean values are reported. Cuts on the CaP-coated alloys were made using the FIB technique (FIB, Helios 5 CX, Thermo Fisher Scientific, Waltham, MA, USA), and SEM (SEM, Helios 5 CX, Thermo Fischer Scientific, Waltham, MA, USA) was used to image the corresponding coating cross section (and to assess its thickness) with an ICE detector at 5 kV acceleration voltage.

X-ray diffraction (XRD), nanoindentation, and electrochemical corrosion studies were performed to study the crystalline phases, mechanical properties, and corrosion behavior, respectively, of the newly synthesized Ti-based alloy and also of the commercial Ti-6Al-4V, for comparison purposes. In the same way, XRD was used to analyze the crystallinity of the CaP-based coatings. XRD patterns were acquired on a Philips X'Pert diffractometer with Cu Kα radiation (λ = 1.5406 Å) in the 2θ range of 10–100° with a step size of 0.026°. Phase identification was performed using the ICDD PDF2 database. Where appropriate, cell parameters, crystal size, and microstrains were determined by Rietveld analysis of the corresponding XRD patterns.

Nanoindentation experiments were carried out on an Anton Paar TriTec Nanoindentation Tester NHT^2 equipped with a Berkovich pyramidal-shaped diamond tip. A maximum load of 250 mN was applied to a 5 × 5 array on the surface of the sample. The reduced Young's modulus (E_r) and hardness (H) were derived from the initial part of the unloading-displacement curve by applying the method of Oliver and Pharr [28]. The adhesion of the CaP coatings to the base alloy was evaluated by means of scratching tests on a Nanoindenter XP from MTS equipped with a Berkovich tip, and the results were compared with those for coatings on commercial Ti-6Al-4V alloy.

For the electrochemical corrosion studies, a disk of 2.5 mm height and 4 mm diameter constructed from the new Ti alloy (TiMoNbTa) was used as working electrode. The disk was embedded in epoxy resin and ground gradually with SiC papers up to 2500 grit, being ultrasonicated in water after each grinding procedure for a few minutes, to finally rinse it with ethanol and pure water and leave it to dry in air. For comparison, a disk of 2.5 mm height and 3 mm diameter constructed from Ti-6Al-4V was employed as a reference material. The electrochemical measurements were then carried out in a double-wall glass cell in a three-electrode configuration. Thus, a large concave platinum net was used as counter electrode, while a saturated calomel electrode (SCE = 0241 V vs. SHE) inserted in a Luggin capillary served as reference electrode. Within the cell, a phosphate-buffered saline (PBS) electrolyte (140 mM NaCl, 10 mM phosphate buffer, and 3 mM KCl; Merck Millipore, Burlington, MA, USA) with pH 7.4 at 25 °C, was maintained at 37 ± 1 °C with the aid of a thermostat coupled to the outer wall of the cell. The whole setup, comprising the cell and the electrodes, was controlled by the Solartron SI 1287 Electrochemical Interface. With this, the corrosion behavior of the TiMoNbTa and the Ti-6Al-4V surfaces were evaluated without any external load in PBS for 2 h, and the open circuit potential (OCP) was determined. Taking into account the final OCP, a potentiodynamic polarization scan was subsequently performed on the working electrode, starting at −50 mV vs. OCP till +2 V vs. SCE, with a scan rate of 0.5 mV/s. Measurements were performed in duplicate.

2.3. Cytocompatibility Studies

2.3.1. Cell Culture

Human osteosarcoma-derived Saos-2 cells (ATCC HTB-85) were used for the biocompatibility analyses. The disks (TiAlV, TiAlV/CaP, TiMoNbTa, and TiMoNbTa/CaP) were sterilized with absolute ethanol for 30 min and individually introduced into a 24-well plate. Then, 1×10^5 Saos-2 cells were seeded on top of the disks in each well and cultured in DMEM (ThermoFisher Scientific, Waltham, MA, USA) supplemented with 10% fetal bovine serum (ThermoFisher Scientific) under standard conditions (37 °C and 5% CO_2).

2.3.2. Cell Proliferation and Cell Viability Analysis

Saos-2 cell proliferation was determined by performing an Alamar Blue cell viability test (Thermo Fisher Scientific, USA) at days 1, 3, and 7 after cell seeding. Briefly, after 24 h of cell seeding, disks were moved to a new plate in order to discard cells growing at the bottom of the wells. Fresh medium with 10% Alamar Blue was added, and cells were incubated for 4 h under standard conditions in the dark. Then, the supernatant was collected, and 200 μL of the solution were transferred to a black-bottom Greiner CELLSTAR® 96-well plate (Sigma-Aldrich, Saint Louis, MO, USA). Supernatant's fluorescence was measured at 590 nm wavelength after excitation at 560 nm on a Spark multimode microplate reader (Tecan, Männedorf, Switzerland). Fresh medium was added to the cultures, and the assay was repeated after 3 and 7 days. Fluorescence values at days 3 and 7 were normalized to values obtained at day 1. Experiments were performed in triplicate.

Cell viability was investigated using the Live/Dead Viability/Cytotoxicity kit for mammalian cells (Invitrogen, Waltham, MA, USA). To carry out these tests, cells were first incubated on top of the materials for 3 days (to allow them to grow), and then the assay was performed following the manufacturer's protocol. Images from different regions of the disks were taken using an Olympus IX71 inverted microscope with epifluorescence capability.

2.3.3. Cell Morphology and Cell Adhesion Analysis

For cell morphology analyses, the same samples used for the cell viability assay were processed to be analyzed by SEM. Briefly, cells were washed with 0.1 M Sodium Cacodylate buffer, pH 7.4 (CBS), fixed in 2.5% glutaraldehyde in CBS for 45 min at room temperature (RT), and washed again twice in CBS. Cell dehydration was performed in a series of increasing ethanol concentrations (50, 70, 90, and twice 100%) for 8 min each. Finally, samples were dried using hexamethyldisilazane (HMDS; Electron Microscope Science) for

15 min. Samples were mounted on special stubs and analyzed using a SEM (Zeiss Merlin, Jena, Germany) to observe cell morphology.

Cell adhesion was determined through the analysis of focal contacts by actin filaments and vinculin detection. To perform the immunodetection, cells were washed twice in PBS 72 h after seeding onto the disks and then fixed in 4% paraformaldehyde in PBS for 15 min at RT. After another PBS wash, cells were permeabilized with 0.1% Triton X-100 (Sigma) in PBS for 15 min and blocked for 25 min with 1% bovine serum albumin (BSA; Sigma), and 0.5% Tween 20 (Sigma) in PBS at RT. Samples were then incubated overnight at 4 °C with a mouse anti-vinculin primary antibody (Millipore, MAB3574, Burlington, MA, USA) at 2 µg/mL and washed with blocking buffer. Then, samples were incubated with a mixture of Alexa fluor 594-conjugated phalloidin (Invitrogen), Alexa fluor 488 chicken anti-mouse IgG (Invitrogen), and Hoechst 33258 (Sigma) for 60 min in the dark at RT. Finally, cells were washed in PBS, air dried, and mounted on specific bottom glass dishes (MatTek, Ashland, MA, USA) using ProLong Antifade mounting solution (Life Technologies, Carlsbad, CA, USA). Immunofluorescence evaluation was performed in a confocal laser scanning microscope (Confocal Leica SP5, Leica Microsystems GmbH, Wetzlar, Germany).

2.3.4. Statistical Analysis

All data were quantified with GraphPad Prism 8 (GraphPad Software Inc., San Diego, CA, USA) and are presented as the mean ± standard deviation. The data from cell proliferation assays were statistically compared using one-way analysis of variance (ANOVA) with Tukey-Kramer multiple comparison test. A value of $p < 0.05$ was considered to be significant. Significance is represented in the figures using an alphabetical superscript on top of the columns. Values with different alphabetical superscripts mean that they are significantly different, whereas values with the same alphabetical superscripts are not significantly different.

3. Results and Discussion

3.1. Fabrication and Characterization of the Ti-18Mo-6Nb-5Ta Alloy

In order to fabricate a potentially non-toxic β-Ti alloy, Mo, Nb, and Ta were selected as alloying elements due to their ability to stabilize the β phase, excellent biocompatibility, and low cytotoxicity [5,9,11]. The proposed composition for the new alloy was established according to the so-called molybdenum equivalency (*MoE*), which is a useful parameter for characterizing the β-phase stability [9]. For the selected alloying elements, the *MoE* is reduced to Equation (6):

$$MoE = 1 \cdot (\text{wt\% } Mo) + 0.28 \cdot (\text{wt\% } Nb) + 0.22 \cdot (\text{wt\% } Ta) \qquad (6)$$

Here, Mo is used as a reference, and the constant before the concentration of each element represents the ratio between the minimum concentration of Mo and the corresponding element to stabilize the β phase. The previous relation was optimized to obtain a *MoE* above 10 and, in this way, ensure the stabilization of the β phase. Thus, the final nominal composition was Ti-16Mo-5Nb-4Ta (wt%).

After the corresponding amounts of each element were placed in the arc-melter and a rod of the alloy was obtained by Cu mold suction casting, pieces from the top and bottom of it were cut and analyzed with EDX spectroscopy to confirm the homogeneity of the rod. As shown in Figure 1a, the weight distributions of the elements between the bottom and top parts of the rod are very similar, and they correspond to a Ti-18Mo-6Nb-5Ta (wt%) formulation that is close to the nominal one (Ti-16Mo-5Nb-4Ta). Similarly, several disks (of about 1 mm thickness) were cut from the rod and were subjected to XRD. A representative XRD pattern is shown in Figure 1b, together with the XRD pattern of commercial Ti-6Al-4V. The identified diffraction peaks of TiMoNbTa match well with the β-Ti (bcc) phase, which confirms the successful fabrication of an alloy with the desired crystal structure. No other diffraction peaks were observed, contrary to Ti-6Al-4V, where a mixture of α and β phases were found.

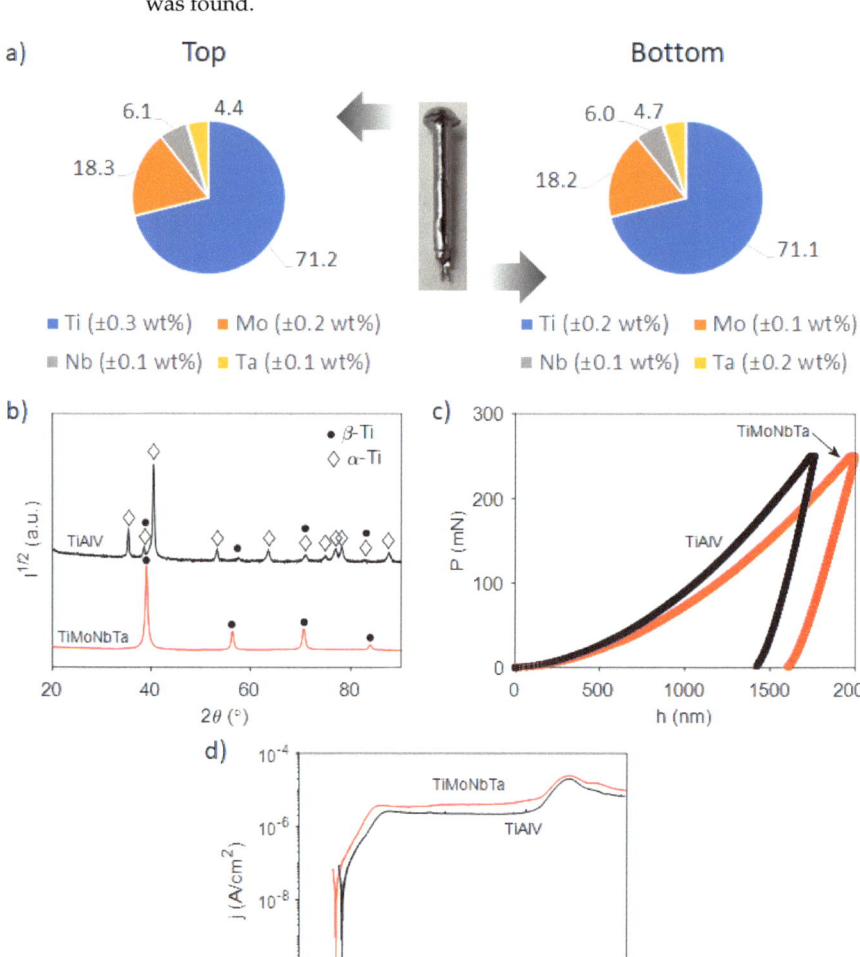

Figure 1. (**a**) Pie charts of the elemental content in wt% at the bottom and top parts of the casted TiMoNbTa rod. (**b**) XRD patterns of TiAlV base alloy (black curve) and TiMoNbTa (red curve). (**c**) Load-unload nanoindentation curves for TiMoNbTa and commercial TiAlV alloys. (**d**) Potentiodynamic polarization curves for TiMoNbTa and TiAlV in PBS solution at 37.5 °C.

Further, the mechanical properties of the new alloy (as well as those of the commercial Ti-6Al-4V) were measured by nanoindentation. Representative curves of the loading and unloading processes recorded during nanoindentation are shown in Figure 1c. Both the reduced Young's modulus (E_r) and hardness (H) values were determined from the obtained data and compared between each other and with those reported in the literature for bone (see Table 1).

Table 1. Reduced Young's modulus, hardness, and H/E_r, H^3/E_r^2, U_{El}/U_{Tot} and U_{Pl}/U_{Tot} ratios experimentally determined by nanoindentation for the newly synthesized Ti-18Mo-6Nb-5Ta and commercial Ti-6Al-4V alloys. Bibliographic data for diaphyseal femoral bone are also given for the sake of comparison.

	Bone [2]	Ti-18Mo-6Nb-5Ta	Ti-6Al-4V
E_r (GPa)	11.4–21.2	79.0 ± 3.0	106.0 ± 1.0
H (GPa)	0.23–0.76	3.40 ± 0.10	4.20 ± 0.1
H/E_r	0.02–0.036	0.044 ± 0.002	0.040 ± 0.001
H^3/E_r^2 (GPa)	0.0001–0.001	0.0070 ± 0.0010	0.0068 ± 0.0005
U_{El}/U_{Tot}	–	0.231 ± 0.010	0.233 ± 0.004
U_{Pl}/U_{Tot}	–	0.77 ± 0.02	0.77 ± 0.01

It can be seen that the new alloy has a considerable hardness (3.4 GPa), albeit one that is lower than that of commercial Ti-6Al-4V (4.2 GPa). The hardness for commercial Ti-6Al-4V is in agreement with the values reported in the literature, namely, 3.6–5.0 GPa, depending on the batch and experimental conditions [1,29]. More importantly, the new alloy also possesses a significantly lower reduced Young's modulus (79 GPa) than Ti-6Al-4V (106.1 GPa), showing a reduction of almost 25% in this property. This is expected from the lower E_r of the β-Ti phase compared with the α-Ti phase. The new elastic modulus is thus closer to the range in which this property varies for the bone. This feature is highly beneficial since it reduces the stress shielding effect caused by a marked difference between the modulus of the bone and the implant, preventing the development of symptoms related to osteoporosis, detachment, and failure of implants, and avoiding revision surgeries [30]. Interestingly, the H/E_r ratio (plasticity index), which is taken as an indirect estimate of the wear resistance of a material [31], is slightly higher for the new alloy. This means that the Ti-18Mo-6Nb-5Ta alloy would offer good resistance to wear and abrasion when subjected to repeated cyclic loads or strains, thereby ensuring its long-term success as an implant [32]. The proxy H^3/E_r^2 represents the resistance to plastic deformation, and, as a first approximation, a material with a larger H^3/E_r^2 is less likely to be plastically deformed and should therefore have higher toughness. According to the values gathered in Table 1, the Ti-18Mo-6Nb-5Ta alloy has slightly higher toughness than commercial Ti-6Al-4V. Finally, the energy ratios are similar for both alloys, in particular the U_{Pl}/U_{Tot}, suggesting that they are able to withstand a similar amount of plastic deformation.

Figure 1d shows the potentiodynamic polarization curves of both the new Ti alloy and commercial Ti-6Al-4V in PBS at 37 °C. Both alloys offer similar corrosion resistance in PBS, yielding close corrosion potential values (−0.341 V and −0.297 V vs. SCE for Ti-6Al-4V and Ti-18Mo-6Nb-5Ta, respectively). Corrosion current densities are well below 1 µA/cm^2, revealing very low free corrosion and metal ion release. In turn, passive current densities are in the range of 2.2–4.0 µA/cm^2. Therefore, it can be concluded that the corrosion resistance of Ti-18Mo-6Nb-5Ta is similar to that of the widely established Ti-6Al-4V material.

3.2. Electrodeposition of CaP Coatings and Their Characterization

Once the base alloy was successfully produced, disk-shaped specimens were coated with CaP by PC deposition. For comparison purposes, the Ti-6Al-4V was also coated with CaP under the same experimental conditions. Figure 2 shows the detail of a potential-time (E-t) transient recorded during the deposition of CaP onto the Ti-18Mo-6Nb-5Ta base alloy. During the t_{on}, the potential (E) shifts towards more negative values, while E relaxes and approaches the open circuit potential during t_{off}. Notice that E at t_{on} = 1 s varies between −1.0 and −1.8 V vs. Ag/AgCl, and hence, reduction of water at values more negative than −1.5 V most likely occurs in our case. According to Eliaz [33], having achieved some nucleation already at −842 mV vs. SCE, potentials below −1.26 V vs. SCE were required to promote HA growth on Ti-6Al-4V substrate when potentiostatic deposition from an electrolyte containing 0.61 mM Ca(NO$_3$)$_2$ and 0.36 mM NH$_4$H$_2$PO$_4$ at pH = 6 was attempted. Although it is not strictly comparable with a galvanostatic pulse deposition scheme, potentiostatic (current-time) transients recorded during the DC deposition of HA on a glassy carbon electrode were associated with the nucleation and growth of deposits on a conducting surface [34], wherein the current density shifted towards more negative values as the applied potential was made more negative.

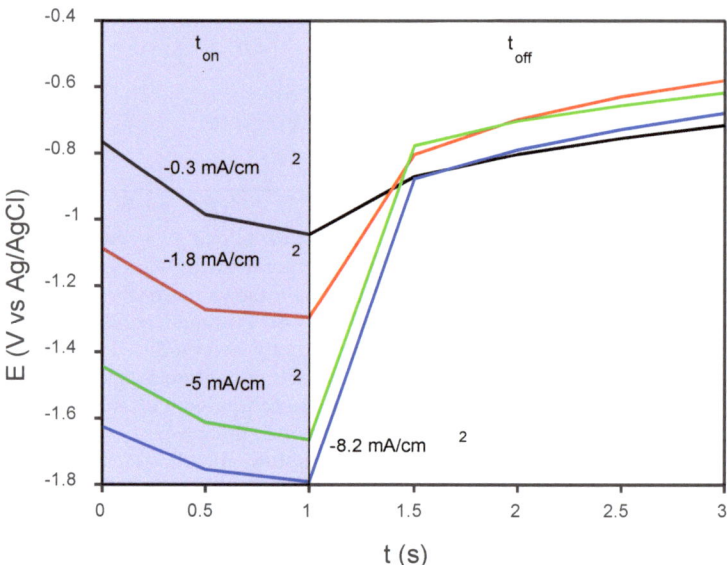

Figure 2. E-t transient recorded during last cycle at the indicated PC densities. Data acquisition was performed in steps of 0.5 s. t_{on} and t_{off} times applied were 1 and 2 s, respectively, and the overall charge (Q) was kept constant for every deposition (0.5 C/cm^2). The electrolyte solution (100 mL) contained 0.042 M Ca(NO$_3$)$_2$·4H$_2$O and 0.025 M NH$_4$H$_2$PO$_4$.

Figure 3 shows the SEM images of the CaP coatings obtained at varying j_c values. The morphology of the coatings changed from belt- to flake-like as j_c was made more negative, i.e., −0.3, −1.8, and −5 mA/cm^2 (Figure 3a–c). The insets show magnified details of the coating morphology, in which the formation of a belt network or grass-like morphology is appreciated. In parallel, the Ca/P ratio increased from 1.2 to 1.3. Further increase of the current density to −8.2 mA/cm^2 caused the precipitation of oxyhydroxides, likely due to local variation of the pH value caused by intensified hydrogen coevolution, without bringing about a notorious increase in the Ca/P ratio. The mean Ca/P ratios varied between 1.28 and 1.40 (Figure 4). A representative EDX spectrum of an as-deposited CaP coating is given in Figure 5a. The pattern shows that both the element peaks originated from the coating layer and the alloy underneath. Considering that the spectra had enough counts and that the Ca and P peaks are isolated, the minimum detectable mass for elements with Z > 11 can ideally be as low as 0.02% wt. In practice, the detection limit of EDX in modern SEMs is about 1–2% wt. for light atoms, as it is in the case of calcium and phosphorous [35,36]. Hence, the estimated Ca/P ratios must be taken with caution, although the trends are meaningful.

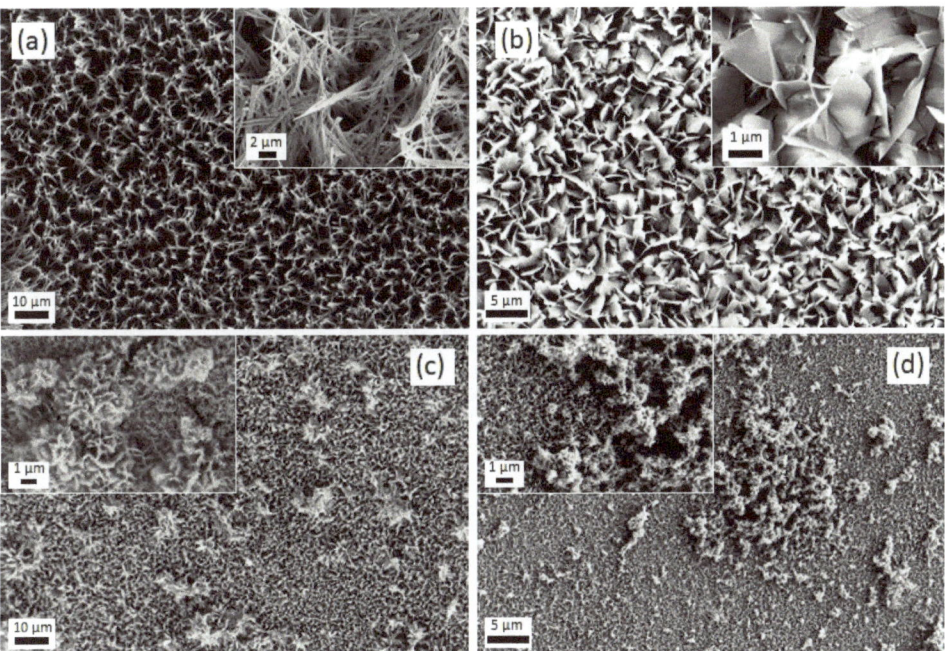

Figure 3. SEM images of CaP coatings deposited on the TiMoNbTa alloy by PC with Q = 0.5 C/cm^2 and (**a**) j$_c$ = −0.3 mA/cm^2, (**b**) j$_c$ = −1.8 mA/cm^2, (**c**) j$_c$ = −5 mA/cm^2, and (**d**) j$_c$ = −8.2 mA/cm^2. t$_{on}$ and t$_{off}$ times applied were 1 and 2 s, respectively. The electrolyte solution (100 mL) contained 0.042 M Ca(NO$_3$)$_2$·4H$_2$O and 0.025 M NH$_4$H$_2$PO$_4$.

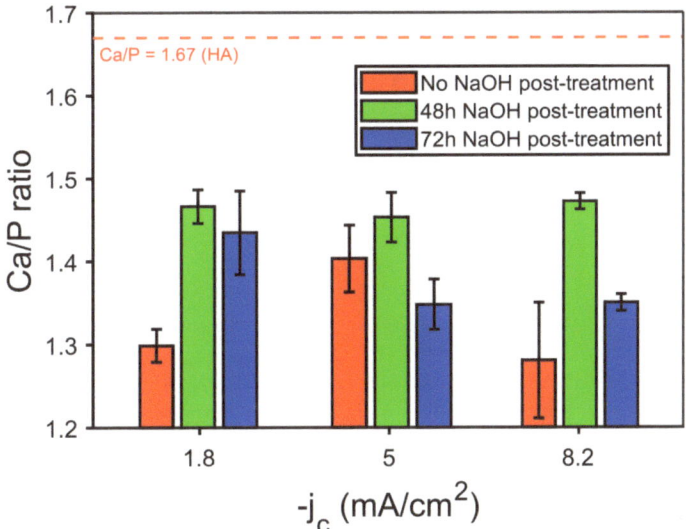

Figure 4. Ca/P atomic ratio in the as-prepared (no NaOH post-treatment) and NaOH-treated coatings produced at different j$_c$ values and similar overall charge (−Q = 0.5–0.8 C/cm^2). The electrolyte solution (100 mL) contained 0.042 M Ca(NO$_3$)$_2$·4H$_2$O and 0.025 M NH$_4$H$_2$PO$_4$.

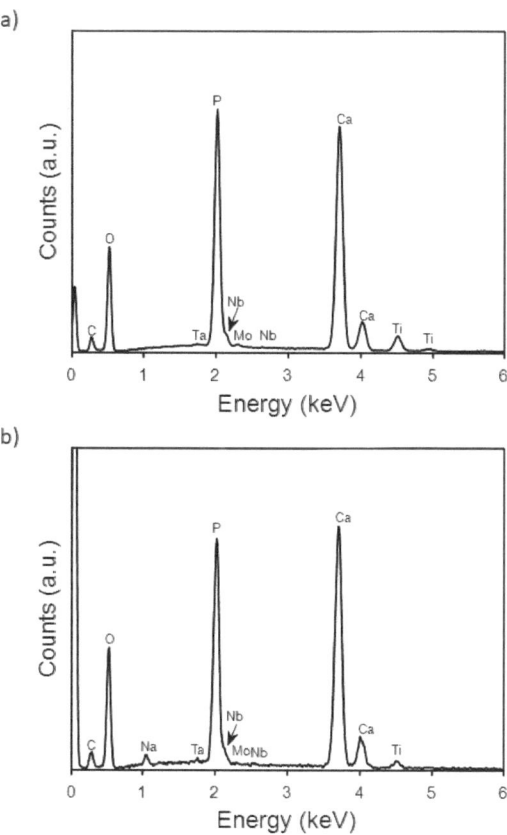

Figure 5. EDX spectra of CaP coatings deposited on the TiMoNbTa alloy by PC at $j_c = -1.8$ mA/cm^2, $t_{on} = 1$ s, $t_{off} = 2$ s, and $Q = 0.8$ C/cm^2 in (**a**) as-deposited state and (**b**) after immersion in 0.1 M NaOH for 48 h.

A powerful strategy that has been considered to promote the formation of HA is an alkaline post-treatment [15], e.g., a post-synthesis immersion in NaOH solution. Since the growth of HA requires an excess of OH$^-$ ions, such alkaline treatment will supply the sufficient OH$^-$ concentration, which in combination with Ca^{2+} and PO_4^{3-} will result in the transformation of intermediate CaP phases into HA [15]. The Ca/P ratios determined by EDX of coatings that were immersed in NaOH 0.1 M for different time periods (48 and 72 h) after electrodeposition at varying current densities are summarized in Figure 4.

It is evident that the post-treatment with NaOH can successfully increase the Ca/P ratio of the as-deposited samples (see Figures 4 and 5). However, higher values of the Ca/P ratio are obtained with 48 h of immersion, while after 72 h this ratio tends to decrease. According to several authors, the increase observed at 48h could be explained by the release of Ca^{2+} and PO_4^{3-} ions from the as-deposited coatings when they are soaked in the alkaline solution. These ions would combine with the OH$^-$ ions present in solution to yield the HA phase [10]. The decline in the Ca/P ratio at higher immersion times suggests a redeposition of the Ca^{2+} and PO_4^{3-} onto the coating. For the samples grown at j_c of -1.8 and -8.2 mA/cm^2 values, the Ca/P ratio can be as high as ~1.47. Notice the relative increase in the peak intensity for Ca, which in turn will increase the Ca/P ratio in Figure 5b. These values match those of CDHA, whose Ca/P can range, according to the literature, from 1.33 up to 1.67 [20]. The CDHA is characterized by higher solubility and consequently more bioactivity than the HA phase [20,37], and it has improved bioresorbable abilities, which stimulate a strong interaction and attachment to the bone thanks to a quick resorption [21]. The increase in Ca/P ratio upon alkalinization treatment for 48 h is accompanied by a change in the on-top morphology of the coatings (Figure 6a,b). Namely, the leaves become wider and adopt a more flower-like structure.

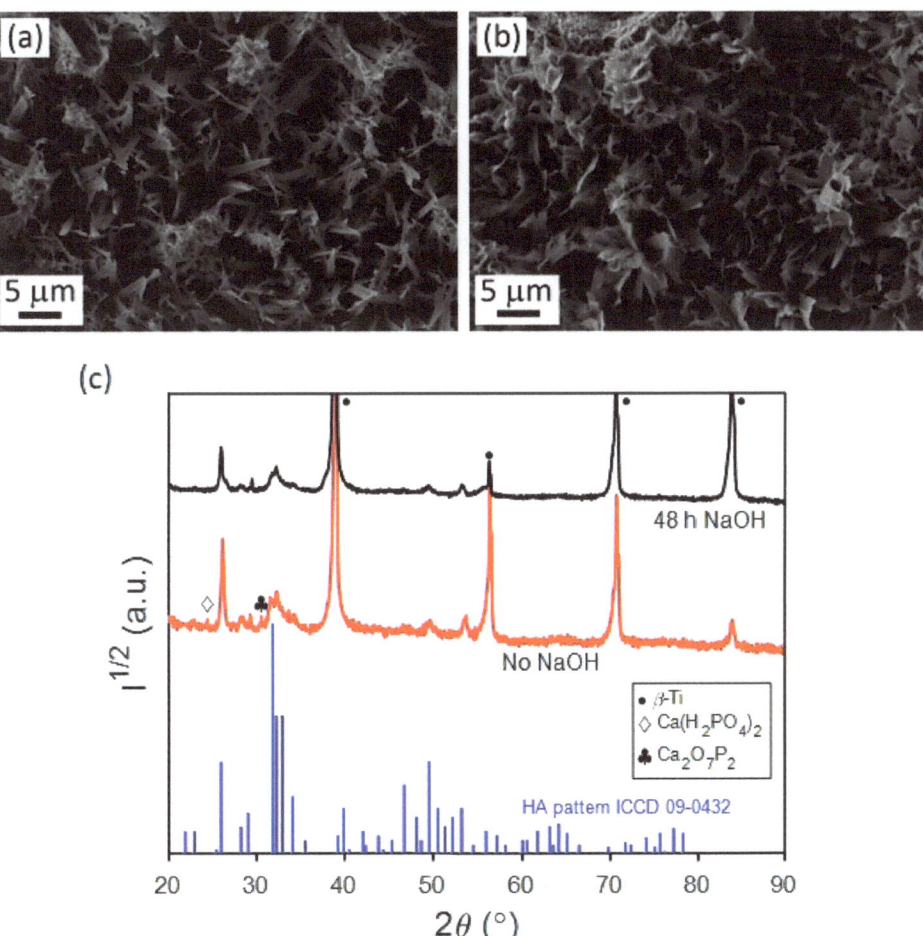

Figure 6. SEM images of CaP coatings deposited on the TiMoNbTa alloy by PC at $j_c = -1.8$ mA/cm^2, t_{on} = 1 s, t_{off} = 2 s, and Q = 0.8 C/cm^2 in (**a**) as-deposited state and (**b**) after immersion in 0.1 M NaOH for 48 h, and (**c**) corresponding XRD patterns. The theoretical pattern for HA is shown in blue at the bottom.

The crystal structure of these coatings was studied by XRD (Figure 6c). The patterns show characteristic peaks for the base alloy and less intense peaks ascribed to the CaP coatings. Although the differences in the patterns for the coatings were quite subtle, on the sample without NaOH post-treatment, the peaks could be attributed to a mixture of different CaP phases: CDHA, monocalcium phosphate monohydrate (Ca(H$_2$PO$_4$)$_2$), and calcium diphosphate (Ca$_2$O$_7$P$_2$). Meanwhile, all the peaks in the pattern of the NaOH-treated coating could be attributed to the CDHA phase. By performing a Rietveld analysis of the XRD pattern of the NaOH-treated coating, the cell parameters, crystal size, and microstrains of both the base alloy and the CDHA coating were determined and are summarized in Table 2.

Table 2. Cell parameters, crystal size, and microstrains determined by Rietveld fitting of the XRD pattern of NaOH-treated coating deposited on the TiMoNbTa substrate shown in Figure 6c.

Material	a (Å)	c (Å)	Crystal Size (nm)	Microstrains
Ti-18Mo-6Nb-5Ta (β phase)	3.2632	–	>200	$5.3 \cdot 10^{-4}$
CDHA (hcp phase)	9.2911	6.8540	16	$11.8 \cdot 10^{-4}$

The cell parameter for the base alloy is a = 3.2632 Å and it is microcrystalline (crystal size well beyond 200 nm). Meanwhile, the CDHA coating is nanocrystalline, and the microstrains are rather low considering the deposition method used. In some reports [38] on stoichiometric hydroxyapatite, a has been determined to be 9.18 Å, which is lower than the values obtained from the Rietveld fitting of the experimental patterns. This difference might be explained by the deficiency of Ca ions in the hexagonal unit cell [38].

In order to determine the thickness of the CDHA coatings on the Ti alloys, specimens were cut using a FIB. The corresponding SEM pictures are shown in Figure 7, where the average thickness is about 5 μm. In the cross-section SEM images, the overall porous structure of the CDHA can be observed, with pore sizes that vary along depth, from a few nm close to the substrate to μm-sized pores at the upper parts of the films, starting at the interface with the alloy with a more compact morphology and evolving toward a more open structure with larger plate-like features [25]. Note that the flaky morphology observed by SEM in Figure 6b corresponds to the top part of the cross-section view in Figure 7.

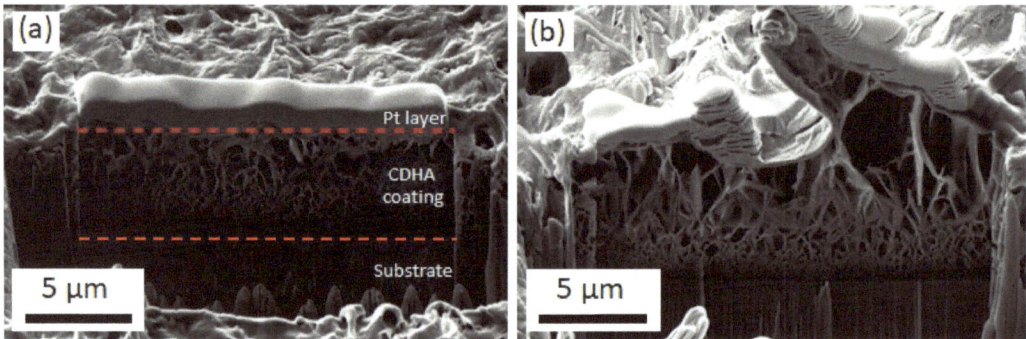

Figure 7. FIB cuts of CDHA coatings grown at j_c = −1.8 mA/cm², t_{on} = 1 s, t_{off} = 2 s, and Q = 0.8 C/cm², followed by an NaOH treatment for 48 h, on (**a**) Ti-18Mo-6Nb-5Ta and (**b**) Ti-6Al-4V alloys.

Scratch tests were also performed on the Ti-6Al-4V and the new alloy, both coated with CDHA, as shown in Figure 8. First, the friction force scales linearly with the scratch distance (Figure 8a). The slightly lower frictional forces for the CDHA-coated TiMoNbTa material are likely not significant considering that the deposited CDHA is similar from the morphological and structural standpoints. Importantly, no critical load was determined to cause adhesive failure of the coatings within the explored force range. The penetration depth increases with the scratch distance and reaches 5 μm at approximately 200 μm of scratch distance (Figure 8b). Further recorded penetration depths of around 8 μm might suggest local failure. Hence, the behavior of the coating on the new alloy upon loading follows quite the same trend as that on the commercially used Ti-6Al-4V, which indicates a similar adhesion of the CDHA to the substrate.

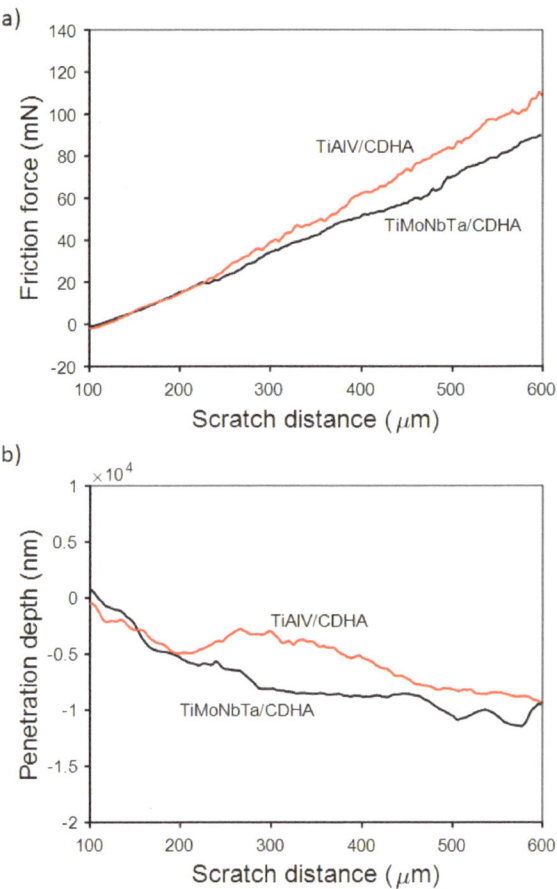

Figure 8. Representative (**a**) friction force and (**b**) penetration depth vs. scratch distance for the CDHA-coated Ti-18Mo-6Nb-5Ta and Ti-6Al-4V alloys using the optimized conditions of Figure 7.

3.3. Cytocompatibility of CaP-Coated and Uncoated Alloys

The cytocompatibility of the materials was assessed by cell viability, cell proliferation, cell morphology, and cell adhesion analyses. Cell viability analysis determines if direct cell contact with the alloys produces a toxic effect (i.e., a decrease in the number of live cells), whereas cell proliferation analysis allows to assess whether cells cultured on top of the alloys can grow (i.e., increase in number over time).

A Live/Dead kit was used to determine the cytotoxicity of four different samples: TiAlV and TiMoNbTa substrates, plus the previously optimized coated ones, namely, TiAlV/CaP (where "CaP" refers to the CDHA coating) and TiMoNbTa/CaP (where "CaP" also refers to the CDHA coating). As shown in Figure 9a, none of the materials proved to be toxic, as a high number of live cells were observed growing on all the disks' surfaces after 3 days in culture. Although no quantitative analyses were performed, images revealed a slightly higher number of cells on top of the TiAlV/CaP and TiMoNbTa/CaP alloys when compared with their counterparts without CaP. Furthermore, in all samples, only a few cells were stained red, meaning that none of the materials is cytotoxic.

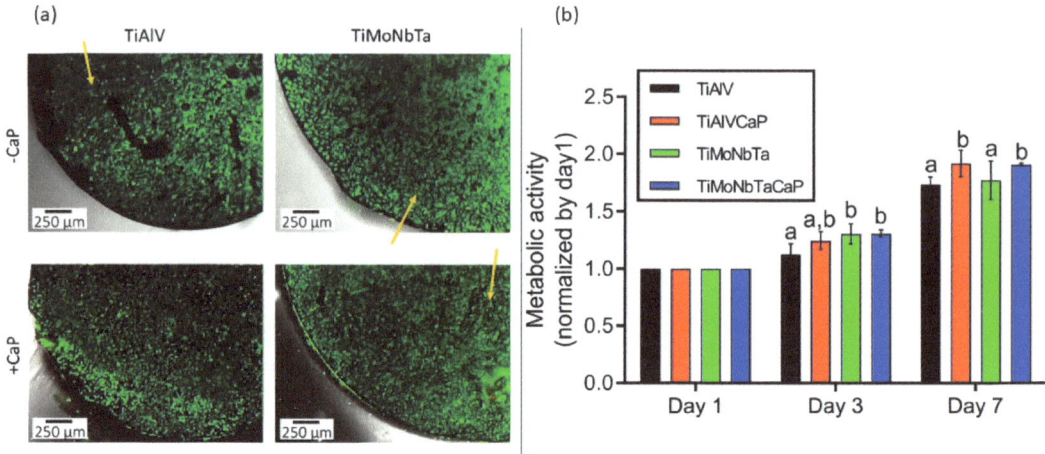

Figure 9. Cytocompatibility of TiAlV, TiAlV/CaP, TiMoNbTa, and TiMoNbTa/CaP alloys. (**a**) Cytotoxicity measured on Saos-2 cells after 3 days in culture using the live/dead kit. Live and dead cells appear in green and red, respectively. Yellow arrows point to dead cells. (**b**) Proliferation of Saos-2 cells grown at 1, 3, and 7 days in culture. Results were normalized by day 1. Different superscripts on top of the columns denote significant differences ($p < 0.05$) among the materials at the same time point.

The proliferation of Saos-2 cells cultured on the different samples was assessed on days 1, 3, and 7 after seeding. Results of the metabolic activity were normalized with respect to the values for day 1 and compared among materials at each time point. As shown in Figure 9b, the proliferation of cells growing on TiMoNbTa (without and with CaP) alloys was higher than that of cells growing on TiAlV at day 3. Furthermore, after 7 days, CaP coating of both TiAlV and TiMoNbTa resulted in higher cell proliferation when compared with alloys without CaP coating. The results of the Alamar Blue seem to be in agreement with the live/dead kit results, confirming that the new TiMoNbTa alloy is good in terms of cell proliferation.

Ti-6Al-4V alloys have long been used as standard bone implant materials because of their excellent reputation for corrosion resistance and biocompatibility. Nonetheless, the long-term performance of these alloys has raised some concerns due to the release of aluminum and vanadium from the alloy [39]. The results here presented are in agreement with those of other authors who have shown that Ti-6Al-4V exhibits good biocompatibility [40]. Furthermore, our results show that the newly developed Ti-18Mo-6Nb-5Ta allows increased cell proliferation while avoiding the use of aluminum and vanadium.

Alternatively, SEM analysis of osteoblasts grown on the different samples showed that cells were randomly distributed on their surfaces after 3 days of culture (Figure 10). Regarding cell morphology, we observed that on all the alloys, cells presented a flattened polygonal morphology with cytoplasmic extensions in different directions. Thus, differences in wettability between uncoated and CaP-coated alloy surfaces do not appear to interfere with cell adhesion.

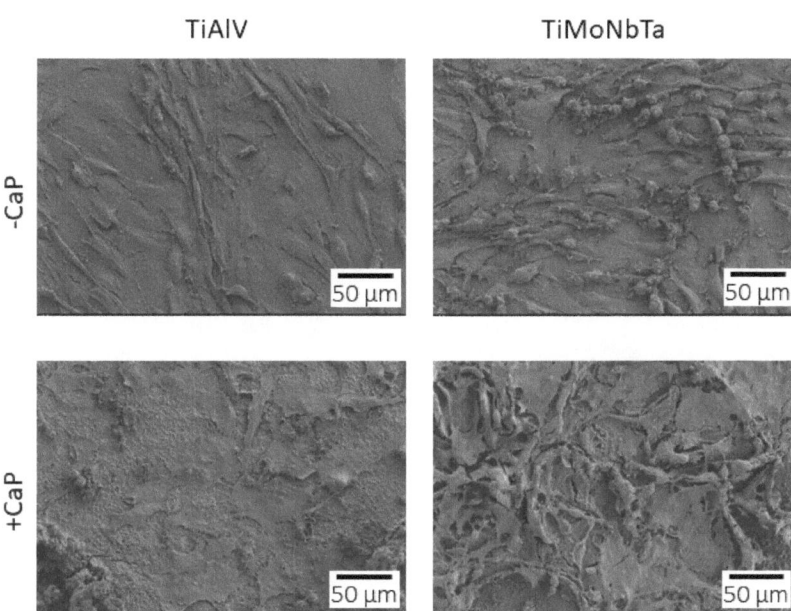

Figure 10. Scanning electron microscope images of human Saos-2 cells cultured for 3 days on TiAlV, TiAlV/CaP, TiMoNbTa, and TiMoNbTa/CaP alloys.

Cell adhesion to the alloys was confirmed through immunofluorescence analysis of vinculin and stress fibers (actin). These analyses showed that Saos-2 cells were completely adhered to the samples' surfaces after 3 days of culture (Figure 11), corroborating the results obtained by SEM, and were able to establish focal contacts. Cells presented well-defined stress fibers, crossing the totality of the cell's cytoplasm and ending in focal contacts. Most of the stress fibers were found in parallel, and some of them were found without a defined orientation.

According to the results obtained in the cell proliferation, viability, morphology, and adhesion analyses, both TiAlV and TiMoNbTa alloys, with and without CaP coating, are biocompatible. As reviewed by Kolli et al. [9], β-Ti derived alloys present good mechanical and biocompatibility properties, which, according to our results, are both true for TiAlV and TiMoNbTa alloys.

In addition, the surface coating with CaP provides better results over both alloys in terms of cell proliferation when compared with the non-coated ones. CaP has been demonstrated to be a good biocoating for enhancing bone growth in osteogenic materials [41,42], and in the present work, we show that both alloys could benefit from the addition of a CaP in the CDHA structure later at the surface. Furthermore, our results suggest that the use of TiMoNbTa/CaP would provide better osseointegration than TiAlV/CaP.

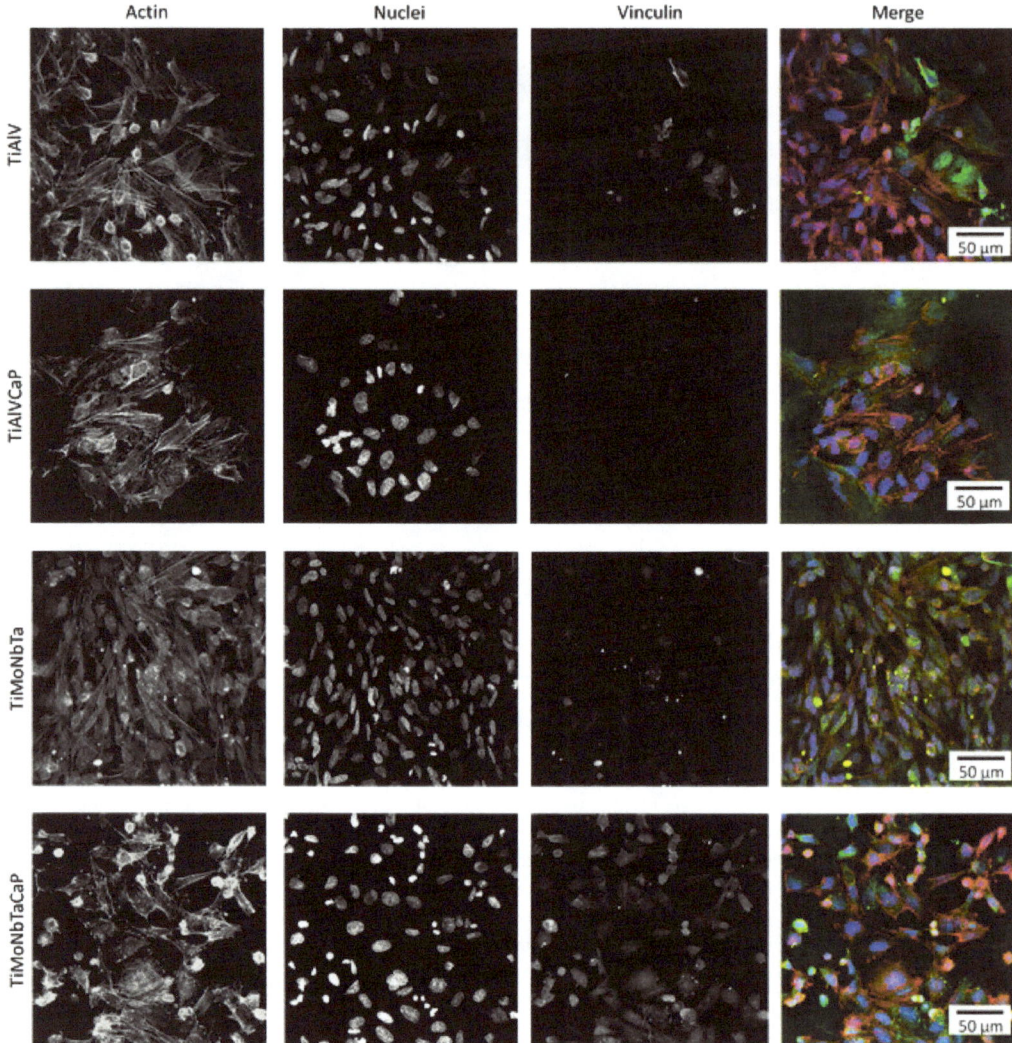

Figure 11. Saos-2 cells adhered to the surfaces of the TiAlV, TiAlV/CaP, TiMoNbTa, and TiMoNbTa/CaP after 3 days in culture. Stress fibers (actin; red), vinculin (green), and nuclei (DNA; blue) can be observed.

4. Conclusions

A new β-Ti alloy (Ti-18Mo-6Nb-5Ta (wt%)) with non-toxic alloying elements was successfully fabricated by arc melting, using the *MoE* as a predictive tool to obtain the desired β-phase. The resulting alloy showed improved mechanical properties, maintaining a high hardness (3.4 GPa) and lowering its reduced Young's modulus (79 GPa) by almost 25% in comparison with the value for the commonly used Ti-6Al-4V (106 GPa). This, in combination with the biological advantages provided by the replacement of Al and V with Mo, Nb, and Ta as documented in literature, pointed out the need for fabrication of a superior β-Ti alloy. Simultaneously, the corrosion behavior of PBS was similar to that of Ti-6Al-4V. The potential integration of the alloy into the bone was further enhanced by the functionalization of its surface with a CaP coating produced by electrodeposition. Structural characterization (SEM, EDX, XRD, FIB, and scratching tests) confirmed the formation of a nanocrystalline CDHA coating with a plate/flower-like morphology, a Ca/P ratio of 1.47, a thickness of 5 μm, and good adhesion to the substrate. Finally, the cytocompatibility of the new alloy, uncoated and coated with CDHA, was also assessed by means of cell viability, cell proliferation, cell morphology, and cell adhesion analyses. The results indicated that the bare alloy and the composite material, i.e., TiMoNbTa/CDHA, do not produce a toxic effect since cells attach well to their surfaces, forming anchor points across the area with good morphology. Additionally, the proliferation analysis showed that the growth rate of cells is higher when cultured on the newly fabricated TiMoNbTa alloy than on the commercial TiAlV, and this rate increased further when the material was coated with CDHA, suggesting a higher capacity to promote bone regeneration and growth.

Author Contributions: Conceptualization, E.P. and J.S.; methodology, M.E., O.C., A.G., E.I., A.B. (Andreu Blanquer), C.N., J.S. and E.P.; formal analysis, M.E. and O.C.; investigation, M.E., O.C., N.F.N., A.B. (Aleksandra Bartkowska), L.A.A., J.F., P.S. and T.G.; resources, A.G., C.N., J.S. and E.P.; writing—original draft preparation, M.E. and O.C; writing—review and editing, N.F.N., L.A.A., J.F., A.G., E.I., A.B. (Andreu Blanquer), C.N., J.S. and E.P.; supervision, C.N., J.S. and E.P.; funding acquisition, J.S. and E.P. All authors have read and agreed to the published version of the manuscript.

Funding: This research was funded by the European Commission within the H2020-MSCA grant agreement No. 861046 (BIOREMIA-ITN). Partial funding was also provided by the Spanish Government (PID2020-116844RB-C21) and the Generalitat de Catalunya (2021-SGR-00651 and 2021-SGR-00122).

Institutional Review Board Statement: Not applicable.

Informed Consent Statement: Not applicable.

Data Availability Statement: Not applicable.

Conflicts of Interest: The authors declare no conflict of interest.

References

1. Tuninetti, V.; Jaramillo, A.F.; Riu, G.; Rojas-Ulloa, C.; Znaidi, A.; Medina, C.; Mateo, A.M.; Roa, J.J. Experimental Correlation of Mechanical Properties of the Ti-6Al-4V Alloy at Different Length Scales. *Metals* **2021**, *11*, 104. [CrossRef]
2. Zysset, P.K.; Edward Guo, X.; Edward Hoffler, C.; Moore, K.E.; Goldstein, S.A. Elastic Modulus and Hardness of Cortical and Trabecular Bone Lamellae Measured by Nanoindentation in the Human Femur. *J. Biomech.* **1999**, *32*, 1005–1012. [CrossRef] [PubMed]
3. Trincă, L.C.; Mareci, D.; Solcan, C.; Fântânariu, M.; Burtan, L.; Hrițcu, L.; Chiruță, C.; Fernández-Mérida, L.; Rodríguez-Raposo, R.; Santana, J.J.; et al. New Ti-6Al-2Nb-2Ta-1Mo Alloy as Implant Biomaterial: In Vitro Corrosion and in Vivo Osseointegration Evaluations. *Mater. Chem. Phys.* **2020**, *240*, 122229. [CrossRef]
4. Verma, R.P. Titanium Based Biomaterial for Bone Implants: A Mini Review. *Mater. Today Proc.* **2020**, *26*, 3148–3151. [CrossRef]
5. Kuroda, D.; Niinomi, M.; Morinaga, M.; Kato, Y.; Yashiro, T. Design and Mechanical Properties of New i Type Titanium Alloys for Implant Materials. *Mater. Sci. Eng. A* **1998**, *243*, 244–249. [CrossRef]
6. Xu, D.; Wang, T.; Wang, S.; Jiang, Y.; Wang, Y.; Chen, Y.; Bi, Z.; Geng, S. Antibacterial Effect of the Controlled Nanoscale Precipitates Obtained by Different Heat Treatment Schemes with a Ti-Based Nanomaterial, Ti–7.5Mo–5Cu Alloy. *ACS Appl. Bio Mater.* **2020**, *3*, 6145–6154. [CrossRef]
7. Guo, S.; Meng, Q.; Zhao, X.; Wei, Q.; Xu, H. Design and Fabrication of a Metastable β-Type Titanium Alloy with Ultralow Elastic Modulus and High Strength. *Sci. Rep.* **2015**, *5*, 14688. [CrossRef]
8. Mohammed, M.T.; Khan, Z.A.; Siddiquee, A.N. Beta Titanium Alloys: The Lowest Elastic Modulus for Biomedical Applications: A Review. *Int. J. Chem. Nucl. Metall. Mater. Eng.* **2014**, *8*, 726–731.
9. Kolli, R.; Devaraj, A. A Review of Metastable Beta Titanium Alloys. *Metals* **2018**, *8*, 506. [CrossRef]
10. Pinotti, V.E.; Plaine, A.H.; Romero da Silva, M.; Bolfarini, C. Influence of Oxygen Addition and Aging on the Microstructure and Mechanical Properties of a β-Ti-29Nb–13Ta–4Mo Alloy. *Mater. Sci. Eng. A* **2021**, *819*, 141500. [CrossRef]
11. Barceloux, D.G.; Barceloux, D. Molybdenum. *J. Toxicol. Clin. Toxicol.* **1999**, *37*, 231–237. [CrossRef] [PubMed]
12. Majumdar, P.; Singh, S.B.; Chakraborty, M. Elastic Modulus of Biomedical Titanium Alloys by Nano-Indentation and Ultrasonic Techniques—A Comparative Study. *Mater. Sci. Eng. A* **2008**, *489*, 419–425. [CrossRef]
13. Sarimov, R.M.; Glinushkin, A.P.; Sevostyanov, M.A.; Konushkin, S.V.; Serov, D.A.; Astashev, M.E.; Lednev, V.N.; Yanykin, D.V.; Sibirev, A.V.; Smirnov, A.A.; et al. Ti-20Nb-10Ta-5Zr Is Biosafe Alloy for Building of Ecofriendly Greenhouse Framework of New Generation. *Metals* **2022**, *12*, 2007. [CrossRef]

14. Gudkov, S.V.; Simakin, A.V.; Konushkin, S.V.; Ivannikov, A.Y.; Nasakina, E.O.; Shatova, L.A.; Kolmakov, A.G.; Sevostyanov, M.A. Preparation, Structural and Microstructural Characterization of Ti–30Nb–10Ta–5Zr Alloy for Biomedical Applications. *J. Mater. Res. Technol.* **2020**, *9*, 16018–16028. [CrossRef]
15. Safavi, M.S.; Walsh, F.C.; Surmeneva, M.A.; Surmenev, R.A.; Khalil-Allafi, J. Electrodeposited Hydroxyapatite-Based Biocoatings: Recent Progress and Future Challenges. *Coatings* **2021**, *11*, 110. [CrossRef]
16. Picard, Q.; Olivier, F.; Delpeux, S.; Chancolon, J.; Warmont, F.; Bonnamy, S. Development and Characterization of Biomimetic Carbonated Calcium-Deficient Hydroxyapatite Deposited on Carbon Fiber Scaffold. *J. Carbon Res.* **2018**, *4*, 25. [CrossRef]
17. Dumelie, N.; Benhayoune, H.; Richard, D.; Laurent-Maquin, D.; Balossier, G. In Vitro Precipitation of Electrodeposited Calcium-Deficient Hydroxyapatite Coatings on Ti6Al4V Substrate. *Mater. Charact.* **2008**, *59*, 129–133. [CrossRef]
18. Li, T.-T.; Ling, L.; Lin, M.-C.; Peng, H.-K.; Ren, H.-T.; Lou, C.-W.; Lin, J.-H. Recent Advances in Multifunctional Hydroxyapatite Coating by Electrochemical Deposition. *J. Mater. Sci.* **2020**, *55*, 6352–6374. [CrossRef]
19. Schmidt, R.; Gebert, A.; Schumacher, M.; Hoffmann, V.; Voss, A.; Pilz, S.; Uhlemann, M.; Lode, A.; Gelinsky, M. Electrodeposition of Sr-Substituted Hydroxyapatite on Low Modulus Beta-Type Ti-45Nb and Effect on in Vitro Sr Release and Cell Response. *Mater. Sci. Eng. C* **2020**, *108*, 110425. [CrossRef]
20. Beaufils, S.; Rouillon, T.; Millet, P.; Le Bideau, J.; Weiss, P.; Chopart, J.-P.; Daltin, A.-L. Synthesis of Calcium-Deficient Hydroxyapatite Nanowires and Nanotubes Performed by Template-Assisted Electrodeposition. *Mater. Sci. Eng. C* **2019**, *98*, 333–346. [CrossRef]
21. Gecim, G.; Dönmez, S.; Erkoc, E. Calcium Deficient Hydroxyapatite by Precipitation: Continuous Process by Vortex Reactor and Semi-Batch Synthesis. *Ceram. Int.* **2021**, *47*, 1917–1928. [CrossRef]
22. Wang, L.; Wang, M.; Li, M.; Shen, Z.; Wang, Y.; Shao, Y.; Zhu, Y. Trace Fluorine Substituted Calcium Deficient Hydroxyapatite with Excellent Osteoblastic Activity and Antibacterial Ability. *CrystEngComm* **2018**, *20*, 5744–5753. [CrossRef]
23. Zhang, H.; Zhang, M. Characterization and Thermal Behavior of Calcium Deficient Hydroxyapatite Whiskers with Various Ca/P Ratios. *Mater. Chem. Phys.* **2011**, *126*, 642–648. [CrossRef]
24. Drevet, R.; Benhayoune, H. Electrodeposition of Calcium Phosphate Coatings on Metallic Substrates for Bone Implant Applications: A Review. *Coatings* **2022**, *12*, 539. [CrossRef]
25. Mokabber, T.; Lu, L.Q.; van Rijn, P.; Vakis, A.I.; Pei, Y.T. Crystal Growth Mechanism of Calcium Phosphate Coatings on Titanium by Electrochemical Deposition. *Surf. Coat. Technol.* **2018**, *334*, 526–535. [CrossRef]
26. Fornell, J.; Feng, Y.P.; Pellicer, E.; Suriñach, S.; Baró, M.D.; Sort, J. Mechanical Behaviour of Brushite and Hydroxyapatite Coatings Electrodeposited on Newly Developed FeMnSiPd Alloys. *J. Alloys Compd.* **2017**, *729*, 231–239. [CrossRef]
27. Vidal, E.; Buxadera-Palomero, J.; Pierre, C.; Manero, J.M.; Ginebra, M.-P.; Cazalbou, S.; Combes, C.; Rupérez, E.; Rodríguez, D. Single-Step Pulsed Electrodeposition of Calcium Phosphate Coatings on Titanium for Drug Delivery. *Surf. Coat. Technol.* **2019**, *358*, 266–275. [CrossRef]
28. Oliver, W.C.; Pharr, G.M. An Improved Technique for Determining Hardness and Elastic Modulus Using Load and Displacement Sensing Indentation Experiments. *J. Mater. Res.* **1992**, *7*, 20. [CrossRef]
29. Hynowska, A.; Blanquer, A.; Pellicer, E.; Fornell, J.; Suriñach, S.; Baró, M.; González, S.; Ibáñez, E.; Barrios, L.; Nogués, C.; et al. Novel Ti–Zr–Hf–Fe Nanostructured Alloy for Biomedical Applications. *Materials* **2013**, *6*, 4930–4945. [CrossRef]
30. Campos-Quirós, A.; Cubero-Sesín, J.M.; Edalati, K. Synthesis of Nanostructured Biomaterials by High-Pressure Torsion: Effect of Niobium Content on Microstructure and Mechanical Properties of Ti-Nb Alloys. *Mater. Sci. Eng. A* **2020**, *795*, 139972. [CrossRef]
31. Leyland, A.; Matthews, A. On the Significance of the H/E Ratio in Wear Control: A Nanocomposite Coating Approach to Optimised Tribological Behaviour. *Wear* **2000**, *246*, 1–11. [CrossRef]
32. Geetha, M.; Singh, A.K.; Asokamani, R.; Gogia, A.K. Ti Based Biomaterials, the Ultimate Choice for Orthopaedic Implants—A Review. *Prog. Mater. Sci.* **2009**, *54*, 397–425. [CrossRef]
33. Eliaz, N.; Eliyahu, M. Electrochemical Processes of Nucleation and Growth of Hydroxyapatite on Titanium Supported by Real-Time Electrochemical Atomic Force Microscopy. *J. Biomed. Mater. Res.* **2007**, *80A*, 621–634. [CrossRef]
34. Vladislavić, N.; Rončević, I.Š.; Buzuk, M.; Buljac, M.; Drventić, I. Electrochemical/Chemical Synthesis of Hydroxyapatite on Glassy Carbon Electrode for Electroanalytical Determination of Cysteine. *J. Solid State Electrochem.* **2021**, *25*, 841–857. [CrossRef]
35. Falcone, R.; Sommariva, G.; Verità, M. WDXRF, EPMA and SEM/EDX Quantitative Chemical Analyses of Small Glass Samples. *Microchim. Acta* **2006**, *155*, 137–140. [CrossRef]
36. Sensitivity/Detection Limit of EDS. Available online: https://www.globalsino.com/EM/page4792.html (accessed on 5 November 2022).
37. Drevet, R.; Velard, F.; Potiron, S.; Laurent-Maquin, D.; Benhayoune, H. In Vitro Dissolution and Corrosion Study of Calcium Phosphate Coatings Elaborated by Pulsed Electrodeposition Current on Ti6Al4V Substrate. *J. Mater. Sci. Mater. Med.* **2011**, *22*, 753–761. [CrossRef]
38. Victor, S.P.; Kumar, T.S.S. Tailoring Calcium-Deficient Hydroxyapatite Nanocarriers for Enhanced Release of Antibiotics. *J. Biomed. Nanotechnol.* **2008**, *4*, 203–209. [CrossRef]
39. Rao, S.; Ushida, T.; Tateishi, T.; Asao, S. Effect of Ti, Al, and V Ions on the Relative Growth Rate of Fibroblasts (L929) and Osteoblasts (MC3T3-El) Cells. *Biomed. Mater. Eng.* **1996**, *6*, 79–86. [CrossRef]
40. Navarro, M.; Michiardi, A.; Castaño, O.; Planell, J.A. Biomaterials in Orthopaedics. *J. R. Soc. Interface* **2008**, *5*, 1137–1158. [CrossRef]

41. Cuijpers, V.M.J.I.; Alghamdi, H.S.; Van Dijk, N.W.M.; Jaroszewicz, J.; Walboomers, X.F.; Jansen, J.A. Osteogenesis Around CaP-Coated Titanium Implants Visualized Using 3D Histology and Micro-Computed Tomography. *J. Biomed. Mater. Res. Part A* **2015**, *103*, 3462–3473. [CrossRef]
42. Surmenev, R.A.; Surmeneva, M.A.; Ivanova, A.A. Significance of Calcium Phosphate Coatings for the Enhancement of New Bone Osteogenesis—A Review. *Acta Biomater.* **2014**, *10*, 557–579. [CrossRef] [PubMed]

Disclaimer/Publisher's Note: The statements, opinions and data contained in all publications are solely those of the individual author(s) and contributor(s) and not of MDPI and/or the editor(s). MDPI and/or the editor(s) disclaim responsibility for any injury to people or property resulting from any ideas, methods, instructions or products referred to in the content.

Article

Human Whole Blood Interactions with Craniomaxillofacial Reconstruction Materials: Exploring In Vitro the Role of Blood Cascades and Leukocytes in Early Healing Events

Viviana R. Lopes [1,2,†], Ulrik Birgersson [3,4,†], Vivek Anand Manivel [5], Gry Hulsart-Billström [2], Sara Gallinetti [1,6], Conrado Aparicio [7,8] and Jaan Hong [5,*]

1. OssDsign AB, SE-754 50 Uppsala, Sweden; vl@ossdsign.com (V.R.L.); sg@ossdsign.com (S.G.)
2. Department of Medicinal Chemistry, Translational Imaging, Uppsala University, SE-751 83 Uppsala, Sweden; gry.hulsart_billstrom@mcb.uu.se
3. Department of Clinical Science, Intervention and Technology, Division of Imaging and Technology, Karolinska Institute, SE-141 52 Huddinge, Sweden; ulrik.birgersson@ki.se
4. Department of Clinical Neuroscience, Neurosurgical Section, Karolinska University Hospital, SE-171 77 Stockholm, Sweden
5. Rudbeck Laboratory, Department of Immunology, Genetics and Pathology (IGP), Uppsala University, SE-751 85 Uppsala, Sweden; vivekanand.manivel@igp.uu.se
6. Department of Engineering Sciences, Applied Materials Science Section, Uppsala University, SE-751 03 Uppsala, Sweden
7. Faculty of Odontology, UIC Barcelona-International University of Catalonia, 08195 Barcelona, Spain; cjaparicio@uic.es
8. IBEC—Institute for Bioengineering of Catalonia, 08028 Barcelona, Spain
* Correspondence: jaan.hong@igp.uu.se
† These authors contributed equally to this work.

Abstract: The present study investigated early interactions between three alloplastic materials (calcium phosphate (CaP), titanium alloy (Ti), and polyetheretherketone (PEEK) with human whole blood using an established in vitro slide chamber model. After 60 min of contact with blood, coagulation (thrombin–antithrombin complexes, TAT) was initiated on all test materials (Ti > PEEK > CaP), with a significant increase only for Ti. All materials showed increased contact activation, with the KK–AT complex significantly increasing for CaP ($p < 0.001$), Ti ($p < 0.01$), and PEEK ($p < 0.01$) while only CaP demonstrated a notable rise in KK-C1INH production ($p < 0.01$). The complement system had significant activation across all materials, with CaP ($p < 0.0001$, $p < 0.0001$) generating the most pronounced levels of C3a and sC5b-9, followed by Ti ($p < 0.001$, $p < 0.001$) and lastly, PEEK ($p < 0.001$, $p < 0.01$). This activation correlated with leukocyte stimulation, particularly myeloperoxidase release. Consequently, the complement system may assume a more significant role in the early stages post implantation in response to CaP materials than previously recognized. Activation of the complement system and the inevitable activation of leukocytes might provide a more favorable environment for tissue remodeling and repair than has been traditionally acknowledged. While these findings are limited to the early blood response, complement and leukocyte activation suggest improved healing outcomes, which may impact long-term clinical outcomes.

Keywords: biomaterials; human whole blood; coagulation; complement; calcium phosphate

1. Introduction

Today, the majority of reconstructive surgical procedures involve the use of various alloplastic materials, ranging from the rigid fixation of bone fractures using different titanium constructs to the filling of bony cavities with calcium-phosphate-based bone cement. The application of these diverse materials is dependent on their interfacial properties and subsequent interactions with cells and biological fluids, such as blood [1–3]. Post implantation, blood proteins and platelets immediately adhere to the material's surface, consequently altering the blood clotting pathway. Coagulation occurs through surface-mediated reactions or tissue factor expression on cells [1,4,5]. Surface-mediated reactions (intrinsic pathway) occur via biomaterial surface interaction between coagulation factor XII (FXII) and prekallikrein, whereas tissue factor (extrinsic pathway) originates from proteins released from damaged tissue or expressed on activated immune (e.g., monocytes) and endothelial cells' surfaces. The two pathways remain distinct until Factor X activation, which directs thrombin and fibrin production, culminating in clot formation [6,7]. Additionally, platelet activation is initiated by platelets adhering to the blood protein layer that develops on the biomaterial surface after implantation [1,6] Existing evidence suggests that specific coagulation factors and activated platelets are also involved in the activation of the complement system, [2,3,7,8], consequently creating a communication link between the coagulation and complement systems [4,5,9–11]. This crosstalk results in blood-mediated thromboinflammation, which has demonstrated beneficial effects on regeneration by promoting the homing of leukocytes near the biomaterial implant. This supports the immune cells, such as monocytes, in switching to pro-inflammatory (M1) or pro-regenerative (M2) phenotypes based on the cellular signals of the surrounding tissue under favorable conditions [12]. The nature and intensity of this acute response are crucial for promoting (or inhibiting) healing. Blood clotting and biomaterial interfacial properties significantly influence compatibility and tissue healing [13–17].

The pivotal function of blood in bone tissue regeneration is highlighted in a study by Thor [18], which demonstrates that blood clots on dental implant surfaces undergo conversion into bone. Nonetheless, the mechanism of action underlying the interaction between whole blood and biomaterials, particularly during the initial stages, has yet to be fully understood.

In this study, we focus on the initial stages of material–blood interaction by exposing three disc-shaped replicas of craniomaxillofacial implants constructed from polyether ether ketone (PEEK), titanium alloy (Ti), and triphasic calcium phosphate (CaP) to freshly collected human whole blood in an established in vitro slide model. The aim is to provide insights into how the early interaction between these materials and whole blood can contribute to their varying capabilities to promote in situ healing.

2. Materials and Methods

2.1. Material Manufacturing

Medical-grade polyether ether ketone (PEEK) (Batch #6190394, ESSDE AB, Uppsala, Sweden), titanium (Ti) medical-grade 5, alloy Ti6Al4V (Lot # 01-958, Livallco stål AB, Stenkullen, Sweden), and triphasic calcium phosphate (CaP) (produced according to the process outlined by Gallinetti et al. [19]) were used. Each material was shaped into discs with a diameter of 16 mm and a height of 5 mm. The discs made from Ti and PEEK discs were cut from rods of the same diameter, polished with silicon carbide paper, and then subjected to an ultrasonic cleaning procedure to remove any residual particles. The discs underwent autoclaving for 20 min at a temperature of 121 °C (Autoclave Nüve OT 23B, Biotechlab, Sofia, Bulgaria). The CaP discs were prepared and supplied by OssDsign AB, Uppsala, Sweden. As experimental controls, polyvinylchloride (PVC) slides were used to ensure adequate heparinization and experimental reproducibility. The PVC slides were sterilized for 60 min at no less than 60 °C using 5% amoniumperoxidesulphate (APS).

2.2. Material Characterization

2.2.1. Surface Morphology

The surface morphology of the materials was visualized using scanning electron microscopy (SEM, Zeiss Merlin, Oberkochen, Germany). To acquire high-quality SEM images, the following steps were performed. First, the samples underwent vacuum drying at 60 °C for 2 h prior to analysis to guarantee their complete dryness. Second, before observation, the samples were secured to the sample holder using carbon tape, silver tape was applied to the sides, and a Au-Pd coating was deposited on the surface to prevent charging during analysis. The sputter coating process utilized a Polaron SC7640 Sputter Coater (Thermo VG Scientific, Waltham, MA, USA) with a voltage of 2 kV and a current of 20 mA, operating for a duration of 60 s.

2.2.2. Wettability

The materials' hydrophilic or hydrophobic properties were assessed using water contact angle measurements, employing both sessile drop tests with ultrapure water and captive bubble methods. A macro contact angle meter (DM-CE1, Kyowa, Japan) with appropriate software (FAMAS, Kyowa, Japan) was used to perform the wettability tests and calculate the contact angles. The contact angle (θ) was defined as the angle between the solid phase and the liquid phase as described [20]. Prior to the contact angle experiments all materials were cleaned in water and dried in a desiccator. For the sessile method, a 2 µL drop of ultrapure water was dispensed on the surface of the tested sample. With the captive bubble method, the sample was immersed in water in a small container and a bubble of air was formed underneath the surface of the tested sample [21]. In both methods, the contact angles were calculated through the first 20 s of contact of the fluid with the tested surface to describe the dynamic wettability response. Three measurements were performed for each type of sample material.

2.3. Study Design

An in vitro slide chamber model developed by Hong et al. [22] was used to investigate the interaction between whole blood and the materials under investigation. In brief, we collected 1.3 mL of blood from seven healthy donors and transferred it to a two-well heparinized slide chamber with the same diameter as the prepared materials' discs. Every material tested was subjected to the blood from each individual donor. The heparin-coating of the chamber surfaces allows blood contact to occur without artificial activation of the coagulation cascade. Subsequently, the materials were placed atop each well chamber pre-filled with human whole blood and secured with a clip. The closed chambers were then incubated for 60 min under rotation on a wheel at 20 rpm in an incubator at 37 °C.

At time point 0 min, 1 mL of fresh human blood was collected from each donor into ethylenediaminetetraacetic acid (EDTA) containing tubes at a final concentration of 4 mM, to serve as the baseline sample. Each experiment was carried out seven times using the two-well setup (i.e., fourteen surfaces tested per material) as depicted in Figure 1. None of the investigated materials underwent pre-treatment or received the addition of exogenous growth factors.

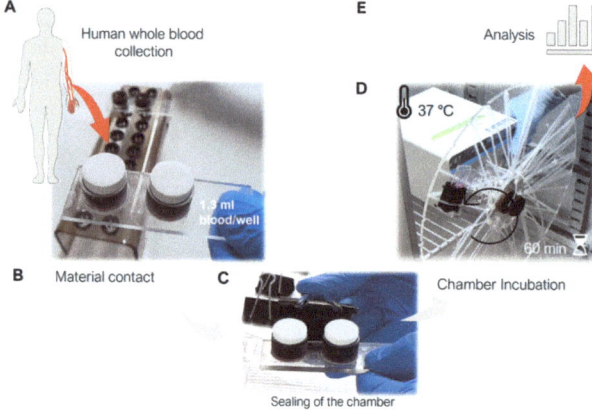

Figure 1. Schematic design setup of the whole blood chamber model. (**A**) Blood was drawn from healthy human donors, followed by the addition of 1.3 mL of fresh blood to each well of 2-well blood

chambers. (**B**) Two samples of each material were introduced to the 2-well chambers just prior to (**C**) securely sealing the chambers using heparinized ethylene propylene o-rings and a clip. (**D**) Thereafter, the chamber was positioned on a rotating wheel at 37 °C for 60 min. (**E**) One mL of blood from each well was immediately collected and analyzed both before and after incubation, while the remaining blood was centrifuged, rapidly frozen, and stored at −80 °C for subsequent analysis.

2.4. Heparinization and Blood Collection

The tubes and tips used in the blood experiments as well as the slide chamber were coated with the Corline heparin surface (Corline Systems AB, Uppsala, Sweden) according to the manufacturer's recommendation. This resulted in a double-layered heparin coating, exhibiting a binding capacity of 12 pmol/cm^2 antithrombin, as previously described by Andersson et al. [23].

Human whole blood was collected from healthy adult donors who had abstained from taking any medication known to impact blood coagulation (e.g., ibuprofen and aspirin) for at least two weeks. During blood collection, the blood was partially heparinized through the use of 50 mL of Falcon tubes with 100 IU/mL heparin (LEO Pharma, Malmö, Sweden), resulting in a final concentration of 0.25 IU of heparin per ml of blood to partially inhibit blood coagulation. Blood was used within 20 min after collection at room temperature.

Informed consent was obtained from all blood donors prior to the experiment. Ethical approval was obtained from the regional ethics committee, with reference number 2008/264.

2.5. Processing of the Blood Samples

For each material, the fluid phase of the blood was collected into tubes containing ethylenediaminetetraacetic acid (EDTA) at a final concentration of 4 mM and subsequently processed for platelet count. The supernatants were analyzed with enzyme-linked immunosorbent assays (ELISA) for myeloperoxidase (MPO) and eosinophil peroxidase (EPX) release, coagulation, and complement and kallikrein–kinin markers.

Due to the porous nature of the CaP material, a reduction in the fluid phase was anticipated for the CaP material; therefore, all measured values were normalized based on the observed volume reduction. No pretreatment of the CaP material, such as saturation with saline solution or an equivalent, was performed prior to the study. This choice was made to mimic clinical use and prevent the introduction of potential confounding factors.

2.6. Analytical Procedures

2.6.1. Macroscopic Visualization of Blood Interactions

Following 60-min incubation in the in vitro model, macroscopic evaluation of blood clotting was conducted for each material surface included in the study. The CaP discs were also sectioned using a scalpel to gross examine the internal structure. The results were captured in photographs.

2.6.2. Platelet Count

Baseline samples were collected at 0 min and the residual whole blood was mixed with EDTA at a final concentration of 4 mM. Platelet count was determined with the use of a hematology analyzer (Sysmex XP-300® Corporation, Kobe, Japan). Thereafter, the samples were centrifuged at 4500× g for 15 min at 4 °C to collect plasma for subsequent analysis. The plasma samples were stored at −70 °C until analysis and were measured in duplicate.

2.6.3. Coagulation and Contact Markers

Thrombin–antithrombin (TAT) was analyzed quantitatively, as an indicator of coagulation activity. Anti-human thrombin pAb was used for capture and HPR-conjugated anti-human antithrombin pAb was used for detection (Enzyme Research Laboratories, South Bend, IN, USA). A standard pooled human serum diluted in working buffer served as a standard for TAT, in an in-house sandwich enzyme-linked immunosorbent assay

(ELISA) for quantification as described earlier [24]. In summary, the capture antibodies were coated on Nunc Maxisorp ELISA plates (Thermo VG Scientific, Roskilde, Denmark) using phosphate buffer saline (PBS) and incubated overnight at 4 °C. Subsequently, the plates were blocked with 1% bovine serum albumin (BSA, Sigma Aldrich, Darmstadt, Germany) and incubated with samples for 60 min while shaking at room temperature. The plates were then washed with PBS-0.05% Tween (non-ionic surfactant), and biotinylated detection antibodies were added, followed by 60-min incubation. Subsequently, the plates were washed for 15 min with streptavidin–horseradish peroxidase (HRP). Tetramethylbenzidine (TMB, Surmodics, Eden Prairie, MN, USA) was then added until a bright signal was obtained, at which point the reaction was stopped by adding 1M H_2SO_4. Absorbance was measured via spectrometry at 450 nm.

2.6.4. Myeloperoxidase (MPO) and EPX Release

The release of heme-containing enzymes, myeloperoxidase (MPO), and eosinophil peroxidase (EPX), was measured using MPO (Invitrogen, ThermoFisher Scientific, Vienna, Austria) and EPX (Mybioscource, San Diego, CA, USA) according to the manufacturer's instructions.

2.6.5. Complement Markers

The complement activation markers, the C3a fragment and soluble complexes C5b-9, were quantified using in-house sandwich ELISA as outlined in Section 2.6.3. Monoclonal antibody (mAb) 4SD17.3 and biotinylated rabbit polyclonal anti-C3a antibody (pAb) Rb-a-Hu were used for capture and detection, respectively. For the sC5b-9 ELISA, mAb anti-neoC9 (Diatec Monoclonals AS, Oslo, Norway) was used for capture, while anti-C5 pAb (Biosite BP373, Täby, Sweden) followed by SA-HRP (GE Healthcare, RPN1231V, Uppsala, Sweden) served as detection. Standards were prepared from Zymosan-activated serum calibrated against commercially available kits (MicroVue, Quidel Corp., Santa Clara, CA, USA).

2.6.6. Kallikrein–Bradykinin (KK) Markers

Kallikrein in complex with the C1-inhibitor and anti-thrombin was quantified using an in-house sandwich ELISA. For the Kallikrein-C1 inhibitor (KK-C1Inh) and Kallikrein- anti thrombin (KK-AT) complex ELISA, sheep anti-human prekallikrein was used for capture and either biotinylated denatured anti-C1Inh antibody (alpha-antitrypsin purified) or HPR-conjugated anti-human antithrombin pAb was used for detection. Premade KK–C1inh complexes diluted in plasma were used as a standard for quantification.

2.7. Statistical Analysis

All statistical analyses were performed with Prism 9.4.1 (458) for Mac OS X, GraphPad Software Inc. (Boston, MA, USA). The results are expressed as mean + standard deviation (SD), unless stated otherwise. Outliers were identified using the nonlinear-regression-based method (ROUT) and removed. Subsequently, the statistical significance was calculated using a one-way analysis of variance (ANOVA), followed by Tukey's post hoc test. The significance levels were set at $p \leq 0.05$.

3. Results

3.1. Surface Topography Visualization

Scanning electron microscopy (SEM) unveiled clear differences in the surface topography among the studied materials (Figure 2A–C).

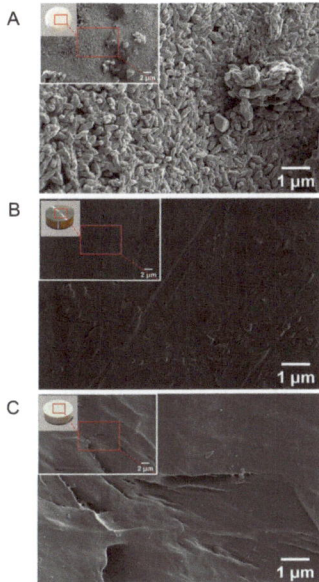

Figure 2. Scanning electron microscopy (SEM) images of the (**A**) calcium phosphate (CaP), (**B**) titanium alloy (Ti), and (**C**) polyetheretherketone (PEEK) disc surfaces examined in this study. Magnification of the top-left SEM images—10.00 kx for (**A**,**B**), and 5.00 kx for (**C**). Magnification of the central images—30.00 kx for (**A**–**C**).

The CaP surface displayed a nanoporous architecture with crystal formations smaller than one micrometer (Figure 2A). In comparison, PEEK (Figure 2B) and Ti (Figure 2C) both demonstrated relatively consistent surfaces, similar to those found in commercial implants with only minor superficial grooves, attributable to the cutting and polishing processes.

3.2. Wettability

The wettability of the tested materials was measured using two different methods: sessile water drop and captive air bubble (Figure 3A,B).

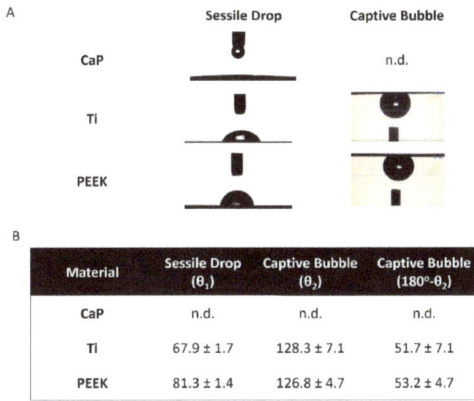

Material	Sessile Drop (θ_1)	Captive Bubble (θ_2)	Captive Bubble ($180°$-θ_2)
CaP	n.d.	n.d.	n.d.
Ti	67.9 ± 1.7	128.3 ± 7.1	51.7 ± 7.1
PEEK	81.3 ± 1.4	126.8 ± 4.7	53.2 ± 4.7

Figure 3. Assessment of wettability for the investigated materials. (**A**) Water contact angle measurements for calcium phosphate (CaP), titanium alloy (Ti), and polyetheretherketone (PEEK) obtained through sessile and captive methods. (**B**) Corresponding contact angle presented as the mean ± SD (n = 3). n.d.—not determined because of being experimentally inaccessible (see the text for further information).

CaP cement is a highly hydrophilic and porous material and thus, measurements of water contact angles of the cement with the sessile drop method were not possible, as anticipated. This is because the dispensed water drop was fully spread out on the CaP surface very quickly after contact (<1 s) and subsequently (in the next 1 s) absorbed in the bulk of the porous cement (Figure 3A).

Similarly, but unexpectedly, in this case, measurements of contact angle with the CaP cement using the captive air bubble method were not possible. This was because the delivered air bubble could not displace the water already in contact with the cement surface, which is strongly indicative of a highly hydrophilic surface (Figure 3A). The inability of obtaining quantitative values of water contact angles for the CaP material is noted as n.d. in the table shown in Figure 3B.

A surface is considered hydrophilic if the water contact angle is lower than 90° (sessile drop method). Under this definition, all tested materials are hydrophilic materials. CaP showed the highest hydrophilicity, as presented in the previous paragraph, followed by Ti and PEEK. The Ti surfaces were mildly hydrophilic ($67.9 \pm 1.7°$) and the PEEK surfaces had water contact angles close to the hydrophilic limit ($81.3 \pm 1.4°$). In the literature, the contact angle of the Ti alloy with different topographic finishes and untreated PEEK is between 30–70° and 70–90°, respectively [25,26].

3.3. Highest Coagulation to Ti

Similar to the conditions in an operating room where implant materials are exposed to the blood of a patient, this study used human whole blood. Following 60-min exposure to the surfaces at 37 °C in rotating chambers (as shown in Figure 1, in the Section 2), the discs of each material were inspected for coagulation reactions.

The macroscopic evaluation of the different material surfaces revealed strong activation and adherence of the blood cells and platelets. A dense blot clot was observed on the Ti surface, whereas PEEK and CaP displayed less dense blood clot formation (Figure 4A).

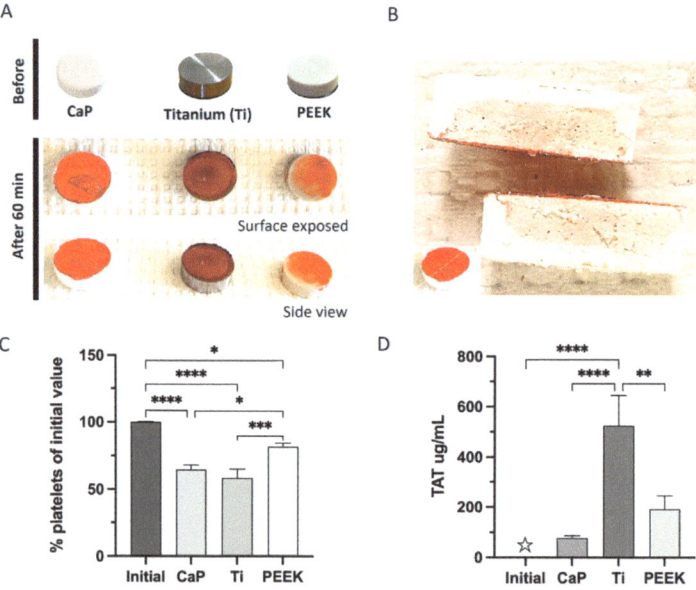

Figure 4. Coagulation activation after 60 min of contact with the calcium phosphate (CaP), titanium alloy (Ti), and polyetheretherketone (PEEK) material discs. (**A**) Representative macroscopic images of adherent blood cells and/or components with the surface of the materials. (**B**) Sectioning of CaP after

60 min of blood contact. (**C**) Percentage of platelets remaining in terms of initial values measured prior to contact at time 0 min. (**D**) The thrombin–antithrombin complex (TAT) compared to the initial amount (☆—the initial values of TAT are above zero, but due to the scale range applied, it is not visible). **** $p < 0.0001$, *** $p < 0.001$, ** $p < 0.01$, * $p < 0.05$; ANOVA with Tukey's multiple comparisons test. Data are the average value of two wells/donor represented as the mean ± SEM ($n = 13$). All comparisons not presented are non-significant.

The residual platelet percentage indicates the level of platelet activation and adherence to the materials. Blood samples were obtained prior to exposure, and the initial number of platelets was determined. All materials demonstrated a significant platelet reduction compared to the initial quantity (Figure 4C). The CaP and Ti surfaces displayed comparable values which were significantly lower than PEEK. Blood clot formation for each material was confirmed by the generation of the thrombin–antithrombin (TAT) complex. All material surfaces initiated TAT, albeit at varying levels (Figure 4D). Ti caused significant production of TAT complexes (525 ug/mL ± 120) compared to the initial values or the other materials. CaP and PEEK showed low generation of TAT complexes but this was not significant compared to the initial amount (TAT levels of 77 ug/mL ± 10 for CaP and 192 ug/mL ± 54 for PEEK).

3.4. Highest Activation of Kallikrein–Bradykinin with CaP

The contact activation system is also linked to the kallikrein–bradykinin (KK) system. Here, the KK–AT complex was significantly induced by CaP (56 ± 9 nM), Ti (47 ± 8 nM), and PEEK (51 ± 10 nM) compared with the initial levels (Figure 5A). All materials exhibited an elevation in KK-C1INH production relative to the initial values; however, only CaP demonstrated a significant increase (144 ± 29 nM) (Figure 5B). Interestingly, the CaP surface showed the most pronounced increase in both complexes.

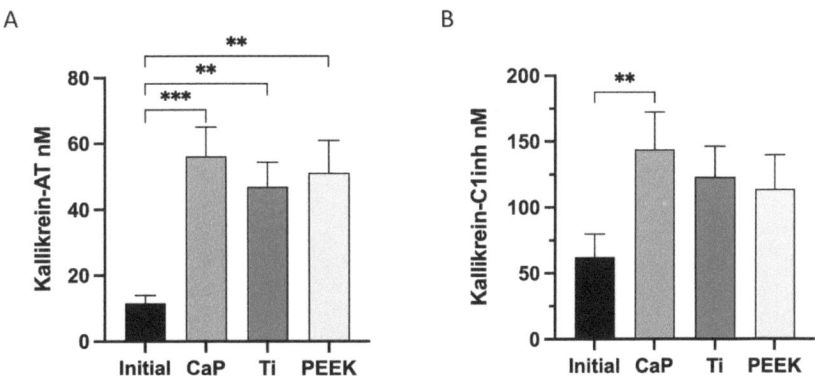

Figure 5. Activation of kallikrein–bradykinin (KK) system mediators after 60-min contact with the calcium phosphate (CaP), titanium alloy (Ti), and polyetheretherketone (PEEK) material discs. (**A**) The kallikrein (KK)–antithrombin (AT) complex. (**B**) The KK–C1–esterase inhibitor complex (C1- C1Inh). *** $p < 0.001$; ** $p < 0.01$. ANOVA with Tukey's multiple comparisons test. The data are the average value of two wells/donor represented as the mean ± SEM ($n = 12$). All comparisons not presented are non-significant.

3.5. Highest Activation of Complement C3a and C5b-9 Complexes in CaP

The activation of the complement system was also significantly increased in the other materials, as shown by the values of C3a (497 ug/L ± 69) and sC5b-9 (304 ug/L ± 54) for Ti and C3a (461 ug/L ± 74) and sC5b-9 (237 ug/L ± 49) for PEEK, as shown in Figure 6A,B, respectively.

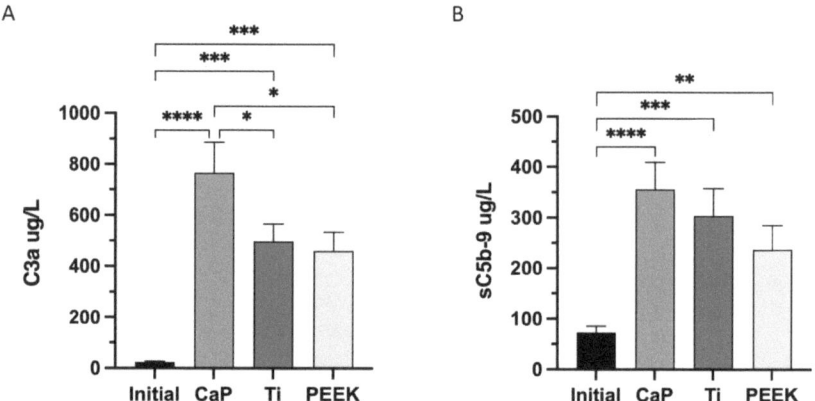

Figure 6. Complement system activation after 60 min of contact with the calcium phosphate (CaP), titanium alloy (Ti), and polyetheretherketone (PEEK) material discs. (**A**) Quantification of complement component 3 (C3a). (**B**) Plasma terminal C5b-9 complement complex. **** $p < 0.0001$, *** $p < 0.001$, ** $p < 0.01$, and * $p < 0.05$. ANOVA with Tukey's multiple comparisons test. Data are the average value of two wells/donor represented as the mean ± SEM ($n = 13$). All comparisons not presented are non-significant.

3.6. Highest Release of MPO with CaP

Leukocytes play an important role during blood coagulation and healing processes; hence, the activation of the most prevalent leukocytes and the release of their granule content were assessed. The focus was placed on two specific types of leukocytes: neutrophils and eosinophils, which are involved in the release of both human peroxidases, myeloperoxidase (MPO) and eosinophil peroxidase (EPX), respectively. Activation of both cell types and release of their granule content was confirmed by the activity of the human peroxidases MPO and EPX, as depicted in Figure 7.

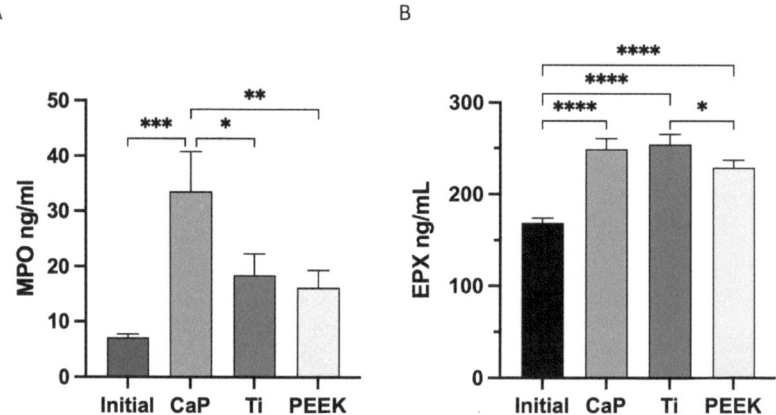

Figure 7. Human peroxidase activity after 60 min of contact with the calcium phosphate (CaP), titanium alloy (Ti), and polyetheretherketone (PEEK) material discs. (**A**) Quantification of myeloperoxidase (MPO) and (**B**) quantification of human eosinophil peroxidase (EPX). **** $p < 0.0001$, *** $p < 0.001$, ** $p < 0.01$, and * $p < 0.05$. ANOVA with Tukey's multiple comparisons test. Data represent the average value of two wells/donor represented as the mean ± SEM ($n = 10$). All comparisons not presented are non-significant.

CaP caused a significant release of MPO (34 ± 7 ng/mL), relative to the initial values, followed by a non-significant increase by Ti (18 ± 4 ng/mL) and PEEK (16 ± 3 ng/mL), as shown in Figure 7A.

All materials triggered the activation of eosinophils with a significant release of EPX compared to the initial values (Figure 7B); however, CaP and Ti exhibited comparable effects.

4. Discussion

In the present study, coagulation is most significantly activated by Ti and PEEK, whereas CaP appears to trigger blood cascades through complement activation. All materials activate leukocytes (EPX and MPO), with CaP inducing the highest neutrophil response (MPO). It has previously been assumed that (i) a strong trigger for coagulation and (ii) subsequent activation of platelets are essential to promote healing following the implantation of metals, alloys, and plastics [6]. These mechanisms were observed for Ti and PEEK; however, the blood response to the CaP material suggests that complement activation, rather than an initial strong coagulation and platelet activation, promotes tissue healing.

Ti exhibited the highest coagulation response of all materials studied (Ti > PEEK > CaP), as demonstrated by thrombin–antithrombin (TAT). CaP had notably low levels of the coagulation marker, TAT, despite an equal reduction in free circulating platelets in the blood as Ti at 60 min. Platelets were likely entrapped in the material's porous structure, which is supported by the blood stain observed in the sectioned CaP material. This aspect is particularly important in the context of healing and repair processes, as CaP could function as a reservoir of growth factors, emphasizing the importance of the blood clot [13]. When confined within the pores, platelets could gradually release their content over time, which encompasses a variety of growth factors such as TGF−β, PDGF, VEGF, and IGF [27]. These growth factors have been shown to influence both angiogenesis and the differentiation of osteoprogenitor cells, promoting healing [28]. Upon activation, platelets have also been shown to regulate the production of monocyte cytokines that counteract inflammatory responses [29].

Beyond porosity, surface wettability influences platelet adherence and, consequently, blood clot formation [6,30,31]. Wettable surfaces (i.e., hydrophilic) are generally more favorable to interactions with blood [32]. This largely accounts for the increased clot formation on Ti and the smaller clot found on PEEK along with the higher and lower release of TAT, respectively, as anticipated given the surface wettability of the two materials. Prior findings by Hong et al. support our data demonstrating elevated levels of TAT associated with Ti surfaces [24,33]. Increasingly hydrophilic Ti surfaces can upregulate the deposition of the fibrin matrix, leading to more substantial blood clots [24]. In the context of bone healing, the hydrophilicity of Ti surfaces has been shown to promote proliferation and osteoblast precursor differentiation, as well as positively regulate angiogenesis, bone mineralization, and bone remodeling [34]. Interestingly, we found less clotting on the CaP material compared to Ti despite its high hydrophilicity. This outcome suggests that factors beyond wettability such as topography, charge, and intrinsic chemical composition influence blood coagulation, especially for CaP materials.

Another blood cascade tightly interlinked with the coagulation systems is the kallikrein–kinin system or the so-called contact activation system [35]. Kallikrein production is regulated by two complexes, kallikrein (KK)–AT and –C1INH, which are ubiquitously present in plasma. Kallikrein has a dual role: first, it interacts with coagulation elements that activate it, and second, it generates bradykinin in the kallikrein–kinin system. Bradykinin promotes tissue regeneration and stimulates the migration of various cell types, including neutrophils, fibroblasts, and endothelial cells among others [36–39]. In our study, CaP exhibited the highest production of both complexes, followed by PEEK and Ti with no significant upregulation observed for Ti or PEEK regarding KK–C1INH complexes.

The complement activation was markedly more pronounced with CaP, followed by Ti and PEEK as indicated by a significant increase in the production of both C3a and C5a (evidenced by sC5b-9) complement proteins. These findings align with those reported by

Klein et al. [40], who demonstrated cleavage of C3 with calcium phosphate powders. Given the reciprocal relationship between contact and complement system activation as shown by Huang et al. [3], the significant complement system activation could explain the low values of the coagulation marker TAT seen for CaP, where fewer protein triggers occur in the clotting cascade. Furthermore, complement system activation has previously been linked with enhanced healing following trauma. The proteolytic cleavage of complement proteins C3 and C5 induces a diverse range of cellular responses, such as chemotaxis, cell activation, and cell adhesion. In preclinical models, C5aR activation by C5a induced strong chemotactic activity in osteoblasts [41]. Bone cells express complement components, including C3, C5, and related receptors, and are responsive to C3a and C5a. During bone healing, these molecules can promote osteoclast formation in a pro-inflammatory environment. Additionally, C5a promotes strong chemotactic activity in osteoblasts [2,42–45].

For instance, Ehrnthaller et al. [41] showed that mice deficient in C3 and C5 exhibited significantly reduced bone formation during the early healing phases, highlighting the complement systems' importance for successful healing. The calcium phosphate (CaP) used in this investigation has demonstrated the ability to undergo resorption and replacement by bone during the healing process in both preclinical and clinical settings. A considerable presence of osteoblasts has been observed at early and late stages in preclinical models. Interestingly, minimal osteoclast activity was observed [19,20,46–48]; instead, a moderate to substantial presence of material-filled macrophages was observed, which are also responsive to the complement proteins C3a, C5a, and bradykinin [49]. Contrarily, titanium meshes [19], solid titanium implants, and PEEK have been found to trigger a foreign body reaction, leading to chronic inflammation that promotes fibrous encapsulation rather than bone formation [19,25,50].

Furthermore, the complement system is known for its ability to induce immune cell activation. Leukocytes, such as neutrophils and eosinophils, express receptors for complements of C3a, C5a, and other complement molecules [49]. Notably, neutrophils have more C5a receptors, while eosinophils possess a higher density of C3a receptors [51].

In response to various stimuli, leukocytes can release a variety of granule proteins, which includes hemeprotein myeloperoxidase (MPO) from neutrophils and human eosinophile peroxidase (EPX) from eosinophils [52,53]. In this study, the CaP induced a significant release of both MPO and EPX relative to the baseline values. Concurrently, both Ti and PEEK induced an increase in MPO and EPX levels, although only the elevation of EPX was statistically significant. In a study by Burkhardt et al. [17] neutrophils by releasing MPO and reactive oxygen species (ROS) were found to facilitate the initial provisional fibrin matrix deposition, as well as the secretion of growth factors and cytokines by the first wave of invading cells, such as blood cells.

Both MPO and EPX peroxidases can stimulate fibroblasts and osteoblasts to migrate, secrete collagen for normal tissue repair, and exhibit angiogenetic properties such as the VEGF factor [54–56]. Additionally, MPO and EPX inhibit osteoclast differentiation and consequently bone resorption [57,58].

Furthermore, MPO and EPX exhibit notable antibacterial properties. Myeloperoxidase (MPO), the primary component of neutrophil granules, plays a significant role in creating and maintaining an alkaline environment. This environment is essential for restricting bacterial proliferation and effectively fighting infections. Neutrophils account for 60 to 70% of immune cells in the blood, emphasizing their importance in this context [57,59].

Both neutrophils and eosinophils possess bactericidal activity through the production of reactive oxygen species (ROS). The ROS generated by EPX can attack and damage various bacterial components, including proteins, lipids, and nucleic acids, ultimately leading to bacterial cell death. By damaging and killing bacteria, EPX plays a vital role in controlling and preventing infections, supporting the immune system's ability to clear invading pathogens [60].

The results of our investigation, emphasizing the antimicrobial characteristics of these molecules, may bear clinical relevance. These findings could provide insights regarding the relatively lower infection rates associated with the use of CaP implants in craniomaxillofacial reconstructive procedures [61,62].

Macrophages play a crucial role in the immune system, tissue repair, and regeneration. Originating from circulating monocytes, macrophages participate in multiple stages of the tissue healing process, ranging from the initial inflammatory response to tissue remodeling. It is essential, in the context of tissue regeneration, for macrophages to transition from the pro-inflammatory M1 phenotype to the anti-inflammatory M2 phenotype. The M1 macrophages are instrumental in the initial inflammatory response to injury or infection, assisting in eliminating pathogens and damaged tissue. However, a prolonged M1 response may result in chronic inflammation and subsequent tissue damage. The polarization of macrophages into an anti-inflammatory M2 phenotype helps ensure that the initial inflammatory response is followed by a resolution of inflammation and the facilitation of tissue repair and regeneration. Dysregulation of macrophage polarization can lead to impaired healing, chronic inflammation, or excessive scarring [63–65].

In addition, monocytes and macrophages, which are essential during inflammation, also respond to the complement proteins C3a, C5a, and bradykinin [51,66]. Considering the significant upregulation of C3a especially, as well as C5a in particular, eosinophils are likely to accumulate at the material surfaces, especially for CaP, which showed the most significant activation [66]. Eosinophils have been found to induce macrophage polarization from M1 into M2 [57,59], thereby playing a vital role in modulating the pro-inflammatory milieu into a regenerative environment. PEEK implants generally elicit chronic inflammation (M1) that causes fibrous encapsulation rather than bone formation and integration [63,64]. Conversely, Ti seems to be able to stimulate the body's response in various ways depending on the anatomical location and distance to native bone. In a non-osseous environment or far from the bone defect (e.g., critical size defects), it generally causes fibrous encapsulation with no or minimal bone formation, whereas places in a bony cavity, such as in dental applications, are fully osseointegrated over time [64]. The CaP composition studied here has been shown to have a higher healing capacity in craniomaxillofacial reconstruction where it has been shown to be resorbed and replaced by bone during the healing process both in preclinical and clinical settings without signs of chronic inflammation [19,20,46,48,67–70]. This response seems to be clearly coupled with the material–macrophage interaction over time [19,20,46,47]. Taken together we, therefore, propose that both the complement and contact system activating properties of this CaP with blood at implantation initiate both the activation and recruitment of leukocytes and osteoblasts and inhibit osteoclasts at the site of implantation.

Considering that implantable materials are temporarily exposed to blood, it is highly probable that the complement activation is transient and subsides during the first days of implantation, potentially resulting in reduced neutrophil recruitment. Schmidt-Bleek et al. [71], who studied immune cell subpopulations of a bone hematoma, found that the composition of neutrophils exceeded that of peripheral blood 60 min post-surgery. Nonetheless, the neutrophil numbers returned to levels comparable to those in the circulating blood after 4 h. While neutrophils play an important role in the initial response to implantable materials, it is equally important to regulate this response over time as prolonged recruitment of active neutrophils may extend the initial beneficial inflammatory response and ultimately impair tissue regeneration [58,72].

In summary (Figure 8), this study revealed that upon short contact with circulating human whole blood (i) CaP primarily activated the complement system and leukocytes instead of coagulation cascades, (ii) Ti induced coagulation and fewer other blood cascades and leukocytes, and (iii) PEEK exerted an intermediate influence on all studied systems. These findings suggest an inverse activation relationship between the coagulation system and the complement system during the early stages with CaP and Ti materials.

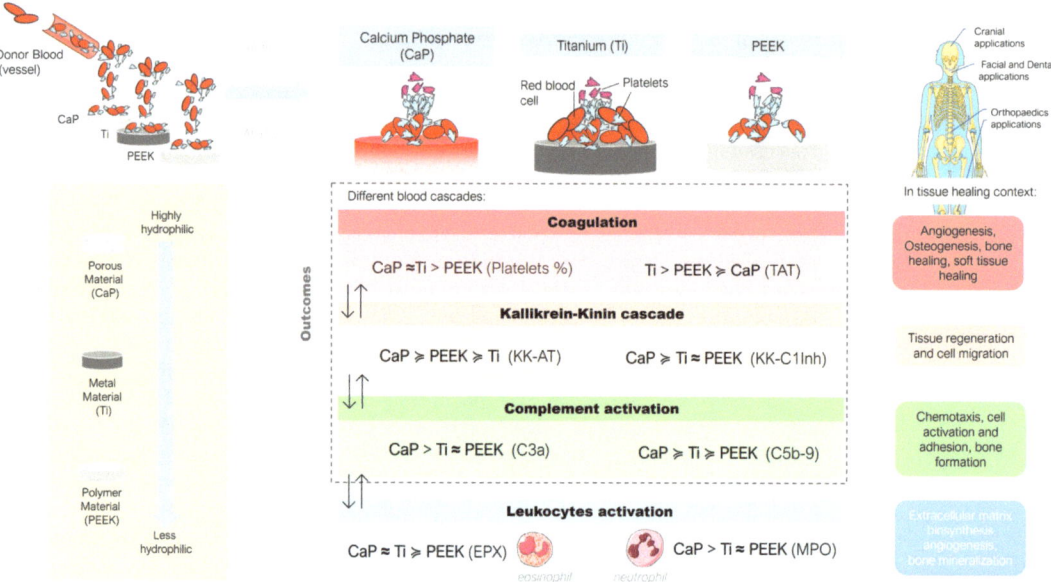

Figure 8. Schematic overview of local interactions between the different components of blood cascades and leukocyte responses during the early stages of contact with calcium phosphate (CaP), titanium alloy (Ti), and polyetheretherketone (PEEK) materials. Further details can be found through the results and discussion. (The symbol almost equal to ($\approx$) by itself or associated with the symbol greater ($\geq$) indicates that the values were significant between the data groups).

The limitations of the study are the lack of inclusion of the tissue factor (extrinsic pathway) and that only one early time point was studied. Furthermore, this study only includes two types of white cells, neutrophils and eosinophils, as a result of the complement activation outcomes. Nevertheless, the primary objective of the study is to investigate the early interactions that arise from the contact between biomaterials and whole blood. Future studies should aim to explore how the inflammatory response to implantable materials is modulated over time, concentrating on the recruitment of various specific cells and their differentiation at various time points.

5. Conclusions

In the current study, coagulation was notably activated by Ti and PEEK, while with CaP, the complement is the main blood cascade triggered. Contrary to previous concepts emphasizing strong coagulation and platelet activation for healing following the implantation of metals, alloys, and plastics, the blood response of CaP material suggests that complement activation may be crucial for tissue healing as an initial event. These findings offer insights into blood–material interactions and could guide the further development and optimization of existing implantable materials used in today's clinical practice. By modulating the initial immune response, the healing process could be optimized, possibly resulting in fewer complications and better osteointegration of implant materials.

Author Contributions: V.R.L., conceptualization, data curation, project administration, formal analysis, visualization, writing—original draft, and writing—review and editing. U.B., conceptualization, data curation, project administration, formal analysis, visualization, writing—original draft, and writing—review and editing. V.A.M., data curation, formal analysis, visualization, and writing—review and editing. G.H.-B., conceptualization and writing—review and editing. S.G., conceptualization and writing—review and editing. C.A., data curation, formal analysis, and writing—review and editing. J.H., conceptualization, data curation, formal analysis, project administration, supervision, visualization, writing—original draft, and writing—review and editing. All authors have read and agreed to the published version of the manuscript.

Funding: This research received no external funding.

Data Availability Statement: Due to confidentiality agreements, data can be made available subject to a non-disclosure agreement and from the corresponding author upon request.

Acknowledgments: The authors wish to acknowledge David Eikram for his help in conducting part of the ELISA tests.

Conflicts of Interest: Vivek Anand Manivel, Conrado Aparicio, and Jaan Hong declare no competing interests. Ulrik Birgersson and Gry Hulsart Billström report personal fees from OssDsign during the conduct of the study. Viviana R Lopes is employed by OssDsign. Sara Gallinetti works as a consultant for OssDsign.

References

1. Weber, M.; Steinle, H.; Golombek, S.; Hann, L.; Schlensak, C.; Wendel, H.P.; Avci-Adali, M. Blood-Contacting Biomaterials: In Vitro Evaluation of the Hemocompatibility. *Front. Bioeng. Biotechnol.* **2018**, *6*, 99. [CrossRef] [PubMed]
2. Mödinger, Y.; Teixeira, G.; Neidlinger-Wilke, C.; Ignatius, A. Role of the Complement System in the Response to Orthopedic Biomaterials. *Int. J. Mol. Sci.* **2018**, *19*, 3367. [CrossRef] [PubMed]
3. Huang, S.; Engberg, A.E.; Jonsson, N.; Sandholm, K.; Nicholls, I.A.; Mollnes, T.E.; Fromell, K.; Nilsson, B.; Ekdahl, K.N. Reciprocal Relationship between Contact and Complement System Activation on Artificial Polymers Exposed to Whole Human Blood. *Biomaterials* **2016**, *77*, 111–119. [CrossRef] [PubMed]
4. Dzik, S. Complement and Coagulation: Cross Talk Through Time. *Transfus. Med. Rev.* **2019**, *33*, 199–206. [CrossRef]
5. Markiewski, M.M.; Nilsson, B.; Ekdahl, K.N.; Mollnes, T.E.; Lambris, J.D. Complement and Coagulation: Strangers or Partners in Crime? *Trends Immunol.* **2007**, *28*, 184–192. [CrossRef] [PubMed]
6. Xu, L.-C.; Bauer, J.W.; Siedlecki, C.A. Proteins, Platelets, and Blood Coagulation at Biomaterial Interfaces. *Colloids Surf. B Biointerfaces* **2014**, *124*, 49–68. [CrossRef]
7. Gorbet, M.B.; Sefton, M.V. Biomaterial-Associated Thrombosis: Roles of Coagulation Factors, Complement, Platelets and Leukocytes. *Biomaterials* **2004**, *25*, 5681–5703. [CrossRef]
8. Gorbet, M.; Sperling, C.; Maitz, M.F.; Siedlecki, C.A.; Werner, C.; Sefton, M.V. The Blood Compatibility Challenge. Part 3: Material Associated Activation of Blood Cascades and Cells. *Acta Biomater.* **2019**, *94*, 25–32. [CrossRef]
9. Kolev, M.; Le Friec, G.; Kemper, C. Complement—Tapping into New Sites and Effector Systems. *Nat. Rev. Immunol.* **2014**, *14*, 811–820. [CrossRef]
10. Eriksson, O.; Mohlin, C.; Nilsson, B.; Ekdahl, K.N. The Human Platelet as an Innate Immune Cell: Interactions Between Activated Platelets and the Complement System. *Front. Immunol.* **2019**, *10*, 1590. [CrossRef]
11. Ekdahl, K.N.; Huang, S.; Nilsson, B.; Teramura, Y. Complement Inhibition in Biomaterial- and Biosurface-Induced Thromboinflammation. *Semin. Immunol.* **2016**, *28*, 268–277. [CrossRef] [PubMed]
12. Koh, T.J.; DiPietro, L.A. Inflammation and Wound Healing: The Role of the Macrophage. *Expert. Rev. Mol. Med.* **2011**, *13*, e23. [CrossRef] [PubMed]
13. Balaguer, T.; Boukhechba, F.; Clavé, A.; Bouvet-Gerbettaz, S.; Trojani, C.; Michiels, J.-F.; Laugier, J.-P.; Bouler, J.-M.; Carle, G.F.; Scimeca, J.-C.; et al. Biphasic Calcium Phosphate Microparticles for Bone Formation: Benefits of Combination with Blood Clot. *Tissue Eng. Part. A* **2010**, *16*, 3495–3505. [CrossRef]
14. Bouler, J.M.; Pilet, P.; Gauthier, O.; Verron, E. Biphasic Calcium Phosphate Ceramics for Bone Reconstruction: A Review of Biological Response. *Acta Biomater.* **2017**, *53*, 1–12. [CrossRef] [PubMed]
15. Liu, Y.; Lin, D.; Li, B.; Hong, H.; Jiang, C.; Yuan, Y.; Wang, J.; Hu, R.; Li, B.; Liu, C. BMP-2/CPC Scaffold with Dexamethasone-Loaded Blood Clot Embedment Accelerates Clinical Bone Regeneration. *Am. J. Transl. Res.* **2022**, *14*, 2874–2893. [PubMed]
16. Zuardi, L.R.; Silva, C.L.A.; Rego, E.M.; Carneiro, G.V.; Spriano, S.; Nanci, A.; de Oliveira, P.T. Influence of a Physiologically Formed Blood Clot on Pre-Osteoblastic Cells Grown on a BMP-7-Coated Nanoporous Titanium Surface. *Biomimetics* **2023**, *8*, 123. [CrossRef]
17. Burkhardt, M.A.; Waser, J.; Milleret, V.; Gerber, I.; Emmert, M.Y.; Foolen, J.; Hoerstrup, S.P.; Schlottig, F.; Vogel, V. Synergistic Interactions of Blood-Borne Immune Cells, Fibroblasts and Extracellular Matrix Drive Repair in an in Vitro Peri-Implant Wound Healing Model. *Sci. Rep.* **2016**, *6*, 21071. [CrossRef]
18. Thor, A. Porous Titanium Granules and Blood for Bone Regeneration around Dental Implants: Report of Four Cases and Review of the Literature. *Case Rep. Dent.* **2013**, *2013*, 410515. [CrossRef]
19. Gallinetti, S.; Linder, L.K.B.; Åberg, J.; Illies, C.; Engqvist, H.; Birgersson, U. Titanium Reinforced Calcium Phosphate Improves Bone Formation and Osseointegration in Ovine Calvaria Defects: A Comparative 52 Weeks Study. *Biomed. Mater.* **2021**, *16*, 035031. [CrossRef]

20. Billström, G.H.; Lopes, V.R.; Illies, C.; Gallinetti, S.; Åberg, J.; Engqvist, H.; Aparicio, C.; Larsson, S.; Linder, L.K.B.; Birgersson, U. Guiding Bone Formation Using Semi-onlay Calcium Phosphate Implants in an Ovine Calvarial Model. *J. Tissue Eng. Regen. Med.* **2022**, *16*, 435–447. [CrossRef]
21. Ye, Z.; Kobe, A.C.; Sang, T.; Aparicio, C. Unraveling Dominant Surface Physicochemistry to Build Antimicrobial Peptide Coatings with Supramolecular Amphiphiles. *Nanoscale* **2020**, *12*, 20767–20775. [CrossRef] [PubMed]
22. Hong, J.; Nilsson Ekdahl, K.; Reynolds, H.; Larsson, R.; Nilsson, B. A New in Vitro Model to Study Interaction between Whole Blood and Biomaterials. Studies of Platelet and Coagulation Activation and the Effect of Aspirin. *Biomaterials* **1999**, *20*, 603–611. [CrossRef] [PubMed]
23. Andersson, J.; Sanchez, J.; Ekdahl, K.N.; Elgue, G.; Nilsson, B.; Larsson, R. Optimal Heparin Surface Concentration and Antithrombin Binding Capacity as Evaluated with Human Non-Anticoagulated Bloodin Vitro. *J. Biomed. Mater. Res.* **2003**, *67*, 458–466. [CrossRef]
24. Hong, J.; Kurt, S.; Thor, A. A Hydrophilic Dental Implant Surface Exhibit Thrombogenic Properties In Vitro. *Clin. Implant. Dent. Relat. Res.* **2013**, *15*, 105–112. [CrossRef]
25. Zhang, J.; Tian, W.; Chen, J.; Yu, J.; Zhang, J.; Chen, J. The Application of Polyetheretherketone (PEEK) Implants in Cranioplasty. *Brain Res. Bull.* **2019**, *153*, 143–149. [CrossRef]
26. Yan, Y.; Chibowski, E.; Szcześ, A. Surface Properties of Ti-6Al-4V Alloy Part I: Surface Roughness and Apparent Surface Free Energy. *Mater. Sci. Eng. C* **2017**, *70*, 207–215. [CrossRef]
27. Rendu, F.; Brohard-Bohn, B. The Platelet Release Reaction: Granules' Constituents, Secretion and Functions. *Platelets* **2001**, *12*, 261–273. [CrossRef]
28. Anitua, E.; Andia, I.; Ardanza, B.; Nurden, P.; Nurden, A. Autologous Platelets as a Source of Proteins for Healing and Tissue Regeneration. *Thromb. Haemost.* **2004**, *91*, 4–15. [CrossRef]
29. Gudbrandsdottir, S.; Hasselbalch, H.C.; Nielsen, C.H. Activated Platelets Enhance IL-10 Secretion and Reduce TNF-α Secretion by Monocytes. *J. Immunol.* **2013**, *191*, 4059–4067. [CrossRef]
30. Fernandes, K.R.; Zhang, Y.; Magri, A.M.P.; Renno, A.C.M.; van den Beucken, J.J.J.P. Biomaterial Property Effects on Platelets and Macrophages: An in Vitro Study. *ACS Biomater. Sci. Eng.* **2017**, *3*, 3318–3327. [CrossRef]
31. Hong, J.; Andersson, J.; Ekdahl, K.N.; Elgue, G.; Axén, N.; Larsson, R.; Nilsson, B. Titanium Is a Highly Thrombogenic Biomaterial: Possible Implications for Osteogenesis. *Thromb. Haemost.* **1999**, *82*, 58–64. [CrossRef]
32. Rodrigues, S.N.; Gonçalves, I.C.; Martins, M.C.L.; Barbosa, M.A.; Ratner, B.D. Fibrinogen Adsorption, Platelet Adhesion and Activation on Mixed Hydroxyl-/Methyl-Terminated Self-Assembled Monolayers. *Biomaterials* **2006**, *27*, 5357–5367. [CrossRef]
33. Hulsart-Billström, G.; Janson, O.; Engqvist, H.; Welch, K.; Hong, J. Thromboinflammation as Bioactivity Assessment of H2O2-Alkali Modified Titanium Surfaces. *J. Mater. Sci. Mater. Med.* **2019**, *30*, 66. [CrossRef] [PubMed]
34. Calciolari, E.; Hamlet, S.; Ivanovski, S.; Donos, N. Pro-Osteogenic Properties of Hydrophilic and Hydrophobic Titanium Surfaces: Crosstalk between Signalling Pathways in in Vivo Models. *J. Periodontal Res.* **2018**, *53*, 598–609. [CrossRef] [PubMed]
35. Schmaier, A.H. The Contact Activation and Kallikrein/Kinin Systems: Pathophysiologic and Physiologic Activities. *J. Thromb. Haemost.* **2016**, *14*, 28–39. [CrossRef] [PubMed]
36. Da Soley, B.S.; Morais, R.L.T.d.; Pesquero, J.B.; Bader, M.; Otuki, M.F.; Cabrini, D.A. Kinin Receptors in Skin Wound Healing. *J. Dermatol. Sci.* **2016**, *82*, 95–105. [CrossRef]
37. Ehrenfeld, P.; Millan, C.; Matus, C.E.; Figueroa, J.E.; Burgos, R.A.; Nualart, F.; Bhoola, K.D.; Figueroa, C.D. Activation of Kinin B1 Receptors Induces Chemotaxis of Human Neutrophils. *J. Leukoc. Biol.* **2006**, *80*, 117–124. [CrossRef]
38. Sheng, Z.; Yao, Y.; Li, Y.; Yan, F.; Huang, J.; Ma, G. Bradykinin Preconditioning Improves Therapeutic Potential of Human Endothelial Progenitor Cells in Infarcted Myocardium. *PLoS ONE* **2013**, *8*, e81505. [CrossRef]
39. Bekassy, Z.; Lopatko Fagerström, I.; Bader, M.; Karpman, D. Crosstalk between the Renin–Angiotensin, Complement and Kallikrein–Kinin Systems in Inflammation. *Nat. Rev. Immunol.* **2022**, *22*, 411–428. [CrossRef]
40. Klein, C.P.; de Groot, K.; van Kamp, G. Activation of Complement C3 by Different Calcium Phosphate Powders. *Biomaterials* **1983**, *4*, 181–184. [CrossRef]
41. Ehrnthaller, C.; Huber-Lang, M.; Nilsson, P.; Bindl, R.; Redeker, S.; Recknagel, S.; Rapp, A.; Mollnes, T.; Amling, M.; Gebhard, F.; et al. Complement C3 and C5 Deficiency Affects Fracture Healing. *PLoS ONE* **2013**, *8*, e81341. [CrossRef]
42. Ignatius, A.; Schoengraf, P.; Kreja, L.; Liedert, A.; Recknagel, S.; Kandert, S.; Brenner, R.E.; Schneider, M.; Lambris, J.D.; Huber-Lang, M. Complement C3a and C5a Modulate Osteoclast Formation and Inflammatory Response of Osteoblasts in Synergism with IL-1β. *J. Cell Biochem.* **2011**, *112*, 2594–2605. [CrossRef] [PubMed]
43. Ignatius, A.; Ehrnthaller, C.; Brenner, R.E.; Kreja, L.; Schoengraf, P.; Lisson, P.; Blakytny, R.; Recknagel, S.; Claes, L.; Gebhard, F.; et al. The Anaphylatoxin Receptor C5aR Is Present During Fracture Healing in Rats and Mediates Osteoblast Migration In Vitro. *J. Trauma Inj. Infect. Crit. Care* **2011**, *71*, 952–960. [CrossRef] [PubMed]
44. Matsuoka, K.; Park, K.; Ito, M.; Ikeda, K.; Takeshita, S. Osteoclast-Derived Complement Component 3a Stimulates Osteoblast Differentiation. *J. Bone Miner. Res.* **2014**, *29*, 1522–1530. [CrossRef]

45. Huber-Lang, M.; Kovtun, A.; Ignatius, A. The Role of Complement in Trauma and Fracture Healing. *Semin. Immunol.* **2013**, *25*, 73–78. [CrossRef]
46. Omar, O.; Engstrand, T.; Kihlström Burenstam Linder, L.; Åberg, J.; Shah, F.A.; Palmquist, A.; Birgersson, U.; Elgali, I.; Pujari-Palmer, M.; Engqvist, H.; et al. In Situ Bone Regeneration of Large Cranial Defects Using Synthetic Ceramic Implants with a Tailored Composition and Design. *Proc. Natl. Acad. Sci. USA* **2020**, *117*, 26660–26671. [CrossRef] [PubMed]
47. Malmberg, P.; Lopes, V.R.; Billström, G.H.; Gallinetti, S.; Illies, C.; Linder, L.K.B.; Birgersson, U. Targeted ToF-SIMS Analysis of Macrophage Content from a Human Cranial Triphasic Calcium Phosphate Implant. *ACS Appl. Bio Mater.* **2021**, *4*, 6791–6798. [CrossRef]
48. Guillet, C.; Birgersson, U.; Engstrand, T.; Åberg, J.; Lopes, V.R.; Thor, A.; Engqvist, H.; Forterre, F. Bone Formation beyond the Skeletal Envelope Using Calcium Phosphate Granules Packed into a Collagen Pouch-a Pilot Study. *Biomed. Mater.* **2023**, *18*, 035007. [CrossRef]
49. Vandendriessche, S.; Cambier, S.; Proost, P.; Marques, P.E. Complement Receptors and Their Role in Leukocyte Recruitment and Phagocytosis. *Front. Cell Dev. Biol.* **2021**, *9*, 624025. [CrossRef]
50. Thien, A.; King, N.K.K.; Ang, B.T.; Wang, E.; Ng, I. Comparison of Polyetheretherketone and Titanium Cranioplasty after Decompressive Craniectomy. *World Neurosurg.* **2015**, *83*, 176–180. [CrossRef]
51. Zwirner, J.; Götze, O.; Begemann, G.; Kapp, A.; Kirchhoff, K.; Werfel, T. Evaluation of C3a Receptor Expression on Human Leucocytes by the Use of Novel Monoclonal Antibodies. *Immunology* **1999**, *97*, 166–172. [CrossRef]
52. Jhunjhunwala, S. Neutrophils at the Biological–Material Interface. *ACS Biomater. Sci. Eng.* **2018**, *4*, 1128–1136. [CrossRef]
53. Arnhold, J. The Dual Role of Myeloperoxidase in Immune Response. *Int. J. Mol. Sci.* **2020**, *21*, 8057. [CrossRef] [PubMed]
54. DeNichilo, M.O.; Shoubridge, A.J.; Panagopoulos, V.; Liapis, V.; Zysk, A.; Zinonos, I.; Hay, S.; Atkins, G.J.; Findlay, D.M.; Evdokiou, A. Peroxidase Enzymes Regulate Collagen Biosynthesis and Matrix Mineralization by Cultured Human Osteoblasts. *Calcif. Tissue Int.* **2016**, *98*, 294–305. [CrossRef] [PubMed]
55. DeNichilo, M.O.; Panagopoulos, V.; Rayner, T.E.; Borowicz, R.A.; Greenwood, J.E.; Evdokiou, A. Peroxidase Enzymes Regulate Collagen Extracellular Matrix Biosynthesis. *Am. J. Pathol.* **2015**, *185*, 1372–1384. [CrossRef]
56. Panagopoulos, V.; Zinonos, I.; Leach, D.A.; Hay, S.J.; Liapis, V.; Zysk, A.; Ingman, W.v.; DeNichilo, M.O.; Evdokiou, A. Uncovering a New Role for Peroxidase Enzymes as Drivers of Angiogenesis. *Int. J. Biochem. Cell Biol.* **2015**, *68*, 128–138. [CrossRef]
57. Panagopoulos, V.; Liapis, V.; Zinonos, I.; Hay, S.; Leach, D.A.; Ingman, W.; DeNichilo, M.O.; Atkins, G.J.; Findlay, D.M.; Zannettino, A.C.W.; et al. Peroxidase Enzymes Inhibit Osteoclast Differentiation and Bone Resorption. *Mol. Cell Endocrinol.* **2017**, *440*, 8–15. [CrossRef] [PubMed]
58. Zhao, X.; Lin, S.; Li, H.; Si, S.; Wang, Z. Myeloperoxidase Controls Bone Turnover by Suppressing Osteoclast Differentiation Through Modulating Reactive Oxygen Species Level. *J. Bone Min. Res.* **2021**, *36*, 591–603. [CrossRef]
59. Liu, L.; Zhang, Y.; Zheng, X.; Jin, L.; Xiang, N.; Zhang, M.; Chen, Z. Eosinophils Attenuate Arthritis by Inducing M2 Macrophage Polarization via Inhibiting the IκB/P38 MAPK Signaling Pathway. *Biochem. Biophys. Res. Commun.* **2019**, *508*, 894–901. [CrossRef]
60. Borelli, V.; Vita, F.; Shankar, S.; Soranzo, M.R.; Banfi, E.; Scialino, G.; Brochetta, C.; Zabucchi, G. Human Eosinophil Peroxidase Induces Surface Alteration, Killing, and Lysis of *Mycobacterium tuberculosis*. *Infect. Immun.* **2003**, *71*, 605–613. [CrossRef]
61. Kwarcinski, J.; Boughton, P.; Ruys, A.; Doolan, A.; van Gelder, J. Cranioplasty and Craniofacial Reconstruction: A Review of Implant Material, Manufacturing Method and Infection Risk. *Appl. Sci.* **2017**, *7*, 276. [CrossRef]
62. Henry, J.; Amoo, M.; Taylor, J.; O'Brien, D.P. Complications of Cranioplasty in Relation to Material: Systematic Review, Network Meta-Analysis and Meta-Regression. *Neurosurgery* **2021**, *89*, 383–394. [CrossRef]
63. Phan, K.; Hogan, J.A.; Assem, Y.; Mobbs, R.J. PEEK-Halo Effect in Interbody Fusion. *J. Clin. Neurosci.* **2016**, *24*, 138–140. [CrossRef]
64. Olivares-Navarrete, R.; Hyzy, S.L.; Slosar, P.J.; Schneider, J.M.; Schwartz, Z.; Boyan, B.D. Implant Materials Generate Different Peri-Implant Inflammatory Factors. *Spine* **2015**, *40*, 399–404. [CrossRef] [PubMed]
65. Hu, Y.; Huang, Y.; Chen, C.; Wang, Y.; Hao, Z.; Chen, T.; Wang, J.; Li, J. Strategies of Macrophages to Maintain Bone Homeostasis and Promote Bone Repair: A Narrative Review. *J. Funct. Biomater.* **2022**, *14*, 18. [CrossRef]
66. Barbasz, A.; Kozik, A. The Assembly and Activation of Kinin-Forming Systems on the Surface of Human U-937 Macrophage-like Cells. *Biol. Chem.* **2009**, *390*, 269–275. [CrossRef] [PubMed]
67. Kuemmerle, J.M.; Oberle, A.; Oechslin, C.; Bohner, M.; Frei, C.; Boecken, I.; von Rechenberg, B. Assessment of the Suitability of a New Brushite Calcium Phosphate Cement for Cranioplasty—An Experimental Study in Sheep. *J. Cranio-Maxillofac. Surg.* **2005**, *33*, 37–44. [CrossRef]
68. Kihlström Burenstam Linder, L.; Birgersson, U.; Lundgren, K.; Illies, C.; Engstrand, T. Patient-Specific Titanium-Reinforced Calcium Phosphate Implant for the Repair and Healing of Complex Cranial Defects. *World Neurosurg.* **2019**, *122*, e399–e407. [CrossRef] [PubMed]
69. Engstrand, T.; Kihlström, L.; Lundgren, K.; Trobos, M.; Engqvist, H.; Thomsen, P. Bioceramic Implant Induces Bone Healing of Cranial Defects. *Plast. Reconstr. Surg. Glob. Open* **2015**, *3*, e491. [CrossRef]
70. Sundblom, J.; Xheka, F.; Casar-Borota, O.; Ryttlefors, M. Bone Formation in Custom-Made Cranioplasty: Evidence of Early and Sustained Bone Development in Bioceramic Calcium Phosphate Implants. Patient Series. *J. Neurosurg. Case Lessons* **2021**, *1*, CASE20133. [CrossRef]

71. Schmidt-Bleek, K.; Schell, H.; Kolar, P.; Pfaff, M.; Perka, C.; Buttgereit, F.; Duda, G.; Lienau, J. Cellular Composition of the Initial Fracture Hematoma Compared to a Muscle Hematoma: A Study in Sheep. *J. Orthop. Res.* **2009**, *27*, 1147–1151. [CrossRef] [PubMed]
72. Eming, S.A.; Koch, M.; Krieger, A.; Brachvogel, B.; Kreft, S.; Bruckner-Tuderman, L.; Krieg, T.; Shannon, J.D.; Fox, J.W. Differential Proteomic Analysis Distinguishes Tissue Repair Biomarker Signatures in Wound Exudates Obtained from Normal Healing and Chronic Wounds. *J. Proteome Res.* **2010**, *9*, 4758–4766. [CrossRef] [PubMed]

Disclaimer/Publisher's Note: The statements, opinions and data contained in all publications are solely those of the individual author(s) and contributor(s) and not of MDPI and/or the editor(s). MDPI and/or the editor(s) disclaim responsibility for any injury to people or property resulting from any ideas, methods, instructions or products referred to in the content.

Article

Femtosecond Laser Irradiation to Zirconia Prior to Calcium Phosphate Coating Enhances Osteointegration of Zirconia in Rabbits

Hirotaka Mutsuzaki [1,2,*], Hidehiko Yashiro [3], Masayuki Kakehata [3], Ayako Oyane [4] and Atsuo Ito [5]

[1] Center for Medical Science, Ibaraki Prefectural University of Health Sciences, 4669-2 Ami, Ibaraki 300-0394, Japan
[2] Department of Orthopaedic Surgery, Ibaraki Prefectural University of Health Sciences Hospital, 4773 Ami, Ibaraki 300-0331, Japan
[3] Research Institute for Advanced Electronics and Photonics, National Institute of Advanced Industrial Science and Technology (AIST), AIST Tsukuba Central 2, 1-1-1 Umezono, Tsukuba, Ibaraki 305-8568, Japan; hidehiko.yashiro@aist.go.jp (H.Y.); kakehata-masayuki@aist.go.jp (M.K.)
[4] Nanomaterials Research Institute, National Institute of Advanced Industrial Science and Technology (AIST), AIST Tsukuba Central 5, 1-1-1 Higashi, Tsukuba, Ibaraki 305-8565, Japan; a-oyane@aist.go.jp
[5] Health and Medical Research Institute, National Institute of Advanced Industrial Science and Technology (AIST), AIST Tsukuba Central 6, 1-1-1 Higashi, Tsukuba, Ibaraki 305-8566, Japan; atsuo-ito@aist.go.jp
* Correspondence: mutsuzaki@ipu.ac.jp; Tel.: +81-29-888-4000; Fax: +81-29-840-2301

Abstract: Calcium phosphate (CaP) coating of zirconia and zirconia-based implants is challenging, due to their chemical instability and susceptibility to thermal and mechanical impacts. A 3 mol% yttrium-stabilized tetragonal zirconia polycrystal was subjected to femtosecond laser (FsL) irradiation to form micro- and submicron surface architectures, prior to CaP coating using pulsed laser deposition (PLD) and low-temperature solution processing. Untreated zirconia, CaP-coated zirconia, and FsL-irradiated and CaP-coated zirconia were implanted in proximal tibial metaphyses of male Japanese white rabbits for four weeks. Radiographical analysis, push-out test, alizarin red staining, and histomorphometric analysis demonstrated a much improved bone-bonding ability of FsL-irradiated and CaP-coated zirconia over CaP-coated zirconia without FsL irradiation and untreated zirconia. The failure strength of the FsL-irradiated and CaP-coated zirconia in the push–out test was 6.2–13.1-times higher than that of the CaP-coated zirconia without FsL irradiation and untreated zirconia. Moreover, the adhesion strength between the bone and FsL-irradiated and CaP-coated zirconia was as high as that inducing host bone fracture in the push-out tests. The increased bone-bonding ability was attributed to the micro-/submicron surface architectures that enhanced osteoblastic differentiation and mechanical interlocking, leading to improved osteointegration. FsL irradiation followed by CaP coating could be useful for improving the osteointegration of cement-less zirconia-based joints and zirconia dental implants.

Keywords: zirconia; calcium phosphate; femtosecond laser; bone formation; bone-bonding ability

1. Introduction

Ceramic implants, such as zirconia and alumina-zirconia composites, are advantageous over metallic implants, owing to their excellent chemical stability, biocompatibility, high wear resistance, and artifact-free magnetic resonance imaging (MRI) [1–6]. Meanwhile, metallic implants have the risk of inducing metal hypersensitivity [7], metallosis [8], and pseudotumors [9], although they show acceptable biocompatibility and favorable mechanical properties. Metallic implants also cause severe MRI artifacts [5,6], which interfere with diagnosis or making appropriate treatment decisions in postoperative evaluation of the implant and surrounding tissues.

However, zirconia and alumina-zirconia composite ceramic implants present a challenge for acquiring bone-bonding ability [10]. While bone-bonding ability is simply obtained on metallic implants by thermal spray coating of osteoconductive calcium phosphate (CaP), the high temperature over 1100 °C used to melt CaP deteriorates the mechanical properties of zirconia, due to crack formation and phase transformation. Thus, zirconia-based implants are used without CaP coatings. If necessary, bone cement is used in clinical practice, although such an approach may induce the development of bone cement implantation syndrome [1,11]. Coating zirconia with CaP using a process with low thermal effect is crucial for improving the bone-bonding ability of zirconia without deteriorating mechanical properties. Zirconia-based implants have previously been coated by CaP without severe heating of substrates using aqueous solution processing [12–18], aerosol deposition [19], magnetron sputtering [20], and ion-beam-assisted deposition [21]. Moreover, the bone-bonding ability of CaP-coated implants improved relative to uncoated ones in animal studies [12,16,22]. However, achieving a high adhesion strength between the CaP coating and zirconia-based implant remains a challenge. A low adhesion strength at the CaP–implant interface eventually causes a low adhesion strength between the bone and implant.

Recently, surface roughening of zirconia at 100-micrometer scale was carried out to improve CaP coating adhesion [23–25]. In this technique, a nano-second Nd:YAG laser was used to produce micro-textures on zirconia, and a CO_2 laser was used to sinter a CaP coating (prepared by dip-coating) on the micro-textured zirconia.

In our previous study, a femtosecond laser (FsL) was used to produce tens-of-micron-sized craters that had linear submicron grooves on their concave surface on zirconia (3 mol% yttrium-stabilized tetragonal zirconia polycrystal: 3Y-TZP), thereby improving CaP coating adhesion [13]. Our previous CaP coating was carried out using a three-step procedure: (i) FsL irradiation on zirconia substrates; (ii) alternate dipping treatment; and (iii) immersion in a supersaturated CaP solution (hereafter, referred to as 'solution process') [13]. In the first step of surface roughening, FsL was employed, since FsL irradiation leads to much less thermal damage to the bulk properties of zirconia compared to nanosecond or longer laser pulses, despite the transient high temperature (over the melting point: 2115 °C) locally at the surface. This is because ultrashort laser pulses make the thermal transport distance negligibly short in contrast to that of nanosecond laser pulses [26,27]. In more detail, the strong electric field or high intensity of FsL pulses causes the nonlinear absorption or multi-photon absorption and ionizes the surface of the zirconia. The ionized materials are ejected from the surface through the mechanism known as Coulomb explosion, which is a nonthermal process. The time required for Coulomb explosion is shorter than that required for the thermal diffusion; therefore, most of the energy absorbed by the nonlinear absorption is transported to the ablation products. Thus, the thermal effect of FsL pulses on the ablated surface is small compared with that of nanosecond or longer laser pulses [27]. Furthermore, the tens-of-micron-sized craters and submicron grooves on the zirconia enhance the osteogenic differentiation of mesenchymal stem cells [28]. In the second step, the FsL-irradiated zirconia was alternately dipped in calcium and phosphate solutions three times for the deposition of a CaP underlayer. In the third step, the thus-treated zirconia was immersed in a supersaturated CaP solution to increase the thickness of the CaP coating.

In our previous CaP coating technique describe above, the tens-of-micron- and submicron-scale surface roughening effectively increased CaP coating adhesion [13], without causing noticeable phase transformation in the zirconia from tetragonal to monoclinic [29]. However, the bone-bonding abilities of FsL-irradiated and CaP-coated zirconia substrates have yet to be clarified. In the present study, we refined our previous CaP coating technique using pulsed laser deposition (PLD) in place of the alternate dipping treatment (in the second step) and employing clinically available infusions and injection fluids as sources for the supersaturated CaP solution (in the third step). The PLD process can precisely control the composition and structure of CaP [30]. In addition, the PLD process can be used for CaP coating on complex-shaped substrates such as zirconia screws [31]. In the present study, the PLD process was carried out at the substrate temperature of 20–25 °C for the deposition of a CaP underlayer. Clinically available infusions and injection fluids were utilized to improve the biosafety of coated CaP for clinical applications of CaP-coated zirconia. The infusions and injections used were water, calcium-containing solutions (two types with different concentrations), phosphorus-containing solutions (two types with different concentrations), a $NaHCO_3$ solution (an alkalizer) to increase pH, and saline to adjust ionic strength. CaP-coated zirconia with and without FsL irradiation (Groups A and B, respectively) along with untreated zirconia (Group C) were implanted in rabbit tibial metaphyses. This study aimed to clarify the effect of FsL irradiation on the bone-bonding ability of CaP-coated zirconia implants in rabbits. After implantation for four weeks, the push-out strength of the implant and bone-to-implant contact ratio were evaluated. We hypothesized that FsL irradiation would improve the bone-bonding ability of CaP-coated zirconia.

2. Materials and Methods

2.1. Preparation of Zirconia Implants

Quadratic prism implants of zirconia were prepared by compacting 3Y-TZP powders (TZ-3YB-E, Tosoh Co., Tokyo, Japan) using a cold isostatic pressing (CIP) technique followed by sintering at 1350 °C for 2 h and cutting. Four rectangular planes with dimensions of 2.4 mm (Y- and Z-directions) × 21 mm (X-direction) were subjected to wet-polishing to mirror quality surfaces (Ra < 0.05 μm). The implants were ultrasonically washed three times with acetone and then dried at room temperature (20–25 °C).

The implants were divided into Groups A, B, and C, where Group A implants were subjected to FsL irradiation followed by a CaP coating with PLD and solution processes, implants of Group B were subjected to the same CaP coating processes (PLD and solution processes) without FsL irradiation, and implants in Group C were not treated (hereafter referred to as 'untreated implants').

2.2. FsL Irradiation for Tens-of-Micron- and Submicron-Scale Surface Roughening

The method of FsL irradiation has been described elsewhere [13]. The implants were irradiated through FsL using a Ti:Sapphire chirped-pulse amplification system with a stamp-scan mode at a peak laser fluence F_{peak} of approximately 4 J/cm^2 per pulse. The laser system generated 810 nm centered 80 fs full width at half maximum (FWHM) pulses at a 570 Hz repetition rate. The direction of laser polarization was set parallel to the Y-direction of the implant, as shown in Figure 1. After each spot was irradiated with 40 pulses, the irradiation position was moved by 60 μm in the X-direction. When the laser reached the end of the X-direction, the irradiation position was moved by 30 and 90 μm in the X- and Y-directions, respectively, to irradiate the next row. The 60 μm stepwise irradiation was then restarted in the X-direction, to fully fill the interstices of elliptical craters. The irradiation was repeated until the length of the irradiated area reached 20 mm in the X-direction. In this laser treatment, five samples were placed side by side and the 12.0 mm (Y) × 20 mm (X) plane was irradiated at the same time. All four rectangular planes of the implant were irradiated. After the laser irradiation, the samples were cleaned with cellulose wiper wetted with ethanol, washed ultrasonically three times with acetone, and then air dried.

FsL irradiation generated elliptical craters with a lateral dimension of 90 μm (Y-direction) × 60 μm (X-direction) and a maximum depth of 6.7–7.0 μm, as measured using a confocal laser microscope (VK-9700, Keyence Co., Osaka, Japan) and a scanning electron microscope (SEM; S-4800, Hitachi High-Tech Corporation, Tokyo, Japan). Linear submicron grooves parallel to the Y-direction with a period of around 840 nm (Figure 1a) were formed on the concave surface of the craters (Figure 1b). Approximately half of the surface area exhibited linear submicron grooves, while the other half showed a non-periodic submicron roughness.

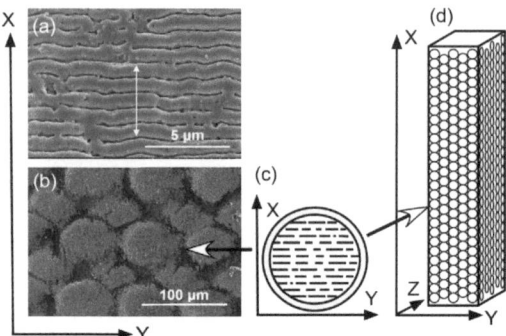

Figure 1. (**a**) High magnification scanning electron microscope (SEM) image of the implant surface, showing linear submicron grooves formed through FsL irradiation. (**b**) Low magnification SEM image of the implant surface. (**c**) Schematic diagram of a crater. (**d**) Schematic diagram of the quadratic prism zirconia implant. The white arrow in (**a**) indicates the length (~4.2 µm) of 5 periods, which means that the length of one period is approximately 840 nm.

2.3. Pulsed Laser Deposition of CaP Underlayer

The zirconia implants with and without FsL irradiation (for Groups A and B, respectively) were subjected to PLD for depositing a CaP underlayer using the 4th harmonics of an Nd:YAG laser system (Quanta-Ray Pro-350, Spectra-Physics, Milpitas, CA, USA, wavelength: 266 nm, pulse duration: 10 ns, repetition rate: 10 Hz). A disk of β-tricalcium phosphate (β-TCP: $Ca_3(PO_4)_2$), 14 mm diameter and 1 mm thick, was used as the target. This experimental configuration for PLD has been described elsewhere [30]. The disk targets were prepared by compacting and sintering β-TCP powders (Olympus Terumo Biomaterials Corp., Tokyo, Japan) at 1100 °C for 1 h in air. The laser beam was focused onto a β-TCP disk target using a planoconvex lens (f = 400 mm) at an incident angle of 45°, and the focused beam size on the target was elliptical with major and minor axis lengths of 270 and 150 µm, respectively. The laser energy was controlled to approximately 19 mJ using a polarizer and a half wave plate, which corresponded to 60 J/cm^2 with a stability within ±5% on the target surface. The target was moved 60 µm after each shot in the raster scan. Ablated particles were ejected intermittently toward the samples placed 20 mm away from the β-TCP target surface. Four samples placed side by side were treated at the same time at room temperature (20–25 °C), and all four rectangular planes of each sample were coated. Water vapor was introduced into the process chamber to control the stoichiometry of the coating, as discussed below. The water vapor pressure was controlled using a water vapor supply system consisting of a water cell at 20 °C, a capacitance manometer (Type 627A01 (1 Torr full scale), MKS instruments Inc., Andover, MA, USA), and a pressure controller (Type 250, MKS instruments Inc., Andover, MA, USA) with a low flow metering valve, as well as a butterfly valve at the inlet port of the vacuum pump. Depending on the water vapor pressure, the process chamber vacuum was in the range from 1×10^{-5} to 0.7 Torr. In a preliminary study using a copper substrate, the Ca/P molar ratio of the CaP coating, measured using an X-ray fluorescence analyzer (SEA5120A, SII NT Co., Chiba, Japan), decreased with increasing water vapor pressure. Based on this preliminary result, the water vapor pressure for the PLD of zirconia implants was set to 0.093 Torr to obtain a CaP underlayer with a Ca/P molar ratio of 1.60–1.67, which is slightly lower than that of stoichiometric hydroxyapatite.

2.4. Solution Processing for CaP Growth

After PLD (deposition of the CaP underlayer), the implants for Groups A and B were subjected to the solution process; they were immersed in a supersaturated CaP solution to grow CaP, as described previously [32]. The supersaturated CaP solution was prepared by mixing clinically available infusions and injection fluids using the same method as in [32]:

7% MEYLON® injection (Otsuka Pharmaceutical Co., Ltd., Tokyo, Japan); water for injection (FUSO Pharmaceutical Industries, Ltd., Osaka, Japan); Klinisalz® (KYOWA CritiCare Co., Ltd., Tokyo, Japan); dibasic potassium phosphate injection 20 mEq kit (TERUMO Co., Tokyo, Japan); Ringer's solution OTSUKA (Otsuka Pharmaceutical Co., Ltd., Tokyo, Japan); calcium chloride corrective injection 1 mEq/mL (Otsuka Pharmaceutical Co., Ltd., Tokyo, Japan); and OTSUKA normal saline (Otsuka Pharmaceutical Co., Ltd., Tokyo, Japan). All samples in Groups A and B were immersed in 200 mL of the solution at 37 °C for 48 h.

2.5. Surface Characterization of the Implants

The implant surface was analyzed using a SEM (S-4800, Hitachi High-Tech Corporation, Tokyo, Japan) equipped with an energy-dispersive X-ray analyzer (EDX; EMAX x-act, HORIBA, Ltd., Kyoto, Japan). The calcium and phosphorus contents of the coating were analyzed using an inductively coupled plasma atomic emission spectrometer, after acid dissolution of the coating (ICP: SPS7800, Seiko Instruments, Inc., Chiba, Japan). The thickness of the CaP underlayer (after PLD) and that of the final coating layer (after subsequent solution process) were measured using a laser confocal microscope (VK-X3000, Keyence Co., Osaka, Japan), only for Group B samples. The measurements could not be carried out on Group A samples because of the craters on the sample surface due to FsL irradiation. In addition, since the coating process was the same for both Groups A and B, the thickness of the coating should have been the same for both groups. Prior to the measurements, straight-line portions of the surface layers were scratched using a knife blade, to expose the zirconia surface. The thicknesses of the layers were measured using 3D surface profiles. Five different points per cross-sectional height profile, and five different cross-sectional height profiles were used to calculate average and standard deviation of the thickness of each layer ($N = 25$).

2.6. Implantation in Rabbit Tibia

Eight implants in Groups A, B, and C were randomly implanted in both proximal tibial metaphysis of twelve male Japanese white rabbits (weight range: 2.5–2.8 kg, 14 weeks old) by a single physician who was blinded to the sample identifications. Three implants in each group were used for histomorphometry analysis and five for radiographical analysis, followed by push-out test.

After an intravenous injection of barbiturate (40 mg/kg body weight), small (10 mm) incisions were aseptically made in the skin at the medial proximal tibia. Bone tunnels 3.5 mm in diameter were then drilled bicortex in both proximal tibial metaphysis. The implants were manually inserted into the tunnels, from the medial to lateral cortex of the tibia (Figure 2). After the implantation, the skin was sutured with a 2-0 nonabsorbable suture. Postoperatively, each animal was allowed free activities in its own cage. All animals were then sacrificed four weeks after the operations.

(a) Implantation in 3.5-mm-diameter bone tunnel in tibia

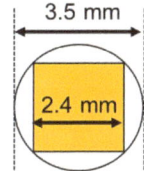

(b)

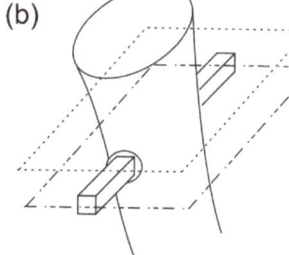

Figure 2. Schematic diagram of implantation of the implant into the proximal tibial metaphysis in a rabbit. (**a**) Front view of the 2.4 mm × 2.4 mm square plane of the implant in the bone tunnel with a diameter of 3.5 mm. (**b**) An overall view of the implant in the proximal tibia.

All the animal experiments and breeding were performed according to conditions approved by the ethics committees of both Ibaraki Prefectural University of Health Sciences and National Institute of Advanced Industrial Science and Technology (AIST), and were in accordance with the National Institutes of Health Guidelines for the Care and Use of Laboratory Animals.

2.7. Radiographical Analysis

Radiographs of proximal tibial metaphyses containing the zirconia implants were recorded on imaging plates (Fuji Film, Tokyo, Japan) using a medical Roentgen diagnostics system (DR-150-1, Hitachi Inc., Tokyo, Japan). An X-ray was irradiated perpendicular to the rectangular plane of the implant using a universally rotating sample holder. The bone-to-implant contact ratio was evaluated for the radiographs by a physician blinded to the sample identifications using Image J ver. 1.48. The bone-to-implant contact ratio was defined as the percentage of the length of direct contact between the bone and the implant surface in the total length of the implant within the bone on X-ray radiographs. The length of direct contact was the bone formed on the surface of the implant placed inside the bone and was the sum of the length that was continuous with the bone cortex and in direct contact with the implant. Moreover, bird's-eye views were prepared from the radiographs by plotting the intensity of X-ray absorption at each X-Y point on the Z-axis.

2.8. Push—Out Tests

After radiographical analysis, the failure load of the bone–implant interface was measured in push-out tests using a universal testing machine (Type EZ-L, Shimazu Co., Kyoto, Japan). The proximal tibial metaphyses containing zirconia implants were fixed on a tilting stage. A compressive load was applied to the implant at a cross-head speed of 2 mm/min in the direction of its X-axis with the use of a square socket with a steel ball. Failure strength was calculated by dividing the failure load (F_{max}) by the area of the implant surface within the bone tunnel (S). The surface area was estimated by measuring the four lengths (in the X-direction) of the implant within the bone tunnel using a caliper and multiplying the sum (sum of the four lengths) by 2.4 mm.

2.9. Alizarin Red Staining of the Zirconia Implant

After measuring the failure load, the zirconia implants were carefully removed from the tibial metaphyses by cutting the bone just above the proximal side of the implants. The removed implants were immersed in 10% phosphate-buffered formalin solution for 10 days to fix the tissue, rinsed twice with calcium- and magnesium-free phosphate-buffered saline (PBS (-)), immersed in a 1% alizarin red solution at pH 6.35 for staining calcium ions with alizarin red, and finally rinsed with PBS (-). Thickness of the alizarin red stained area was measured using a laser confocal microscope (OLS-4100, Olympus, Tokyo, Japan). Images of the stained implants were captured using a stereoscopic microscope (SZX16, Olympus Co., Japan), and analyzed for stained area using Image J ver. 1.48.

2.10. Histomorphometric Analysis through SEM

The proximal tibial metaphyses containing zirconia implants were fixed in 10% phosphate-buffered formalin solution, dehydrated in serial concentrations of ethanol (70, 80, 90, and 99.5 vol%) and acetone, and then embedded in an acrylic resin (poly(methyl methacrylate)) through polymerization. The cured resin specimens were cut parallel to the X-Z plane of the implant and perpendicular to the bone axis using a micro-cutting machine (BS-300CL, EXAKT Advanced Technologies, Norderstedt, Germany). The cutting blade was inserted at a position slightly away from the center line of the Y-Z plane considering the cutting allowance (~0.4 mm), so that the longitudinal rectangular section at the center of the implant (~1.2 mm distant from the X-Z plane) was exposed. The surfaces of the resin specimens including the half-cut implants were polished to mirror quality using a micro-grinding machine (MG-400 CS, EXAKT Advanced Technologies, Norderstedt, Germany) and #4000 polishing paper. The interfaces of the bone and implant in 2 mm regions (four regions per resin specimen) from the outer edges of the cortical bone were observed using SEM in backscattered electron imaging mode, and the captured images were analyzed using Image J. The bone-to-implant contact ratio was calculated as the percentage of the length of the bone-contacting implant surface of the total length of the implant surface in the test region. The bone-contacting implant surface was determined as the implant surface where the gap with the bone was ≤5 µm.

2.11. Statistical Analysis

All the data from each group were analyzed using generalized linear models. Statistical analysis was performed using IBM SPSS 28.0, with a 5% level of statistical significance.

3. Results

3.1. Surface Characterization of the Implants

The implants in Groups A and B were fully coated with CaP layers with similar surface morphology and chemical composition. Figure 3 shows 3D profiles (from a laser confocal microscope) of the partially scratched implant (Group B) surfaces after (a) PLD and (b) subsequent solution processing in false color. The diagonal straight lines with blue color are the scratched portions with exposed zirconia surface on the implants. Representative cross-sectional height profiles of the CaP under-layer (after PLD) and final coating layer (after subsequent solution process) including the scratched portions are shown in Figure 3c. Micron-scale protrusions observed on the final coating layer correspond to the particles (red dots) on its 3D surface profile (Figure 3c). According to the quantitative analysis, the CaP underlayer on the implant deposited through PLD was 204 ± 22 nm thick, which increased to 992 ± 195 nm after the subsequent solution process.

The resulting implants in both Groups A and B showed micro-rough surfaces (Figure 4a,b,d,e), whereas those in Group C showed a flat surface (Figure 4c,f). On the implants in Group A, the linear submicron grooves formed by FsL irradiation (Figure 1a) were invisible due to the overlaid CaP layer, although the elliptical crater-like structures were barely recognizable, as shown in Figure 4d. The CaP-coated crater had a size of about 90 × 60 µm in diameter and 5 µm in depth.

EDX analysis revealed that the surface layers on the Group A and B implants were composed of CaP (Figure 5). Detailed analysis of the peak positions of P and Zr in the range 1.7–2.3 keV confirmed the presence of P on the surfaces of these implants (Figure 5d). No noticeable differences in chemical composition and morphology of the CaP layers were found between Groups A and B (Figure 5a,b); the critical difference was the presence (Group A) or absence (Group B) of surface waviness, which reflected the tens-of-micron-sized craters formed by FsL irradiation. The quantitative analysis of Ca and P using ICP showed Ca and P contents of 132.4 µg and 58.1 µg, respectively, with a Ca/P molar ratio of 1.76 for Group A, and 171.4 µg and 73.3 µg, respectively, with a Ca/P molar ratio of 1.81 for Group B.

3.2. Radiographical Analysis

Typical radiographs of the proximal tibial metaphyses containing the implants in Groups A, B, and C are shown in Figure 6a–c, respectively. In the radiographs of Groups B and C, significant radiolucent lines were observed around the implant, whereas no clear radiolucent line was found around the implants in Group A. Radiolucent lines are often associated with intervening fibrous tissue formation between an implant and bone. The bird's-eye views in Figure 6d–f visualize the relative intensity of X-ray absorption in Figure 6a–c, respectively. The views of Groups B and C show valleys of low intensity in the close vicinity of the implants (corresponding to the radiolucent lines in Figure 6b,c) as shown in Figure 6e,f; on the other hand, no such valley was apparent in the vicinity of the Group A implants (Figure 6d). These results suggest that the Group A implants showed better bone-to-implant contact in the rabbit tibias than the implants in Groups B and C, which was confirmed through quantitative radiographical analysis, as described below.

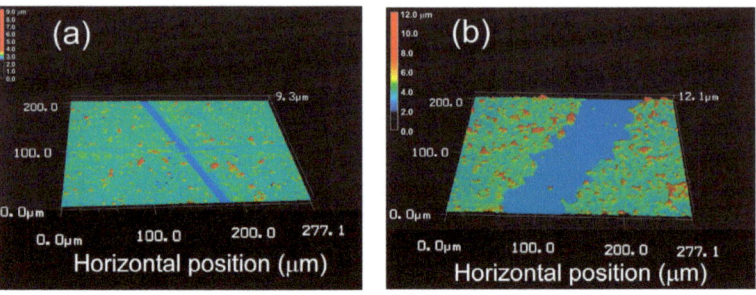

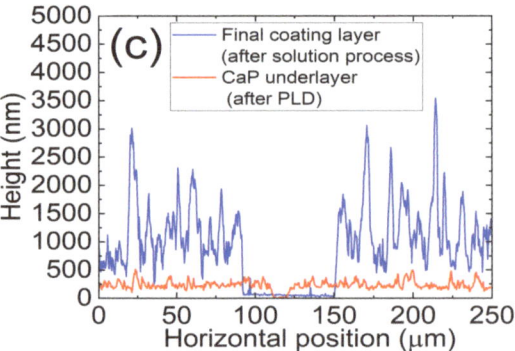

Figure 3. Three-dimensional profiles (from a laser confocal microscope) of the partially scratched implant (Group B) surfaces after (**a**) PLD and (**b**) subsequent solution processing in false color, and (**c**) representative cross-sectional height profiles of the CaP underlayer (after PLD) and those of the final coating layer (after subsequent solution process) including the scratched portions.

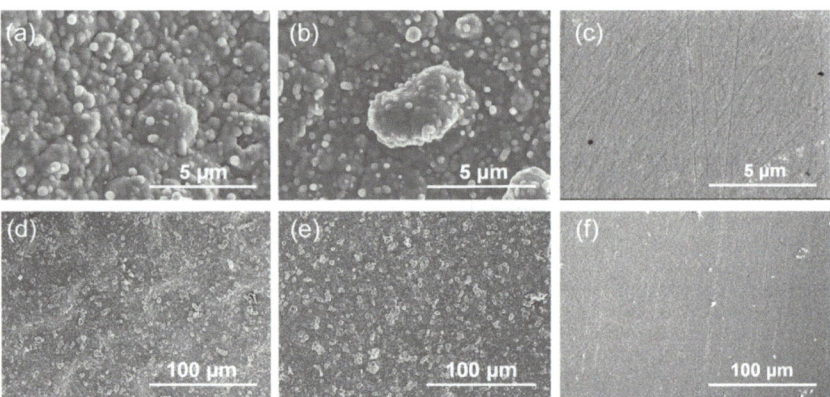

Figure 4. SEM images, with (**a**–**c**) high and (**d**–**f**) low magnifications, of the surfaces of the (**a**,**d**) CaP-coated implant with FsL irradiation (Group A), (**b**,**e**) CaP-coated implant without FsL irradiation (Group B), and (**c**,**f**) untreated implant (Group C).

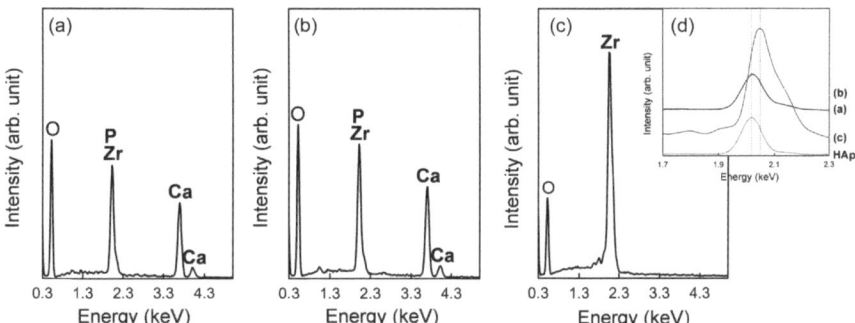

Figure 5. EDX spectra of the surfaces of the (**a**) CaP-coated implant with FsL irradiation (Group A), (**b**) CaP-coated implant without FsL irradiation (Group B), and (**c**) untreated implant (Group C). (**d**) Magnified EDX spectra of (**a**–**c**) and of sintered hydroxyapatite (HAp) in the range 1.7–2.3 keV.

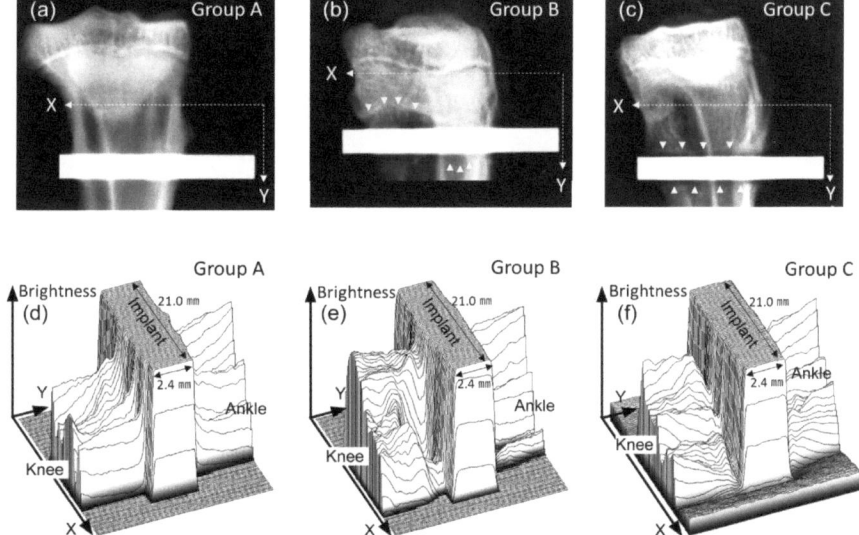

Figure 6. X-ray radiographs of the implants in Groups A (**a**), B (**b**), and C (**c**) in the tibias, and their bird's-eye views, where the brightness (related to the intensity of X-ray absorption) of each X-Y plane in (**a**–**c**) was converted into height in (**d**–**f**), respectively. The yellow arrows are part of the radiolucent line.

According to the radiographical analysis, the bone-to-implant contact ratios were 33.8 ± 8.0, 19.4 ± 13.5, and $17.4 \pm 12.0\%$ for Groups A, B, and C, respectively, as shown in Figure 7. The bone-to-implant contact ratio for Group A was significantly higher than those for Groups B ($p = 0.027$) and C ($p = 0.012$). No significant difference was found between Groups B and C ($p = 0.754$).

3.3. Push−Out Tests

Typical load−displacement curves are shown in Figure 8. A sawtooth pattern invariably only appeared on the curve for Group A. The distance between the two neighboring sawtooth peaks was approximately 56 μm, which is almost the same as the distance of 60 μm between the centers of the elliptical craters generated through FsL irradiation.

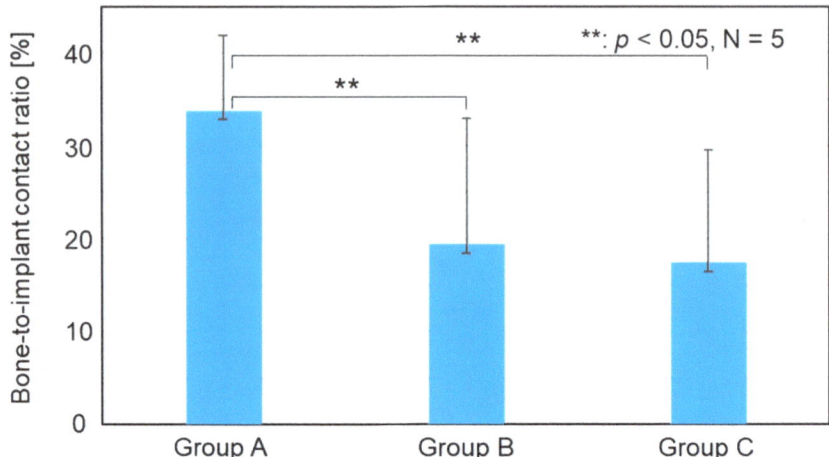

Figure 7. Bone-to-implant contact ratio of Groups A, B, and C measured through radiographical analysis.

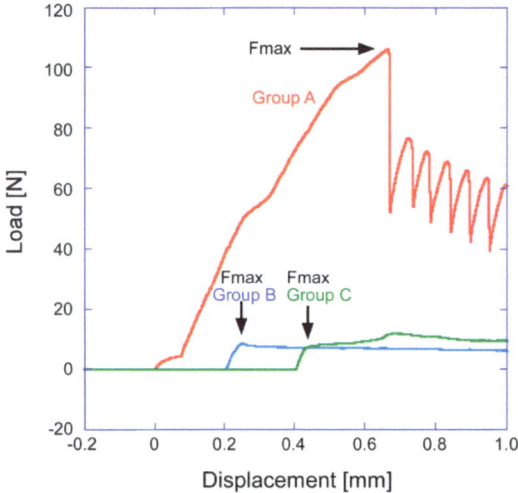

Figure 8. Typical load–displacement curves in the push–out tests for Groups A, B, and C.

The failure strength ($\times 10^4$ Pa) in the push-out test was 51.1 ± 30.6, 8.3 ± 3.7, and 3.9 ± 3.6 for Groups A, B, and C, respectively. The failure strength for Group A was significantly higher than that for Groups B ($p < 0.001$) and C ($p < 0.001$), as shown in Figure 9. No significant difference was found between Groups B and C ($p = 0.667$, Figure 9). These results indicate that the Group A implants showed higher adhesion strength with the bone tissue compared to the implants in Groups B and C.

3.4. Alizarin Red Staining of the Implants after Push-Out Tests

After the push–out tests, the Group A implants removed from the tibial metaphyses were more widely and strongly stained with alizarin red (calcium indicator) than those of Groups B and C (Figure 10). The implants in Group A had a strongly stained region on a large part of the surface (Figure 10 right). The stained portion on the implants in Group A had a thickness from 5 to 10 µm, which was thicker than the estimated initial thickness of the CaP coating layer (approximately 1 µm). Thus, the main component of this stained portion was likely fractured bone tissue. In contrast, the implants in Groups B and C were hardly stained with alizarin red (Figure 10 right), suggesting that the amount of bone tissue on their surfaces was relatively small. The ratio of stained area for Group A was significantly larger than that for Groups B ($p < 0.001$) and C ($p < 0.001$) (Figure 10 left). No significant difference was found between Groups B and C in the ratio of the stained area ($p = 0.144$, Figure 10 left).

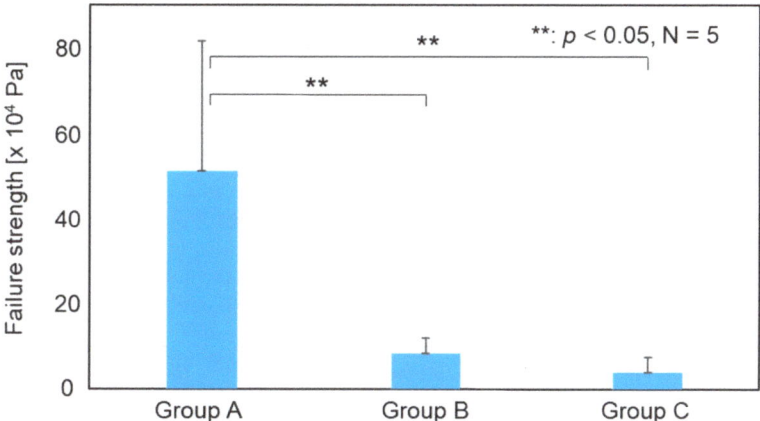

Figure 9. Failure strength of the implants in Groups A, B, and C in the push−out tests.

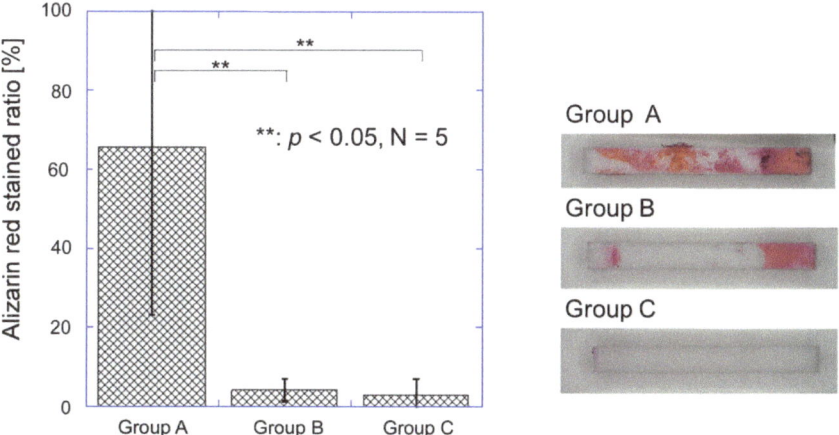

Figure 10. Ratio of the alizarin red stained area to the whole surface area of the implants in Groups A, B, and C after push−out tests (**left**). Typical macroscopic image of alizarin red stained implants in Groups A, B, and C after push-out tests (**right**).

3.5. Histomorphometric Analysis through SEM

Histomorphometric analysis through SEM revealed better contact of the cortical bone with the implant in Group A than for the implants in Groups B and C. Figure 11 shows typical SEM images of the histological sections, showing the interfaces between the cortical bone and zirconia implants. The bone tissue directly contacted the implant surface in a wider region for Group A than for Groups B and C. In Group A, bone tissue was even observed in the elliptical craters on the implant surface created through FsL irradiation. An electron-sparse layer was seen between the zirconia and bone in Groups B and C. As shown in Figure S1, the electron-sparse layer was rich in carbon and contained no calcium. Thus, the electron-sparse layer was that of an intervening soft tissue. Such an electron-sparse layer was scarcely seen between the zirconia and bone in Group A.

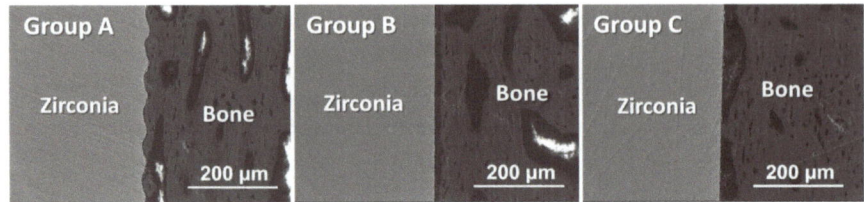

Figure 11. SEM images (backscattered electron images) of the histological sections showing the bone–implant interface of Groups A, B, and C.

According to the SEM and EDX analyses, residual CaP layers were not clearly identified on the implants for either Group A or B. As shown in Figure 12, the bone-to-implant contact ratio of Group A was significantly higher than that of Groups B ($p < 0.001$) and C ($p < 0.001$). The bone-to-implant contact ratio of Group B was also significantly higher than that of Group C ($p < 0.001$).

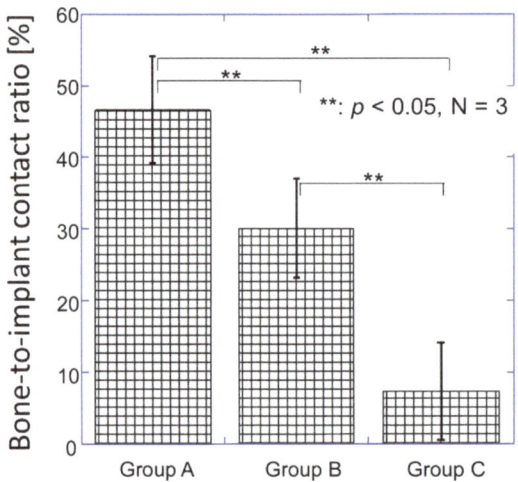

Figure 12. Bone-to-implant contact ratio of Groups A, B, and C.

4. Discussion

FsL irradiation followed by CaP coating on zirconia improved the bone-bonding ability of zirconia. In Group A, FsL irradiation followed by CaP coating generated a micro-ruggedness consisting of partially-overlapped elliptical craters and an approximately 1 μm-thick CaP coating layer on the zirconia. The same CaP coating thickness was deposited on the Group B implants without FsL irradiation. The four different in vivo analyses conducted in the present study (radiographical analysis, push–out test, alizarin red staining, and histomorphometric analysis) demonstrated a much improved bone-bonding ability of zirconia in Group A over that in Groups B and C. In the histomorphometric analysis through SEM (Figure 11), the bone-to-implant contact ratio was higher in Group B than in Group C, most likely because of the CaP coating layer. However, no significant differences between Groups B and C were observed in the radiographical analysis, push-out test, and alizarin red staining. Therefore, FsL irradiation played a critical role in enhancing the bone-bonding ability of CaP-coated zirconia.

FsL irradiation generated a surface architecture that enhanced the osteogenic differentiation of mesenchymal stem cells (MSCs). Studies have shown that surface architectures and roughness similar to that of an osteoclast resorption pit are sensed by MSCs in contact with proteins adsorbed on an implant surface, causing MSCs to undergo osteogenic differentiation [28]. An osteoclast resorption pit has an area of approximately 700–7800 μm^2, lateral diameter of 30–100 μm, and depth of ~10 μm [33]. Elliptical craters formed through FsL irradiation on zirconia under conditions similar to those of the present study had an area of approximately 5000 μm^2, lateral diameter of ~80 μm, and depth of ~10 μm, enhancing the osteogenic differentiation of MSCs [28]. In the present study, the CaP-coated craters on zirconia had an area of approximately 4200 μm^2, lateral diameter of 90 × 60 μm, and depth of ~5 μm.

The surface architecture generated through FsL irradiation supported the osteoconduction of the CaP layer, which resulted in an improvement in the osteointegration of the zirconia in the present animal model. Osteointegration is defined as the formation of direct contact between the implant surface and bone tissue without intervening fibrous tissue, being a key factor for the long-term success of the implant [34]. In the animal experiment, a quadratic prism implant was inserted into a cylindrical bone tunnel (see Figure 2). Thus, the implant was not fixed tightly in the bone tunnel at the time of implantation; the gap between the implant and the inner wall of the bone tunnel was from 0.05 mm to 0.55 mm. Even under such loose-contact conditions, the zirconia in Group A showed significantly higher failure strengths than those in Groups B and C in the push-out test 4 weeks after implantation. This was because the zirconia in Group A was integrated more firmly into the newly formed bone. Although it is unclear whether the CaP layer remained unresorbed and whether osteoconduction was involved four-weeks after implantation, enhanced osteointegration was achieved on the zirconia in Group A. In contrast, the zirconia in Groups B and C exhibited limited osteoconduction or osteointegration, as evidenced by the radiolucent line around the implant in the radiographical analysis, and the lower bone-to-implant contact ratios and electron-sparse layer in the histomorphometric analysis. A radiolucent line and the electron-sparse layer are often associated with intervening fibrous tissue-formation, which means no osteoconduction occurred where the fibrous tissue was formed.

FsL irradiation of zirconia also improved the mechanical interlocking, to enhance the adhesion strength between the zirconia and the CaP layer, as well as between the zirconia and bone. FsL irradiation of zirconia not only induced tens-of-micron-sized craters and submicron grooves but also surface wetting via ablation plasma, enabling the zirconia to completely and strongly bond to the CaP layer [13]. The strong adhesion may also be attributed to the mechanical interlocking effects due to the surface architecture generated by FsL irradiation [13]. Moreover, new bone was formed in the elliptical craters in Group A (Figure 11). The resulting mechanical interlocking structure contributed to the improved adhesion strength between the zirconia and bone, as evidenced by the higher failure strength in the push-out test. Putative schematic diagrams of the failure mode in the push-out tests for Groups A, B, and C are illustrated in Figure 13. In Group A, cracks were generated both in the newly formed bone and at the bone–implant interface. The resulting protruding bone and the craters on the implant caused the sawtooth pattern in the load–displacement curve after fracture, due to improved mechanical interlocking (Figure 8). However, in Groups B and C, the cracks were mostly generated at the bone–implant interface, as shown in the righthand diagram of Figure 13. In the subsequent alizarin red staining, the zirconia in Group A had a larger stained area than those in Groups B and C. Furthermore, the stained portion of zirconia in Group A was 5–10-times thicker than the CaP layer coated on the zirconia. Therefore, the stained portion in Group A mainly consisted of bone tissue detached from the host bone because of fractures within the host bone. The bone-to-zirconia adhesion strength in Group A was as high as that inducing host bone fracture in the push-out test. This enhanced bone-to-zirconia adhesion strength was partly due to the improved mechanical interlocking caused by the surface architecture generated through FsL irradiation.

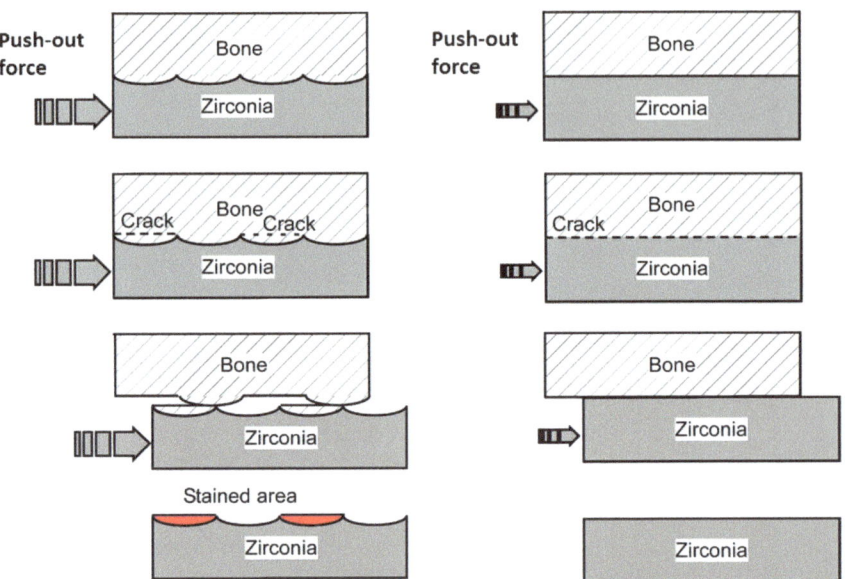

Figure 13. Putative schematic diagrams of failure mode in the push−out tests (upper three rows) and subsequent alizarin red staining (bottom row) for Group A (left column) and Groups B and C (right column).

Clinically, a zirconia knee prosthesis requires cementation for fixation with the bone tissue. However, fixation with bone cements involves the risk of complications (bone cement implantation syndrome) such as hypoxia, hypotension, consciousness disturbance, and death in the worst case [11]. The incidence of bone cement implantation syndrome is 24% in total hip arthroplasty (THA) and 28% in total knee arthroplasty (TKA) [11]. Various CaP coating techniques have been proposed to improve the bone-bonding ability of zirconia-based implants, thereby realizing cement-less zirconia joints [12–21]. Compared to previous techniques, the present surface modification technique utilizing FsL irradiation followed by CaP coating has the advantages of strong coating adhesion [13], supporting osteogenic cell-differentiation [28], and a mechanical interlocking effect, ensuring firm osteointegration without using any toxic reagents such as hydrofluoric acid [12]. The disadvantage of FsL irradiation is the negative impact on the mechanical properties of zirconia [35]. However, this disadvantage can be overcome by improving the manufacturing process of zirconia, reaching mechanical and crystallographic properties that meet the ISO 13556:2015 standard after FsL irradiation (Tables S1 and S2, and Figures S2–S5). As verified in the push-out tests (Figures 9 and 10), the Group A implants showed strong adhesion with the bone tissue, even without cementation. Strong bone-to-implant adhesion is essential for reducing the risk of loosening and ensuring long-term survival. Taken together, application of the present surface modification technique to the bone-contacting region of zirconia-based implants might successfully lead to the development of cement-less zirconia-based joints and zirconia dental implants.

This study had several limitations. Animal experiments were carried out on a small number of rabbits. In the future, the number of animals must be increased, and the experiments must be conducted on larger animals prior to clinical trials. Moreover, this study was limited to a short follow-up period; long-term evaluation is necessary in the future. Third, implants with FsL irradiation and without CaP coating were not considered. In this study, CaP-coated zirconia with and without FsL irradiation was tested to verify the effect of FsL irradiation. The effect of FsL irradiation alone could be a topic for future study.

5. Conclusions

FsL irradiation was carried out on zirconia to produce micro-/submicron surface architectures, followed by CaP coating using PLD and solution processes. The FsL-irradiated and CaP-coated zirconia (Group A) showed improved osteointegration and stronger adhesion with the bone tissue in rabbits than the untreated zirconia (Group C) and CaP-coated zirconia without FsL irradiation (Group B). The bone-to-zirconia adhesion in Group A was so strong that host bone fracture occurred in the push-out test. The present surface modification technique could be useful for developing cement-less zirconia-based joints and zirconia dental implants.

Supplementary Materials: The following supporting information can be downloaded at: https://www.mdpi.com/article/10.3390/jfb15020042/s1, Figure S1: SEM image and EDX line scan profiles (carbon and calcium signals) of the histological section for the bone-implant interface of Group B; Figure S2: Four-point bending strength for zirconia compacted using the additional HIP technique with and without subsequent FsL irradiation, and using the cold isostatic pressing (CIP) technique with and without subsequent FsL irradiation. The compacted temperature was set at 1350 °C for CIP in ambient air and at 1300 °C for the additional HIP treatment in a 147-MPa Ar atmosphere. The four-point bending strength was measured in accordance with ISO 14704 standard (left). FsL irradiation was made on one of the rectangular surfaces. The FsL-irradiated surface was set on the tension side in the bending test (left). Four-point bending strength of zirconia compacted using the additional HIP technique with subsequent FsL irradiation meets the ISO 13556:2015 standard requirement for zirconia implants (right). Reprinted and slightly modified from [36] Copyright 2017, and [37] Copyright 2019 with permission from Japan Laser Processing Society; Figure S3: Four-point bending strength for zirconia compacted using the additional HIP technique and subjected to accelerated aging in hot water (HIP+HW) and that compacted using HIP and subjected to FsL irradiation and accelerated aging in hot water (HIP+FsL+HW). Set-up of specimens and four-point bending test were the same as those in Figure S2. The accelerated aging in hot water was performed in accordance with ISO 13556:2015 standard. After accelerated aging, four-point bending strength of zirconia compacted using the additional HIP technique and subjected to FsL irradiation meets the requirement of ISO 13556:2015 standard for zirconia implants; Figure S4: X-ray diffraction (XRD) patterns for zirconia compacted using the additional HIP technique (HIP), and that subsequently subjected to only FsL irradiation (HIP+FsL) or both FsL irradiation and accelerated aging in hot water (HIP+FsL+HW). The labels "t" and "m" indicate the tetragonal and monoclinic crystal phases, respectively. The monoclinic and tetragonal phases were identified based on the data from The International Centre for diffraction data (ICDD) 01-070-8379 (Baddeleyite, syn) and ICDD 01-081-1544 (Zirconium Oxide), respectively; Figure S5: Monoclinic crystal phase ratio Rm for zirconia compacted using the additional HIP technique (HIP), and that subsequently subjected to only Fs irradiation (HIP+FsL) or Fs irradiation plus accelerated aging in HW (HIP+FsL+HW). Each Rm was calculated from the XRD peak areas for monoclinic and tetragonal phases using ref. [38]. The peak areas for monoclinic and tetragonal phases were analyzed using XRD analysis software (PDXL, Rigaku Co., Tokyo, Japan) with ICDD 01-070-8379 (Baddeleyite, syn) and ICDD 01-081-1544 (Zirconium Oxide), respectively. The Rm data for HIP+FsL were reproduced from [36]. The Rm values meet the ISO 13556:2015 standard requirement for zirconia implants; Table S1: Weibull parameters of four-point bending strength in Figure S2 for zirconia compacted using the additional hot isostatic pressing (HIP), and those subsequently subjected to FsL irradiation (HIP+FsL). Weibull parameters for "HIP+FsL" meet the requirements of ISO 13556:2015 standard for zirconia implants; Table S2: Number of fractured specimens after cyclic fatigue test in accordance with ISO 13556:2015 standard for zirconia implants reproduced from [37].

Author Contributions: Conceptualization, H.M., H.Y., M.K., A.O. and A.I.; Methodology, H.M., H.Y., M.K., A.O. and A.I.; Software, H.M., H.Y., M.K., A.O. and A.I.; Validation, H.M., H.Y., M.K., A.O. and A.I.; Formal Analysis, H.M., H.Y., M.K. and A.O.; Investigation, H.M., H.Y., M.K. and A.O.; Resources, H.M., H.Y., M.K. and A.O.; Data curation, H.M., H.Y., M.K. and A.O.; Writing Original Draft Preparation, H.M. and H.Y.; Writing Review and Editing, H.M., H.Y., M.K., A.O. and A.I.; Visualization, H.Y. and M.K.; Supervision, H.M., H.Y., M.K., A.O. and A.I.; Project Administration, H.M., H.Y., M.K., A.O. and A.I.; Funding Acquisition, H.Y., M.K. and A.O. All authors have read and agreed to the published version of the manuscript.

Funding: This study was supported by JSPS KAKENHI Grant Numbers JP15H00906, JP22H05148, JP22K04977, and AMED A16-81, Japan.

Institutional Review Board Statement: All animal experiments and breeding were performed according to the conditions approved by the ethics committees of both the Ibaraki Prefectural University of Health Sciences (approval code: H26-25; approval date: 23 July 2014, and approval code: H27-1; approval date: 25 March 2015) and the National Institute of Advanced Industrial Science and Technology (AIST) (approval code: DO2014-225; approval date: 24 September 2014, and approval code: DO2015-225; approval date: 25 June 2015), which are in accordance with the National Institutes of Health Guidelines for the Care and Use of Laboratory Animals (https://grants.nih.gov/grants/olaw/Guide-for-the-Care-and-use-of-laboratory-animals.pdf, accessed on 6 February 2024).

Data Availability Statement: The datasets generated and/or analyzed during the current study are available from the corresponding author upon reasonable request.

Acknowledgments: We thank Masahiko Monma for his technical support during radiography. We also thank Ikuko Sakamaki for her technical support and Kenji Torizuka for his scientific advice on laser engineering.

Conflicts of Interest: The authors declare no conflict of interest.

References

1. Meier, E.; Gelse, K.; Trieb, K.; Pachowsky, M.; Hennig, F.F.; Mauerer, A. First clinical study of a novel complete metal-free ceramic total knee replacement system. *J. Orthop. Surg. Res.* **2016**, *11*, 21. [CrossRef] [PubMed]
2. Innocenti, M.; Vieri, B.; Melani, T.; Paoli, T.; Carulli, C. Metal hypersensitivity after knee arthroplasty: Fact or fiction? *Acta Biomed.* **2017**, *88*, 78–83. [CrossRef] [PubMed]
3. Hanawa, T. Zirconia versus titanium in dentistry: A review. *Dent. Mater. J.* **2020**, *39*, 24–36. [CrossRef] [PubMed]
4. Sivaraman, K.; Chopra, A.; Narayan, A.I.; Balakrishnan, D. Is zirconia a viable alternative to titanium for oral implant? A critical review. *J. Prosthodont. Res.* **2018**, *62*, 121–133. [CrossRef]
5. Jungmann, P.M.; Agten, C.A.; Pfirrmann, C.W.; Sutter, R. Advances in MRI around metal. *J. Magn. Reson. Imaging* **2017**, *46*, 972–991. [CrossRef]
6. Duttenhoefer, F.; Mertens, M.E.; Vizkelety, J.; Gremse, F.; Stadelmann, V.A.; Sauerbier, S. Magnetic resonance imaging in zirconia-based dental implantology. *Clin. Oral. Implants Res.* **2015**, *26*, 1195–1202. [CrossRef]
7. Bao, W.; He, Y.; Fan, Y.; Liao, Y. Metal allergy in total-joint arthroplasty: Case report and literature review. *Medicine* **2018**, *97*, e12475. [CrossRef]
8. Sahan, I.; Anagnostakos, K. Metallosis after knee replacement: A review. *Arch. Orthop. Trauma. Surg.* **2020**, *140*, 1791–1808. [CrossRef]
9. Kurmis, A.P.; Herman, A.; McIntyre, A.R.; Masri, B.A.; Garbuz, D.S. Pseudotumors and high-grade aseptic lymphocyte-dominated vasculitis-associated lesions around total knee replacements identified at aseptic revision surgery: Findings of a large-scale histologic review. *J. Arthroplast.* **2019**, *34*, 2434–2438. [CrossRef] [PubMed]
10. Schünemann, F.H.; Galárraga-Vinueza, M.E.; Magini, R.; Fredel, M.; Silva, F.; Souza, J.C.M.; Zhang, Y.; Henriques, B. Zirconia surface modifications for implant dentistry. *Mater. Sci. Eng. C Mater. Biol. Appl.* **2019**, *98*, 1294–1305. [CrossRef] [PubMed]
11. Rassir, R.; Schuiling, M.; Sierevelt, I.N.; van der Hoeven, C.W.P.; Nolte, P.A. What are the frequency, related mortality, and factors associated with bone cement implantation syndrome in arthroplasty surgery? *Clin. Orthop. Relat. Res.* **2021**, *479*, 755–763. [CrossRef]
12. Takemoto, M.; Fujibayashi, S.; Neo, M.; Suzuki, J.; Kokubo, T.; Nakamura, T. Bone-bonding ability of a hydroxyapatite coated zirconia-alumina nanocomposite with a micorporous surface. *J. Biomed. Mater. Res.* **2006**, *78*, 693–701. [CrossRef] [PubMed]
13. Oyane, A.; Kakehata, M.; Sakamaki, I.; Pyatenko, A.; Yashiro, H.; Ito, A.; Torizuka, K. Biomimetic apatite coating on yttria-stabilized tetragonal zirconia utilizing femtosecond laser surface processing. *Surf. Coat. Technol.* **2016**, *296*, 88–95. [CrossRef]
14. Zain, N.M.; Hussain, R.; Kadir, M.R.A. Quinone-rich polydopamine functionalization of yttria stabilized zirconia for apatite biomineralization: The effects of coating temperature. *Appl. Surf. Sci.* **2015**, *346*, 317–328. [CrossRef]
15. Desante, G.; Pudełko, I.; Krok-Borkowicz, M.; Pamuła, E.; Jacobs, P.; Kazek-Kęsik, A.; Nießen, J.; Telle, R.; Gonzalez-Julian, J.; Schickle, K. Surface multifunctionalization of inert ceramic implants by calcium phosphate biomimetic coating doped with nanoparticles encapsulating antibiotics. *ACS Appl. Mater. Interfaces* **2023**, *15*, 21699–21718. [CrossRef] [PubMed]
16. Teng, F.; Zheng, Y.; Wu, G.; Beekmans, B.; Wismeijer, D.; Lin, X.; Liu, Y. Bone tissue responses to zirconia implants modified by biomimetic coating incorporated with BMP-2. *Int. J. Periodontics Restor. Dent.* **2019**, *39*, 371–379. [CrossRef] [PubMed]
17. Goldschmidt, G.-M.; Krok-Borkowicz, M.; Zybała, R.; Pamuła, E.; Telle, R.; Conrads, G.; Schickle, K. Biomimetic in situ precipitation of calcium phosphate containing silver nanoparticles on zirconia ceramic materials for surface functionalization in terms of antimicrobial and osteoconductive properties. *Dent. Mater.* **2021**, *37*, 10–18. [CrossRef] [PubMed]

18. Zamin, H.; Yabutsuka, T.; Takai, S.; Sakaguchi, H. Role of magnesium and the effect of surface roughness on the hydroxyapatite-forming ability of zirconia induced by biomimetic aqueous solution treatment. *Materials* **2020**, *13*, 3045. [CrossRef] [PubMed]
19. Cho, Y.; Hong, J.; Ryoo, H.; Kim, D.; Park, J.; Han, J. Osteogenic responses to zirconia with hydroxyapatite coating by aerosol deposition. *J. Dent. Res.* **2015**, *94*, 491–499. [CrossRef]
20. Kozelskaya, A.I.; Bolbasov, E.N.; Golovkin, A.S.; Mishanin, A.I.; Viknianshchuk, A.N.; Shesterikov, E.V.; Ashrafov, A.; Novikov, V.A.; Fedotkin, A.Y.; Khlusov, I.A.; et al. Modification of the ceramic implant surfaces from zirconia by the magnetron sputtering of different calcium phosphate targets: A comparative study. *Materials* **2018**, *11*, 1949. [CrossRef]
21. Pae, A.; Lee, H.; Noh, K.; Woo, Y.-H. Cell attachment and proliferation of bone marrow-derived osteoblast on zirconia of various surface treatment. *J. Adv. Prosthodont.* **2014**, *6*, 96–102. [CrossRef] [PubMed]
22. Aboushelib, M.N.; Shawky, R. Osteogenesis ability of CAD/CAM porous zirconia scaffolds enriched with nano-hydroxyapatite particles. *Int. J. Implant. Dent.* **2017**, *3*, 21. [CrossRef] [PubMed]
23. Faria, D.; Henriques, B.; Souza, A.C.; Silva, F.S.; Carvalho, O. Laser-assisted production of HAp-coated zirconia structured surfaces for biomedical applications. *J. Mech. Behav. Biomed. Mater.* **2020**, *112*, 104049. [CrossRef] [PubMed]
24. Carvalho, O.; Sousa, F.; Madeira, S.; Silva, F.S.; Miranda, G. HAp-functionalized zirconia surfaces via hybrid laser process for dental applications. *Opt. Laser Technol.* **2018**, *106*, 157–167. [CrossRef]
25. Mesquita-Guimarães, J.; Detsch, R.; Souza, A.C.; Henriques, B.; Silva, F.S.; Boccaccini, A.R.; Carvalho, O. Cell adhesion evaluation of laser-sintered HAp and 45S5 bioactive glass coatings on micro-textured zirconia surfaces using MC3T3-E1 osteoblast-like cells. *Mater. Sci. Eng. C Mater. Biol. Appl.* **2020**, *109*, 110492. [CrossRef] [PubMed]
26. Chichkov, B.N.; Momma, C.; Nolte, S.; von Alvensleben, F.; Tünnermann, A. Femtosecond, picosecond and nanosecond laser ablation of solids. *Appl. Phys. A* **1996**, *63*, 109. [CrossRef]
27. Delgado-Ruíz, R.A.; Calvo-Guirado, J.L.; Moreno, P.; Guardia, J.; Gomez-Moreno, G.; Mate-Sánchez, J.E.; Ramirez-Fernández, P.; Chiva, F. Femtosecond laser microstructuring of zirconia dental implants. *J. Biomed. Mater. Res. B Appl. Biomater.* **2011**, *96*, 91–100. [CrossRef]
28. Hashimoto, S.; Yasunaga, M.; Hirose, M.; Kakehata, M.; Yashiro, H.; Yamazaki, A.; Ito, A. Cell attachment area of rat mesenchymal stem cells correlates with their osteogenic differentiation level on substrates without osteoconductive property. *Biochem. Biophys. Res. Commun.* **2020**, *525*, 1081–1086. [CrossRef]
29. Kakehata, M.; Yashiro, H.; Oyane, A.; Ito, A.; Torizuka, K. Femtosecond laser-induced periodic surface structures on yttria-stabilized zirconia. In Proceedings of the 7th International Congress on Laser Advanced Materials Processing (LAMP2015), Kitakyushu, Japan, 26–29 May 2015, #15-017. This paper is available upon request to Japan Laser Processing Society. Available online: http://www.jlps.gr.jp/eng/symposium/index.html (accessed on 4 August 2015).
30. Yashiro, H.; Kakehata, M.; Umebayashi, N.; Ito, A. Crystalline hydroxyapatite coating by hydrolysis using β-tricalcium phosphate target by pulsed-laser deposition. *Jpn. J. Appl. Phys.* **2021**, *60*, 065501. [CrossRef]
31. Li, M.; Komasa, S.; Hontsu, S.; Hashimoto, Y.; Okazaki, J. Structural characterization and osseointegrative properties of pulsed laser-deposited fluorinated hydroxyapatite films on nano-zirconia for implant applications. *Int. J. Mol. Sci.* **2022**, *23*, 2416. [CrossRef]
32. Mutsuzaki, H.; Sogo, Y.; Oyane, A.; Ito, A. Improved bonding of partially osteomyelitic bone to titanium pins owing to biomimetic coating of apatite. *Int. J. Mol. Sci.* **2013**, *14*, 24366–24379. [CrossRef]
33. Boyan, B.D.; Berger, M.B.; Nelson, F.R.; Donahue, H.J.; Schwartz, Z. The biological basis for surface-dependent regulation of osteogenesis and implant osseointegration. *J. Am. Acad. Orthop. Surg.* **2022**, *30*, e894–e898. [CrossRef] [PubMed]
34. Parithimarkalaignan, S.; Padmanabhan, T.V. Osseointegration: An update. *J. Indian. Prosthodont. Soc.* **2013**, *13*, 2–6. [CrossRef] [PubMed]
35. Li, W.; Ding, Q.; Sun, F.; Liu, B.; Yuan, F.; Zhang, L.; Bao, R.; Gu, J.; Lin, Y. Fatigue behavior of zirconia with microgrooved surfaces produced using femtosecond laser. *Lasers Med. Sci.* **2023**, *38*, 33. [CrossRef] [PubMed]
36. Kakehata, M.; Oyane, A.; Yashiro, H.; Ito, A.; Okazaki, Y.; Torizuka, K. Bending strength and cyclic fatigue tests of yttria-stabilized zirconia ceramics modified with femtosecond-laser induced periodic surface structures for medical implants. In Proceedings of the 8th International Congress on Laser Advanced Material Processing (LAMP2019), Hiroshima, Japan, 21–24 May 2019; 2019, #19-063. This paper is available upon request to Japan Laser Processing Society. Available online: http://www.jlps.gr.jp/symposium/information/index.html (accessed on 3 September 2019).
37. Kakehata, M.; Ito, A.; Yashiro, H.; Oyane, A.; Torizuka, K. Effect of femtosecond laser surface treatment on bending strength of yttria-stabilized zirconia ceramics. In Proceedings of the 18th International Symposium on Laser Precision Microfabrication (LPM2017), Toyama, Japan, 5–8 June 2017; 2017, #17-76. This paper is available upon request to Japan Laser Processing Society. Available online: http://www.jlps.gr.jp/symposium/information/index.html (accessed on 28 July 2017). In the proceedings, correct parameters as b = 4 mm, d = 3 mm in Equation(1).
38. Toraya, H.; Yoshimura, M.; Somiya, S. Calibration curve for quantitative analysis of the monoclinic-tetragonal ZrO_2 system by X-ray diffraction. *J Am Ceram Soc* **1984**, *67*, C119. [CrossRef]

Disclaimer/Publisher's Note: The statements, opinions and data contained in all publications are solely those of the individual author(s) and contributor(s) and not of MDPI and/or the editor(s). MDPI and/or the editor(s) disclaim responsibility for any injury to people or property resulting from any ideas, methods, instructions or products referred to in the content.

Journal of Functional Biomaterials

Article

Characterization of Trabecular Bone Microarchitecture and Mechanical Properties Using Bone Surface Curvature Distributions

Pengwei Xiao [1,2,*], Caroline Schilling [3] and Xiaodu Wang [1]

1. Department of Mechanical Engineering, University of Texas at San Antonio, San Antonio, TX 78249, USA
2. Department of Orthopedic Surgery, Massachusetts General Hospital, Harvard Medical School, 55 Fruit St., Boston, MA 02114, USA
3. Department of Biomedical Engineering, University of Texas at San Antonio, San Antonio, TX 78249, USA
* Correspondence: pxiao2@mgh.harvard.edu

Abstract: Understanding bone surface curvatures is crucial for the advancement of bone material design, as these curvatures play a significant role in the mechanical behavior and functionality of bone structures. Previous studies have demonstrated that bone surface curvature distributions could be used to characterize bone geometry and have been proposed as key parameters for biomimetic microstructure design and optimization. However, understanding of how bone surface curvature distributions correlate with bone microstructure and mechanical properties remains limited. This study hypothesized that bone surface curvature distributions could be used to predict the microstructure as well as mechanical properties of trabecular bone. To test the hypothesis, a convolutional neural network (CNN) model was trained and validated to predict the histomorphometric parameters (e.g., BV/TV, BS, Tb.Th, DA, Conn.D, and SMI), geometric parameters (e.g., plate area PA, plate thickness PT, rod length RL, rod diameter RD, plate-to-plate nearest neighbor distance NND_{PP}, rod-to-rod nearest neighbor distance NND_{RR}, plate number PN, and rod number RN), as well as the apparent stiffness tensor of trabecular bone using various bone surface curvature distributions, including maximum principal curvature distribution, minimum principal curvature distribution, Gaussian curvature distribution, and mean curvature distribution. The results showed that the surface curvature distribution-based deep learning model achieved high fidelity in predicting the major histomorphometric parameters and geometric parameters as well as the stiffness tenor of trabecular bone, thus supporting the hypothesis of this study. The findings of this study underscore the importance of incorporating bone surface curvature analysis in the design of synthetic bone materials and implants.

Keywords: surface curvature; trabecular bone; histomorphometric parameters; stiffness tensor; geometric parameter; deep learning; convolution neural network

Citation: Xiao, P.; Schilling, C.; Wang, X. Characterization of Trabecular Bone Microarchitecture and Mechanical Properties Using Bone Surface Curvature Distributions. *J. Funct. Biomater.* **2024**, *15*, 239. https://doi.org/10.3390/jfb15080239

Academic Editor: Kunyu Zhang

Received: 4 August 2024
Revised: 19 August 2024
Accepted: 20 August 2024
Published: 22 August 2024

Copyright: © 2024 by the authors. Licensee MDPI, Basel, Switzerland. This article is an open access article distributed under the terms and conditions of the Creative Commons Attribution (CC BY) license (https://creativecommons.org/licenses/by/4.0/).

1. Introduction

Trabecular bone, characterized by a sponge-like structure, comprises a network of plates and rods at the microstructural level [1]. The microstructure of trabecular bone is essentially important for determining bone's resistance to fractures [2]. Currently, the major method for clinical assessment of bone microstructure is mainly based on dual-energy X-ray absorptiometry (DXA)-based trabecular bone score (TBS) [3,4], an indirect method based on bone mineral density known for its limited accuracy. Recent advancements in micro-CT technology have led to the development of micro-CT image-based reconstruction methods for evaluating trabecular bone microstructure [5,6]. This method encompasses a full set of histomorphometric parameters, including degree of anisotropy (DA), bone volume fraction (BV/TV), connectivity density (Conn.D), bone surface area (BS), structure model index (SMI), and trabecular thickness (Tb.Th), offering an overall evaluation of bone microstructure [7]. However, the histomorphometric parameters are scalar and averaged measures of trabecular microarchitecture at global levels, thus could not fully capture the variance of microarchitectural properties and their influence on mechanical properties of trabecular bone [8]. In addition, Columbia University has developed an Individual Trabeculae Segmentation ITS technique to characterize the geometric parameters of trabecular bone, allowing for the segmentation of trabecular bone into individual plates and rods [9]. Consequently, the description of bone microstructure could encompass parameters related to size (plate area PA, plate thickness PT, rod length RL, rod diameter RD), spatial arrangement (plate-to-plate nearest neighbor distance NND_{PP}, rod-to-rod nearest neighbor distance NND_{RR}), trabeculae number (plate number PN, rod number RN), and orientation, thus providing more microarchitectural features of bone microstructure that contribute to the mechanical competence of trabecular bone. However, these geometric parameters only provide detailed information about individual trabeculae but interpreting these parameters in the context of bone mechanical competence can be challenging. Consequently, the fundamental microarchitectural characteristics of trabecular bone remain to be fully explored.

Recently, a novel methodology utilizing surface curvatures has been proposed for the comprehensive characterization of cancellous microstructure [10]. Various surface curvatures, such as maximum principal curvature, minimum principal curvature, Gaussian curvature, and mean curvature, offer a direct means of assessing the local geometry of bone in terms of convexity and concavity. Moreover, research has demonstrated a strong correlation between surface curvatures, SMI, and Euler number (which can be used to quantify the connectivity of trabecular bone) [11], underscoring the significance of bone surface curvatures as a pivotal metric for delineating both local and global bone geometry and effectively capturing diverse spatial structural aspects. Given the intimate relationship between bone structure and its mechanical properties, the implications of surface curvatures on bone mechanical behavior are noteworthy. Hence, understanding bone surface curvatures is crucial for bone material design as well as prediction of the mechanical behavior and functionality of bone structures. Nevertheless, to the best of our knowledge, few studies have been conducted to investigate how the surface curvatures are quantitatively related to the bone microarchitecture as well as its mechanical properties.

In order to characterize the surface geometry of trabecular bone, it is essential to quantify bone surface curvature distributions [11]. This leads to the technical question of how to describe the surface curvature distributions by utilizing specific parameters and establishing the relationships between bone surface curvature distributions and bone microstructure. Our previous studies [12–14] have shown that a 2D projection image can be used to describe its 3D bone microstructure and mechanical properties by using a deep learning (DL) approach. Inspired by the aforementioned applications, we proposed to describe the curvature spatial distributions through a 2D projection image-based approach and employ a DL model to establish relationships between surface curvature distributions and bone microstructure as well as its mechanical properties.

In this study, we hypothesized that there exists a strong correlation between bone surface curvature distributions and the microstructure as well as mechanical properties of trabecular bone, and hence bone surface curvature distributions could be used as a holistic indicator for prediction of both bone microstructure and mechanical behavior. To test the hypothesis, the spatially distributed surface curvatures across the surface of trabecular bone were projected onto a two-dimensional plane, and then a deep learning model was developed to predict the histomorphometric parameters, geometric parameters, as well as mechanical properties of trabecular bone using the above two-dimensional projections of bone surface curvature distributions.

2. Materials and Methods

2.1. Preparation of Trabecular Bone Specimens and Micro-CT Image-Based Reconstruction

A total of six cadaveric proximal femurs were collected from six different donors (three males and three females, with a mean age of 48.5 ± 24 years) with Institutional Biosafety Committee (IBC) approval (IBC#B94-01-21). All proximal femurs were scanned using a micro-CT system (Sky-Scan 1173, Bruker, Billerica, MA, USA) with a resolution of 35 µm, which was able to capture the trabecular microstructure. Then, a total of eight hundred and sixty-eight trabecular cubes, each with the dimensions of 6 mm × 6 mm × 6 mm, were dissected out from the micro-CT images of the six proximal femurs to serve as representative volume elements (RVEs). It should be noted that trabecular cubes with low BV/TV and/or minimal trabeculae, as well as those containing cortical bone, were excluded from this study. Finally, all trabecular cubes were constructed digitally using STL format.

2.2. Calculation of Bone Surface Curvatures

Surface curvatures of trabecular bone were computed utilizing the STL format of the reconstruction of trabecular bone. A previous study demonstrated that bone surface curvatures, including the principal curvatures, mean curvatures, and Gaussian curvatures, could distinguish bone microstructure across different locations, suggesting their potential for predicting bone failure [10]. Therefore, the maximum and minimum principal curvatures, as well as the mean and Gaussian curvatures (Figure 1), were employed to define the surface curvature of the trabecular bone in this study, as these surface curvatures could capture the most fundamental shape details [10]. Specifically, the maximum and minimum principal curvatures (K_1, K_2) of the trabecular bone surface were computed on the triangle meshes of the trabecular bone surface using a finite-differences approach [15]. Subsequently, Gaussian curvature K and mean curvature H were defined as follows:

$$K = K_1 K_2 \tag{1}$$

$$H = (K_1 + K_2)/2 \tag{2}$$

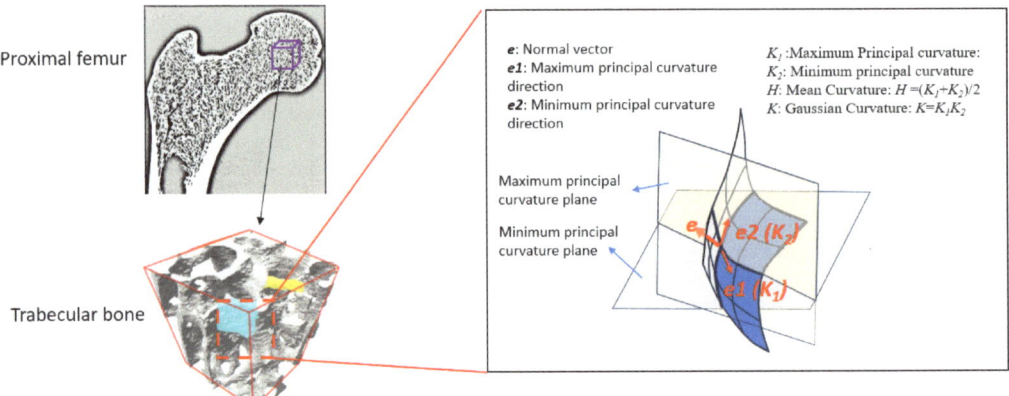

Figure 1. The schematic represents trabecular bone surface curvature using maximum principal curvature (K_1), minimum principal curvature (K_2), mean curvature (H), and Gaussian curvature (K).

The above four types of surface curvatures were computed for each vertex of the triangle meshes in each trabecular cube using MATLAB R2023a (The MathWorks, Inc., Natick, MA, USA).

2.3. Characteristics of Trabecular Microarchitecture

To describe the microarchitecture of trabecular bone globally, six histomorphometric parameters were measured from the micro-CT images of the trabecular bone cubes using ImageJ (1.52 h) and BoneJ (https://bonej.org/). These six histomorphometric parameters included bone volume fraction (BV/TV), bone surface area (BS), trabecular thickness (Tb.Th), structure model index (SMI), the degree of anisotropy (DA), and connectivity density (Conn.D). Using these six histomorphmetric parameters, trabecular bone mass, trabecular size, number, structure types (either plate-like or rod-like), and trabecular orientation can be defined accurately.

Moreover, the geometric parameters of trabecular bone cubes were also assessed utilizing a novel individual trabecula segmentation (ITS) technique [9]. With recent advancements in biomedical image processing technologies, the microarchitecture of trabecular bone can be segmented into individual trabecular plates and rods, allowing describing the microarchitecture of trabecular bone using trabecular number (plate number PN, rod number RN), trabecular size (mean plate thickness PT, mean plate area PA, mean rod diameter RD, mean rod length RL), trabecular arrangement (mean plate-to-plate nearest neighbor distance NND_{PP}, mean rod-to-rod nearest neighbor distance NND_{RR}), and trabecular orientation [8,16]. Using the above parameters, the geometry of trabecular bone could be precisely defined.

2.4. Determination of the Mechanical Properties of Trabecular Bone Using the Micro-FE Method

The anisotropic mechanical behavior of trabecular cubes can be described in terms of the apparent stiffness tensor. The stiffness tensor of trabecular cubes, which is a fourth-rank tensor and is considered elastically orthotropic with three mutually perpendicular planes of symmetry [17,18], can be simplified as

$$C = \begin{bmatrix} C_{11} & C_{12} & C_{13} & 0 & 0 & 0 \\ C_{21} & C_{22} & C_{23} & 0 & 0 & 0 \\ C_{31} & C_{32} & C_{33} & 0 & 0 & 0 \\ 0 & 0 & 0 & C_{44} & 0 & 0 \\ 0 & 0 & 0 & 0 & C_{55} & 0 \\ 0 & 0 & 0 & 0 & 0 & C_{66} \end{bmatrix} \quad (3)$$

where $C_{12} = C_{21}$, $C_{13} = C_{31}$, and $C_{23} = C_{32}$. In this study, trabecular cubes with the dimension of 6 mm × 6 mm × 6 mm dissected out from the various anatomic regions of the femurs, such as the femur head, neck, and greater trochanter regions, were used as RVEs, and the stiffness tensor of trabecular cubes was assessed using micro-CT-based finite element (FE) simulations. The FE analysis was conducted using Abaqus 2021/Standard software package. Specifically, a direct voxel conversion method was employed to transform each voxel of the digitized trabecular cubes into first-order tetrahedral elements (C3D4), generating approximately 0.5 to 2.5 million tetrahedral elements for each trabecular cube. Trabecular bone was assumed to be homogeneous, linearly elastic, and isotropic material, with a Young's modulus of 15 GPa and a Poisson's ratio of 0.3 [19]. Then, six uniform boundary conditions, including three uniaxial compression tests along the three orthogonal coordinate axes and three pure shear tests in the three orthogonal planes, were applied to the FE model sequentially to assess the stiffness matrix. Finally, the stiffness tensor was obtained by rotating the fabric coordinate axes of the stiffness matrix to its principal axes [20] using the MSAT (a toolkit for the analysis of elastic and seismic anisotropy) in a MATLAB environment.

2.5. Development of DL Model

2.5.1. Characterization of Bone Surface Curvature Distributions Using a 2D Projection Image-Based Approach

In this study, the spatial distributions of bone surface curvatures on each vertex of the triangle mesh of trabecular bone were characterized by projecting the surface curvatures onto a 2D plane (Figure 2). Our previous studies have shown that the 2D projection of properties effectively captures their 3D spatial distribution and could be effectively learned by the DL model. Thus, we projected the bone surface curvatures, including maximum principal curvature, minimum principal curvature, Gaussian curvature, and mean curvature, onto four different 2D planes for each trabecular cube using custom MATLAB scripts (MathWorks, Natick, MA, USA). Briefly, the 2D plane was meshed at the resolution of 172 pixels (bins) × 172 pixels (bins), matching the resolution of the micro-CT images for each trabecular cube. Next, the trabecular cubes were meshed along the projection direction with a thickness of one voxel, generating n = 172 planer layers. Then, the curvature values at each vertex in each planer layer of the trabecular cubes were projected to the 2D plane. The curvature values at each bin were obtained by summing all the curvature values falling onto the bin using the following equation:

$$K(x,y) = \frac{1}{n}\sum_{z=1}^{n} k(x,y,z) \quad (4)$$

where, K is the curvature value of the bin at the location (x, y) on the 2D projection plane; k is the summation of the curvature values at the location (x, y, z) in the trabecular cube; n is the number of plane layers of the trabecular cube in the projection direction (n = 172). Finally, the 2D projection plane was converted into a 2D image. In this study, the 2D projection images of maximum principal curvature, minimum principal curvature, Gaussian curvature, and mean curvature were used as input to train the DL model.

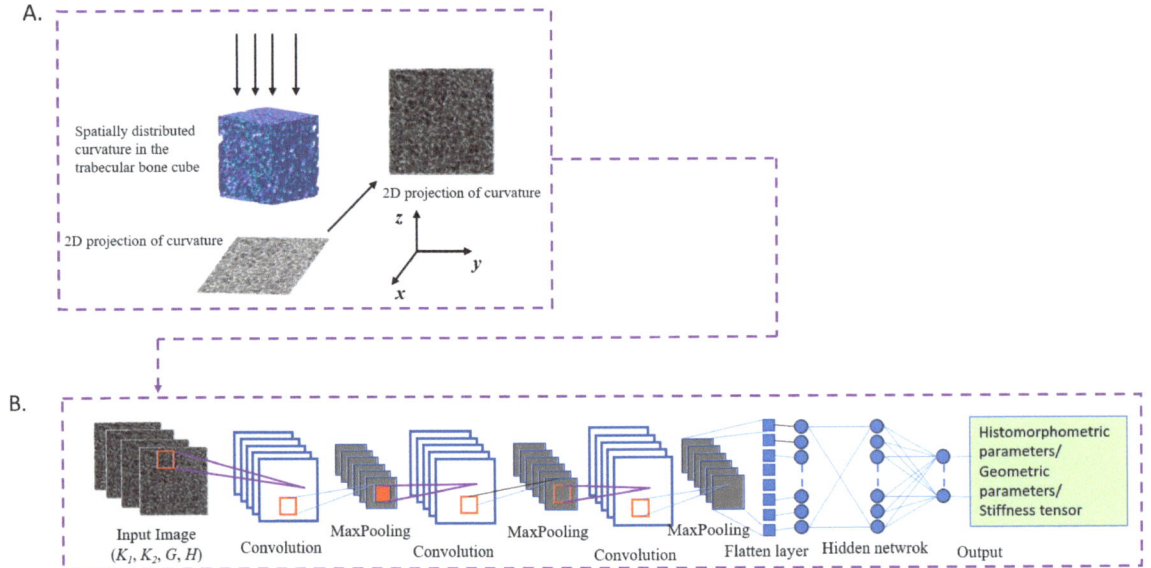

Figure 2. The schematic framework DL model based on 2D projections of trabecular surface curvatures (maximum principal curvature (K_1), minimum principal curvature (K_2), mean curvature (H), and Gaussian curvature (K)). (**A**). Projection of trabecular surface curvatures onto a 2D plane. (**B**). The architecture of the CNN model with the 2D projection images of curvatures as input and the histomorphometric parameters/geometric parameters/stiffness tensor as output.th.

2.5.2. Convolutional Neural Network (CNN) Modeling

In order to explore the correlations between bone surface curvature distributions and the microstructure and mechanical properties of trabecular bone, one CNN model was developed and trained in this study to predict the histomorphometric parameters, geometric parameters, as well as mechanical properties of trabecular bone based on the 2D projection images of bone surface curvatures. The architecture of the proposed CNN model is illustrated in Figure 2, consisting of multiple convolutional layers, max-pooling layers, and a fully connected neural network followed by the outputs. During the training process, the 2D projection images of bone surface curvatures were used as input, while the histomorphometric parameters, geometric parameters, and the apparent stiffness tensor were used as output, respectively. The mean square error (MSE) was utilized as a loss function throughout the training process. Furthermore, hyperparameter optimization was conducted to meticulously refine the architecture of the CNN model to achieve optimal performance across the training process. The parameters assessed in the CNN architecture comprised the number of hidden layers, the number of filters, the number of convolutional layers, kernel size, the optimizer functions, the learning rates, the number of epochs, and the dropout rate. Finally, the details of the optimized architecture of the CNN model were shown in Table 1.

Table 1. Optimized architecture of CNN model in prediction of histomorphometric parameters, geometric parameters, and stiffness tensor using bone surface curvature distributions.

Models	Input	Kernel Size	Pool Size	Convolutional Layers	# of Hidden Layers	Learning Rates	No. of Epochs	No. of Filters	Drop-Out	Output
#1	Bone surface curvature distributions	3 × 3	2 × 2	(8, 16, 32)	3	0.0001	200	128 × 64 × 6	0.3	Histomorphometric parameters
#2		5 × 5	2 × 2	(16, 16, 64)	3	0.0001	300	128 × 64 × 8	0.4	Geometric parameters
#3		3 × 3	2 × 2	(16, 32, 64)	3	0.0001	250	128 × 64 × 9	0.5	Stiffness tensor

In addition, in order to minimize the effect of different scales on the results, all output parameters were normalized by using rescaling (min-max normalization) before training the CNN model using the following formula:

$$x' = \frac{x - \min(x)}{\max(x) - \min(x)} \quad (5)$$

where x is the original value, x' is the normalized value.

In this study, 80% of the datasets were randomly selected as training datasets, while the remaining 20% were used as testing datasets. The CNN model was programmed in Python using the Keras library with a TensorFlow backend and was trained on a Dell desktop computer (XPS 8930, Intel Core i9-9900k 8-Core Processor, 64 GB Memory, NVIDIA R GeForce® GTX 1080 with 8 GB GDDR5X Graphic Memory, Dell, Round Rock, TX, USA).

2.6. Data Analysis

The correlations between the distributions of bone surface curvature and the histomorphometric parameters, geometric parameters, as well as the mechanical properties of trabecular bone were evaluated by quantifying the prediction accuracy of the DL model in predicting those parameters. By performing the linear regression analyses, the prediction accuracy of the DL model was assessed using the Pearson correlation coefficient (R^2), with significance determined at $p < 0.05$. All the statistical analyses were performed using IBM SPSS software (version 29, IBM, Chicago, IL, USA).

3. Results

3.1. Correlation between Bone Aurface Curvature Distributions and Histmorphometric Parameters of Trabecular Bone

The linear regression analyses were performed to assess the prediction accuracy of the surface curvature-based DL model in predicting the histomorphometric parameters (Figure 3). The results showed that the histomorphometric parameters predicted by the surface curvature-based DL model were consistent with those measured directly from micro-CT images. The Pearson correlation coefficients (R^2) were 0.96, 0.94, 0.90, 0.79, 0.57, and 0.11 for bone surface (BS), bone volume fraction (BV/TV), trabecular thickness (Tb.Th), structural model index (SMI), connectivity density (Conn.D), and the degree of anisotropy (DA), respectively, with all the p-values < 0.0001. Employing R^2 as an indicator of predictive accuracy of the surface curvature-based DL model, the results suggest that bone surface curvature distributions were significantly correlated with BS, BV/TV, Tb.Th, SMI, and Conn.D, with the exception of DA. Moreover, bone surface curvature distributions exhibited the highest Pearson correlation coefficient with BS among the histomorphometric parameters, whereas a weak correlation was observed between bone surface curvature distributions and DA.

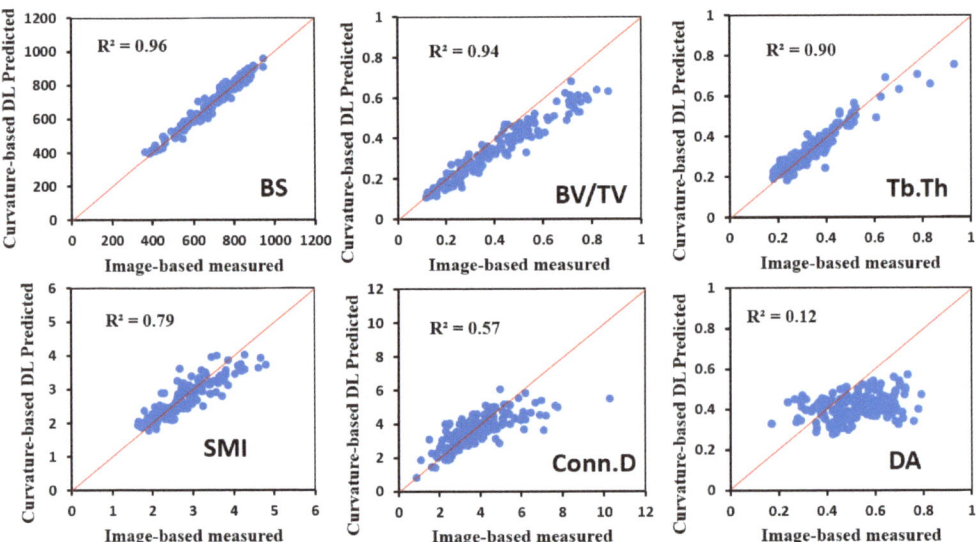

Figure 3. Regression plots of the microstructural parameters predicted by the curvature-based DL model vs. measured by micro-CT images.

3.2. Correlation between Bone Surface Curvature Distributions and Geometric Parameters of Trabecular Bone

The linear regression analyses were also used to assess the prediction accuracy of the geometric parameters, including trabecular size, spatial arrangement, and trabecular number (Figure 4). The results indicated the surface curvature-based DL model exhibited reasonably high accuracy in predicting plate area (PA), plate thickness (PT), and rod length (RL), with Pearson correlation coefficients R^2 of 0.80, 0.63, and 0.79, respectively. However, the model demonstrated lower prediction accuracy for rod diameter (RD), plate-to-plate nearest neighbor distance (NND_{PP}), and rod-to-rod nearest neighbor distance (NND_{RR}), with R^2 values of 0.36, 0.36, and 0.10, respectively. These findings indicated strong correlations between bone surface curvature distributions and PA, PT, and RL but weak correlations with RD, NND_{PP}, and NND_{RR}. Additionally, the Pearson correlation coefficients R^2 were 0.83 for plate number PN and 0.34 for rod number RN, suggesting a strong correlation between bone surface curvature distributions and plate number but a weak correlation with rod number.

3.3. Correlation between Bone Surface Curvature Distributions and Apparent Stiffness Tensor of Trabecular Bone

The prediction accuracy of the bone surface curvature-based DL model in predicting the constant components of the apparent stiffness tensor was assessed by comparing the

DL-predicted stiffness tensor with the ground-true values measured using FE simulations. The Pearson correlation coefficients (R^2) were 0.89, 0.89, 0.88, 0.90, 0.90, 0.90, 0.87, 0.87, and 0.86 for the apparent stiffness tensor constants C_{11}, C_{22}, C_{33}, C_{44}, C_{55}, C_{66}, C_{12}, C_{13}, and C_{23}, respectively, with all the *p*-values < 0.001 (Figure 5), suggesting high correlations between bone surface curvature distributions and the apparent stiffness tenor of trabecular cube. These findings imply that bone surface curvature distributions could be used to predict the anisotropic mechanical behavior of trabecular bone with high accuracy.

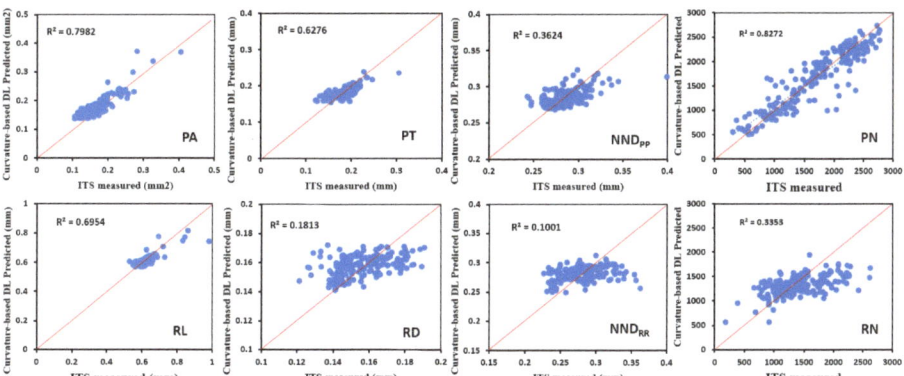

Figure 4. Regression plots of the geometric parameters predicted by the curvature-based DL model vs. measured by micro-CT images.

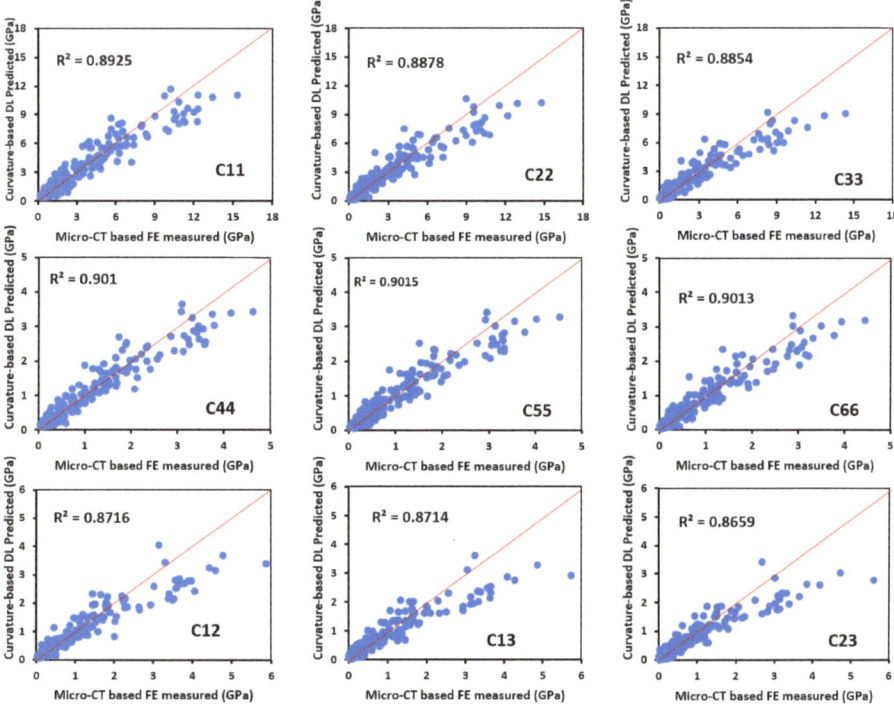

Figure 5. Regression plots of the stiffness tensor predicted by the curvature-based DL model vs. measured by micro-CT images.

4. Discussion

This study investigated the correlations between bone surface curvature distributions and trabecular microstructure as well as mechanical properties using a DL approach. A surface curvature-based CNN model was developed and trained to predict the histomorphometric parameters, geometric parameters, and the apparent stiffness tensor of trabecular bone. The results demonstrated that bone surface curvature distributions were not only highly correlated with the major histomorphometric parameters and geometric parameters, but also with the apparent stiffness tensor of trabecular bone. These findings supported the hypothesis that bone surface curvature distributions can serve as a holistic parameter for predicting bone microstructure and mechanical behavior with reasonably high accuracy, underscoring the significance of incorporating bone surface curvature analysis in the design of synthetic bone materials and implants.

Previous studies have indicated that bone surface curvatures primarily capture the local geometry of trabecular bone [11]. However, our study reveals that bone surface curvature distributions could be used to effectively predict the histomorphometric parameters of trabecular bone using the DL model, suggesting that bone surface curvature distributions can be used to evaluate the overall changes of bone microstructure. The results in this study demonstrated bone surface curvature distributions exhibited the strongest correlation coefficients with bone BS ($R^2 = 0.96$) and Tb.Th ($R^2 = 0.90$) among the histomorphometric parameters. Previous studies have shown that BS is a function of the integration of bone surface curvature [11]. Given the strong correlation between bone surface curvatures and BS as well as Tb.Th, it is reasonable to infer a similarly strong correlation with bone volume fraction (BV/TV) ($R^2 = 0.94$). Indeed, by plotting the distributions of bone curvature of trabecular cubes with different BV/TV values (Figure 6), BV/TV can be clearly identified by the distributions of bone surface curvatures. In addition, serval studies have demonstrated the correlation between bone surface curvatures and SMI [11]. This study is the first attempt to predict SMI using bone surface curvature distributions with high accuracy ($R^2 = 0.79$). Furthermore, this study also finds a reasonable correlation between bone surface curvature distributions and Conn.D. but a low correlation between bone surface curvatures and DA. Nonetheless, this study demonstrated bone surface curvature distributions could accurately predict the major histomorphometric parameters of trabecular bone with reasonable accuracy, suggesting the significance of bone surface curvature distributions in assessing the overall microstructural changes of trabecular bone.

This study also investigated the correlations between bone surface curvature distributions and the geometric parameters measured using individual trabecula segmentation (ITS) analysis, such as trabecular size, spatial arrangement, and trabecular number. These parameters describe the local microstructural features of trabecular bone as well as its mechanical properties [8]. High correlations were found between bone surface curvature distributions and parameters such as PA, PT, and RL, with R^2 values of 0.80, 0.63, and 0.70, respectively, suggesting that bone surface curvature distributions are sensitive to the changes in trabecular plate area, plate thickness, and rod length. However, a low correlation was observed between bone surface curvature distributions and rod diameter (RD) ($R^2 = 0.18$), implying that bone surface curvature distributions might not be able to accurately capture the changes in rod diameter. In addition, it is interesting to find that there were strong correlations between bone surface curvature distributions and plate number ($R^2 = 0.83$), an important parameter closely related to the onset of osteoporosis. It has been observed that plate-like trabeculae were seriously depleted in patients with osteoporotic fractures [21]. The results in this study further demonstrated that bone surface curvature distributions are able to pick up the local changes in bone microstructure during the onset of osteoporosis.

To the best of our knowledge, this study represents the first attempt to predict the mechanical properties of trabecular bone using a DL model based on the distributions of bone surface curvatures. The strong correlations between bone surface curvature distributions and stiffness tensors of trabecular bone ($R^2 = 0.89$–0.91) demonstrated the capability of surface curvature distributions in predicting the anisotropic mechanical properties of trabecular bone. This suggests that bone surface curvature distributions can serve as a novel parameter for governing the mechanical behavior of trabecular bone. Previous studies have shown that the structure-function of trabecular bone is mainly attributed to a full set of histomorphometric parameters [14,22], such as BV/TV, SMI, Conn.D, Tb.Th, BS, and DA. However, the changes in individual histomorphometric parameters might not be able to fully reflect the changes in mechanical behavior. Instead of using a full set of histomorphometric parameters, the high correlations between bone surface curvature distributions and the microstructure as well as mechanical properties allow bone surface curvature distributions to be a holistic parameter in capturing the subtle changes in bone microstructure as well as bone mechanical properties.

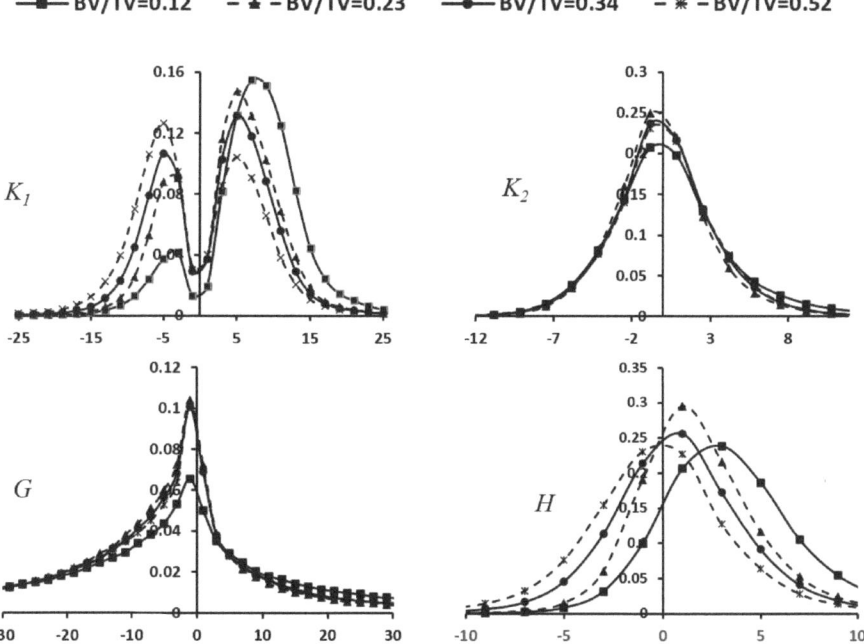

Figure 6. Probability distribution of surface curvatures (maximum principal curvature K_1, minimum principal curvature K_2, Gaussian curvature K, and mean curvature H) vs. bone volume fraction (BV/TV).

Indeed, the spatial distributions of bone surface curvatures could reveal various geometric characteristics of bone microstructures (Figure 7). By examining the spatial distributions of bone surface curvatures, the maximum principal curvature and mean curvature seem to effectively characterize the overall framework of bone microstructure, while the minimum principal curvature and Gaussian curvature are more appropriate for capturing local topological characteristics. Moreover, previous studies have shown that mean curvature describes the local convexity or concavity of a surface, whereas Gaussian curvature delineates different surface types, including saddle-shaped regions ($K < 0$), intrinsically flat regions ($K = 0$), and sphere-shaped regions ($K > 0$) [10]. Additionally, this study investigated the correlations between individual surface curvature distributions and the histomorphometric parameters of trabecular bone using DL models. The results (Table 2) showed that DL models based on maximum principal curvature distribution and mean curvature distribution demonstrated higher prediction accuracy in predicting bone surface area, whereas Gaussian curvature-based DL models demonstrated higher prediction accuracy in predicting trabecular thickness, thus suggesting that maximum principal curvature and mean curvature are more closely correlated with bone surface area while Gaussian curvature is more strongly correlated with trabecular thickness. Furthermore, the results indicated that the prediction accuracy of DL models using individual surface curvature as input is comparable to that of DL models using all four surface curvatures as input, implying that each surface curvature distribution contains the major geometric characteristics regarding bone microstructure.

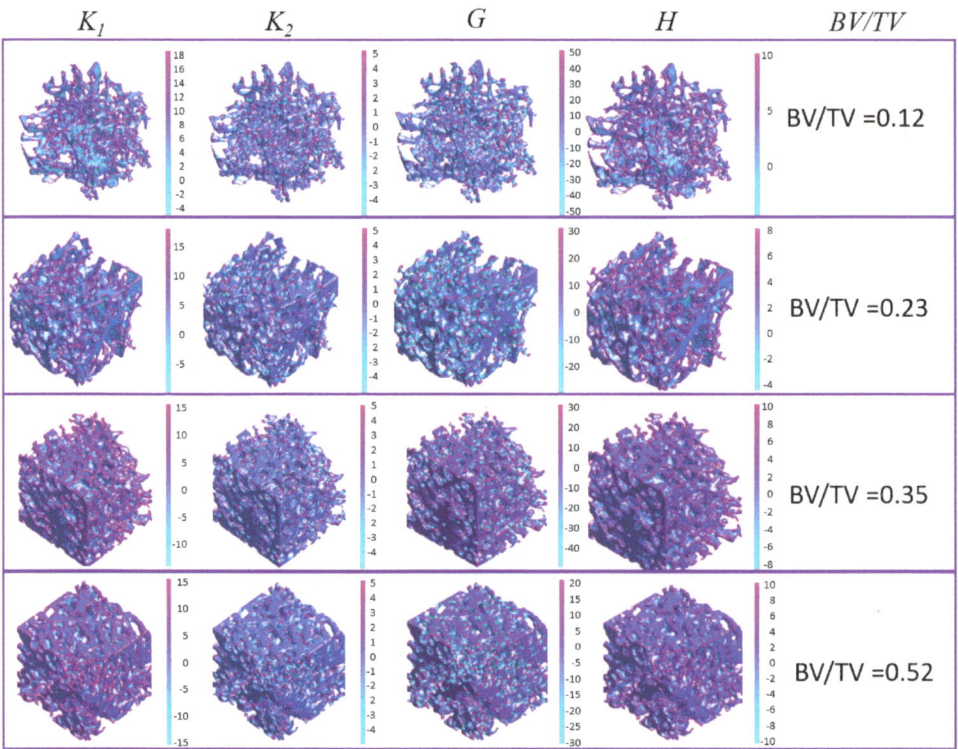

Figure 7. Plots of surface curvatures (maximum principal curvature K_1, minimum principal curvature K_2, Gaussian curvature K, and mean curvature H) over trabecular surface.

Table 2. Comparison of prediction accuracies of microstructural parameters of trabecular bone using different inputs for DL model.

Inputs for DL Model	Prediction Accuracy (R^2)					
	BS	BV/TV	Tb.Th	SMI	Conn.D	DA
K_1	0.94	0.92	0.86	0.77	0.42	0.11
K_2	0.91	0.91	0.85	0.76	0.44	0.11
G	0.84	0.92	0.91	0.77	0.37	0.08
H	0.94	0.91	0.88	0.76	0.47	0.06
K_1, K_2, G, H	0.96	0.94	0.90	0.79	0.57	0.12

Moreover, surface curvatures have been extensively applied in various fields. Several studies have been conducted on applying bone surface curvatures for segmenting and labeling bone surface regions due to their reliable detection of geometric features [23–26]. Furthermore, researchers have applied bone surface curvature to fabricate tissue scaffolds [27], indicating that bone surface curvature allows to create a library of mathematically designed scaffolds, showing its potential for regeneration of anisotropic bone structure. Guo et al., proposed a deep learning approach for designing structures with targeted surface curvature [28], which could be designed to promote mechanical behavior. Researchers further examined the effects of the curvature of the femur and tibia on biomechanical behavior during unloaded uphill locomotion [29], highlighting the strong correlation between bone curvature and locomotor function, as well as underlying skeletal structure. These findings demonstrated the significant role of bone surface curvature in governing the structure and mechanical behavior of bone.

Several limitations should be acknowledged in this study. Firstly, training a robust DL model typically necessitates a comprehensive dataset. However, this study included only six proximal femurs from six distinct donors, which might not represent the general population's bone surface curvatures. Nonetheless, the findings of this study are still valid to support the hypothesis of this study. Secondly, the projection of bone surface curvature distributions onto a 2D plane might not fully capture the spatial distributions of surface curvature by DL model. Future studies could investigate the prediction of bone microstructure as well as mechanical behavior using 3D surface curvature distributions as input for the DL model.

5. Conclusions

This study is the first to quantitatively correlate bone surface curvature distributions with both the microstructure and the mechanical behavior of trabecular bone using a deep learning (DL) model, demonstrating that bone surface curvature distributions can serve as a novel parameter governing the microstructure and mechanical behavior of trabecular bone. The DL model based on the surface curvature distributions demonstrated a high fidelity in predicting the microstructure as well as the mechanical properties of trabecular bone, thus verifying the hypothesis of this study. In addition, the following conclusions could be achieved: Firstly, the maximum principal curvature and mean curvature could effectively capture the overall framework of bone microstructure, whereas the minimum principal curvature and Gaussian curvature are better suited for capturing the local topological features. Secondly, each surface curvature distribution contains the major geometric characteristics regarding bone microstructure. Finally, bone surface curvature could serve as a holistic parameter in describing the bone microstructure and mechanical behavior. The findings of this study underscore the significance of incorporating bone surface curvature analysis in the design of synthetic bone materials and implants.

Author Contributions: Conceptualization, P.X. and X.W.; methodology, P.X.; software, C.S.; validation, P.X., C.S. and X.W.; formal analysis, P.X.; investigation, P.X.; resources, X.W.; data curation, C.S.; writing—original draft preparation, P.X.; writing—review and editing, C.S., X.W.; visualization, P.X.; supervision, X.W.; project administration, X.W. All authors have read and agreed to the published version of the manuscript.

Funding: This research received no external funding.

Data Availability Statement: The original contributions presented in the study are included in the article, further inquiries can be directed to the corresponding author.

Conflicts of Interest: The authors declare no conflicts of interest.

References

1. Gibson, L.J.; Ashby, M.F. *Cellular Solids: Structure and Properties*; Cambridge University Press: Cambridge, UK, 1997.
2. Seeman, E.; Delmas, P.D. Mechanisms of disease—Bone quality—The material and structural basis of bone strength and fragility. *N. Engl. J. Med.* **2006**, *354*, 2250–2261. [CrossRef] [PubMed]
3. Harvey, N.; Glüer, C.; Binkley, N.; McCloskey, E.; Brandi, M.-L.; Cooper, C.; Kendler, D.; Lamy, O.; Laslop, A.; Camargos, B.; et al. Trabecular bone score (TBS) as a new complementary approach for osteoporosis evaluation in clinical practice. *Bone* **2015**, *78*, 216–224. [CrossRef] [PubMed]

4. Pothuaud, L.; Carceller, P.; Hans, D. Correlations between grey level variations on 2D DXA-images (TBS) and 3D microarchitecture in human cadaver bone samples. *Bone* **2007**, *40*, S248.
5. Ohs, N.; Collins, C.J.; Atkins, P.R. Validation of HR-pQCT against micro-CT for morphometric and biomechanical analyses: A review. *Bone Rep.* **2020**, *13*, 100711. [CrossRef]
6. Steiner, L.; Synek, A.; Pahr, D.H. Comparison of different microCT-based morphology assessment tools using human trabecular bone. *Bone Rep.* **2020**, *12*, 100261. [CrossRef] [PubMed]
7. Odgaard, A. Three-dimensional methods for quantification of cancellous bone architecture. *Bone* **1997**, *20*, 315–328. [CrossRef]
8. Haque, E.; Xiao, P.; Ye, K.; Wang, X. Probability-based approach for characterization of microarchitecture and its effect on elastic properties of trabecular bone. *J. Mech. Behav. Biomed. Mater.* **2022**, *131*, 105254. [CrossRef]
9. Liu, X.S.; Cohen, A.; Shane, E.; Stein, E.; Rogers, H.; Kokolus, S.L.; Yin, P.T.; McMahon, D.J.; Lappe, J.M.; Recker, R.R.; et al. Individual trabeculae segmentation (ITS)–based morphological analysis of high-resolution peripheral quantitative computed tomography images detects abnormal trabecular plate and rod microarchitecture in premenopausal women with idiopathic osteoporosis. *J. Bone Miner. Res.* **2010**, *25*, 1496–1505. [CrossRef]
10. Callens, S.J.; Betts, D.C.T.N.; Müller, R.; Zadpoor, A.A. The local and global geometry of trabecular bone. *Acta Biomater.* **2021**, *130*, 343–361. [CrossRef]
11. Jinnai, H.; Watashiba, H.; Kajihara, T.; Nishikawa, Y.; Takahashi, M.; Ito, M. Surface curvatures of trabecular bone microarchitecture. *Bone* **2002**, *30*, 191–194. [CrossRef]
12. Xiao, P.; Zhang, T.; Dong, X.N.; Han, Y.; Huang, Y.; Wang, X. Prediction of trabecular bone architectural features by deep learning models using simulated DXA images. *Bone Rep.* **2020**, *13*, 100295. [CrossRef]
13. Xiao, P.W.; Zhang, T.; Haque, E.; Wahlen, T.; Dong, X.N.; Huang, Y.; Wang, X. Prediction of Elastic Behavior of Human Trabecular Bone Using A DXA Image-Based Deep Learning Model. *JOM* **2021**, *73*, 2366–2376. [CrossRef]
14. Xiao, P.; Haque, E.; Zhang, T.; Dong, X.N.; Huang, Y.; Wang, X. Can DXA image-based deep learning model predict the anisotropic elastic behavior of trabecular bone? *J. Mech. Behav. Biomed. Mater.* **2021**, *124*, 104834. [CrossRef] [PubMed]
15. Rusinkiewicz, S. Estimating curvatures and their derivatives on triangle meshes. In Proceedings of the 2nd International Symposium on 3D Data Processing, Visualization and Transmission, 3DPVT 2004, Thessaloniki, Greece, 9 September 2004; IEEE: Piscataway, NJ, USA, 2004.
16. Kirby, M.; Morshed, A.H.; Gomez, J.; Xiao, P.; Hu, Y.; Guo, X.E.; Wang, X. Three-dimensional rendering of trabecular bone microarchitecture using a probabilistic approach. *Biomech. Model. Mechanobiol.* **2020**, *19*, 1263–1281. [CrossRef] [PubMed]
17. Cowin, S.C.; Mehrabadi, M.M. Identification of the Elastic Symmetry of Bone and Other Materials. *J. Biomech.* **1989**, *22*, 503–515. [CrossRef] [PubMed]
18. Yang, G.; Kabel, J.; van Rietbergen, B.; Odgaard, A.; Huiskes, R.; Cowin, S.C. The anisotropic Hooke's law for cancellous bone and wood. *J. Elast.* **1998**, *53*, 125–146. [CrossRef]
19. Goldstein, S.A. The Mechanical-Properties of Trabecular Bone—Dependence on Anatomic Location and Function. *J. Biomech.* **1987**, *20*, 1055–1061. [CrossRef]
20. Walker, A.M.; Wookey, J. MSAT-A new toolkit for the analysis of elastic and seismic anisotropy. *Comput. Geosci.* **2012**, *49*, 81–90. [CrossRef]
21. Wang, J.; Zhou, B.; Parkinson, I.; Thomas, C.D.L.; Clement, J.G.; Fazzalari, N.; Guo, X.E. Trabecular plate loss and deteriorating elastic modulus of femoral trabecular bone in intertrochanteric hip fractures. *Bone Res.* **2013**, *1*, 346–354. [CrossRef]
22. Mittra, E.; Rubin, C.; Gruber, B.; Qin, Y.-X. Evaluation of trabecular mechanical and microstructural properties in human calcaneal bone of advanced age using mechanical testing, μCT, and DXA. *J. Biomech.* **2008**, *41*, 368–375. [CrossRef]
23. Xi, J.; Hu, X.; Jin, Y. Shape Analysis and Parameterized Modeling of Hip Joint. *J. Comput. Inf. Sci. Eng.* **2003**, *3*, 260–265. [CrossRef]
24. da F. Costa, L.; dos Reis, S.F.; Arantes, R.A.T.; Alves, A.C.R.; Mutinari, G. Biological shape analysis by digital curvature. *Pattern Recognit.* **2004**, *37*, 515–524. [CrossRef]
25. Rohr, K. Extraction of 3D anatomical point landmarks based on invariance principles. *Pattern Recognit.* **1999**, *32*, 3–15. [CrossRef]
26. Subburaj, K.; Ravi, B.; Agarwal, M. Automated identification of anatomical landmarks on 3D bone models reconstructed from CT scan images. *Comput. Med. Imaging Graph.* **2009**, *33*, 359–368. [CrossRef]
27. Blanquer, S.B.G.; Werner, M.; Hannula, M.; Sharifi, S.; Lajoinie, G.P.R.; Eglin, D.; Hyttinen, J.; Poot, A.A.; Grijpma, D.W. Surface curvature in triply-periodic minimal surface architectures as a distinct design parameter in preparing advanced tissue engineering scaffolds. *Biofabrication* **2017**, *9*, 025001. [CrossRef]
28. Guo, Y.; Sharma, S.; Kumar, S. Inverse Designing Surface Curvatures by Deep Learning. *Adv. Intell. Syst.* **2024**, *6*, 2300789. [CrossRef]
29. Murray, A.A.; MacKinnon, M.; Carswell, T.M.R.; Giles, J.W. Anterior diaphyseal curvature of the femur and tibia has biomechanical consequences during unloaded uphill locomotion. *Front. Ecol. Evol.* **2023**, *11*, 1220567. [CrossRef]

Disclaimer/Publisher's Note: The statements, opinions and data contained in all publications are solely those of the individual author(s) and contributor(s) and not of MDPI and/or the editor(s). MDPI and/or the editor(s) disclaim responsibility for any injury to people or property resulting from any ideas, methods, instructions or products referred to in the content.

MDPI AG
Grosspeteranlage 5
4052 Basel
Switzerland
Tel.: +41 61 683 77 34

Journal of Functional Biomaterials Editorial Office
E-mail: jfb@mdpi.com
www.mdpi.com/journal/jfb

Disclaimer/Publisher's Note: The title and front matter of this reprint are at the discretion of the Guest Editors. The publisher is not responsible for their content or any associated concerns. The statements, opinions and data contained in all individual articles are solely those of the individual Editors and contributors and not of MDPI. MDPI disclaims responsibility for any injury to people or property resulting from any ideas, methods, instructions or products referred to in the content.

www.ingramcontent.com/pod-product-compliance
Lightning Source LLC
LaVergne TN
LVHW072352090526
838202LV00019B/2525